Student's Solutions Manual

Beverly Fusfield

Mathematical Reasoning for Elementary Teachers
Seventh Edition

Calvin T. Long
Washington State University

Duane W. DeTemple
Washington State University

Richard S. Millman
Georgia Institute of Technology

PEARSON

Boston Columbus Indianapolis New York San Francisco Upper Saddle River
Amsterdam Cape Town Dubai London Madrid Milan Munich Paris Montreal Toronto
Delhi Mexico City São Paulo Sydney Hong Kong Seoul Singapore Taipei Tokyo

The author and publisher of this book have used their best efforts in preparing this book. These efforts include the development, research, and testing of the theories and programs to determine their effectiveness. The author and publisher make no warranty of any kind, expressed or implied, with regard to these programs or the documentation contained in this book. The author and publisher shall not be liable in any event for incidental or consequential damages in connection with, or arising out of, the furnishing, performance, or use of these programs.

Reproduced by Pearson from electronic files supplied by the author.

Copyright © 2015, 2012, 2009 Pearson Education, Inc.
Publishing as Pearson, 75 Arlington Street, Boston, MA 02116.

All rights reserved. No part of this publication may be reproduced, stored in a retrieval system, or transmitted, in any form or by any means, electronic, mechanical, photocopying, recording, or otherwise, without the prior written permission of the publisher. Printed in the United States of America.

ISBN-13: 978-0-321-90102-6
ISBN-10: 0-321-90102-9

1 2 3 4 5 6 OPM 17 16 15 14 13

www.pearsonhighered.com

PEARSON

Contents

Chapter 1	Thinking Critically	1
Chapter 2	Sets and Whole Numbers	14
Chapter 3	Numeration and Computation	25
Chapter 4	Number Theory	36
Chapter 5	Integers	47
Chapter 6	Fractions and Rational Numbers	54
Chapter 7	Decimals, Real Numbers, and Proportional Reasoning	69
Chapter 8	Algebraic Reasoning, Graphing, and Connections with Geometry	79
Chapter 9	Geometric Figures	89
Chapter 10	Measurement: Length, Area, and Volume	98
Chapter 11	Transformations, Symmetries, and Tilings	111
Chapter 12	Congruence, Constructions, and Similarity	121
Chapter 13	Statistics: The Interpretation of Data	131
Chapter 14	Probability	142

Chapter 1 Thinking Critically

Section 1.1
An Introduction to Problem Solving

Problem Set 1.1

1. (a) Using guess and check:
   Guess 14 bikes, 13 trikes. The number of wheels is $14 \times 2 + 13 \times 3 = 67$.
   Too many wheels, too many trikes.
   Guess again: 17 bikes, 10 trikes. The number of wheels is $17 \times 2 + 10 \times 3 = 64$.
   Still too many wheels.
   Guess again: 21 bikes, 6 trikes. The number of wheels is: $21 \times 2 + 6 \times 3 = 60$. O.K.

   (b)

Bikes	Trikes	Bike Wheels	Trike Wheels	Total Wheels
17	10	34	30	64
18	9	36	27	63
19	8	38	24	62
20	7	40	21	61
21	6	42	18	60

   (c) Place two wheels next to each seat and then add a third wheel to as many seats as necessary to make 60 wheels.

   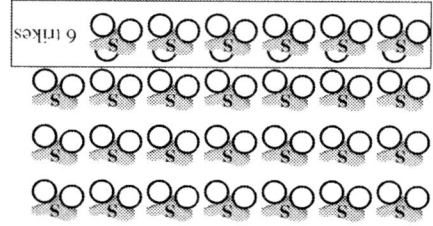

   (d) Yes. If all trikes have one wheel off the ground, there are 54 wheels touching the ground. To make 60 wheels there must be 6 wheels in the air—6 trikes.

3. (a) Suppose all 32 stamps were 18-cent stamps. The total worth of the stamps would be $5.76. $8.07 - $5.76 = $2.31. Since 29-cent stamps cost 11 cents more than 18-cent stamps, the total number of 29-cent stamps is $231 \div 11 = 21$. Mr. Akika has 11 18-cent stamps and 21 29-cent stamps. (Note that this is Jennifer's strategy.)

   (b) Answers will vary. One method is given in (a). Another possibility is to make an educated guess and then adjust the guess upward or downward depending on how the total value of the stamps under your guess compares to $8.07.

5. Use the Make a Table strategy.

Dimes	Nickels	Pennies	Total Value
4	0	5	45¢
4	1	4	49¢

   4 dimes are too many; try 3 dimes.

Dimes	Nickels	Pennies	Total Value
3	1	5	40¢
3	2	4	44¢
3	3	3	48¢

   Xin has 3 dimes, 3 nickels, and 3 pennies.

7. Guess and Check or Make a Table.

Guess	Multiply by 5	Subtract 8 to obtain result
10	50	42
11	55	47
12	60	52 ✓

9. Answers will vary. Two possibilities are given. Who am I? If you multiply me by 4 and subtract 8, the result is 60. Answer: 17
   Sally has 15 nickels and dimes worth a total of $1.10. How many of each coin does she have? Answer: 8 nickels and 7 dimes.

11. (a) Answers will vary. Two possibilities are

    ```
        6              1
      1   5          6   5
     5 4 3 2        2 3 4 5
    ```
    or

    Another possibility is given in part (c).

    (b) Yes.

Copyright © 2015 Pearson Education, Inc.

2 Chapter 1 Thinking Critically

(c) Answers will vary. One method is given. First, choose a number which is to be the sum of the numbers on each side—say, 11. (Several choices are possible.) We need to write 11 as a sum of three numbers (1 through 6) in three different ways. 11 can be written as $6 + 4 + 1$, $6 + 3 + 2$, or $5 + 4 + 2$. Since 2, 4, and 6 each appear in two of the sums, we place 2, 4, and 6 in the "corner" circles and complete the diagram as shown on the next page.

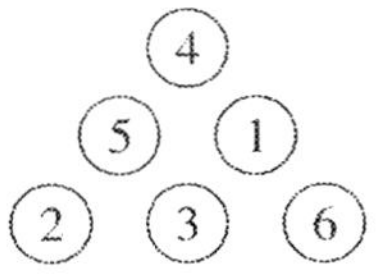

④
⑤ ①
② ③ ⑥

13. Parts (a), (b) and (c) have more than one solution. You can place an arbitrary number in the upper left circle and then complete the rest of the circles.

(a) ⑧   7   15
      11   20

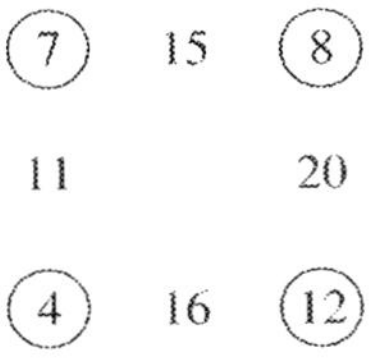

(b) ④   16   ⑫
     ⑦   12   ⑤
     10   21

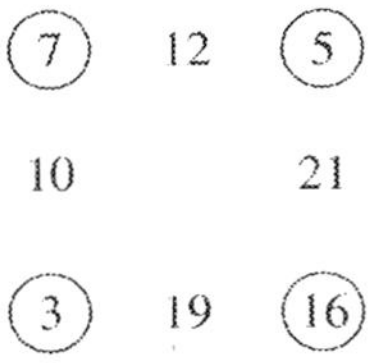

(c) ③   19   ⑯
     ①   7   ⑨
     9   16
     ⑧   18   ⑩

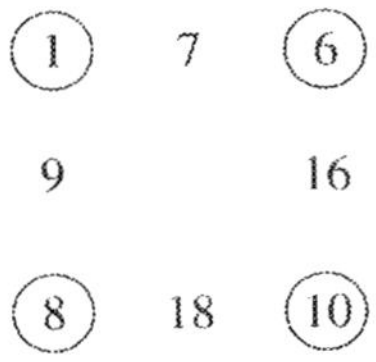

(d) Note that in each of (a), (b), and (c), the sum of the top and bottom numbers given is the sum of the left and right numbers; i.e., $11 + 20 = 15 + 16$, $10 + 21 = 12 + 19$, and $9 + 16 = 7 + 18$. For such a problem to have a solution, this must always be the case. Thus, there is no solution.

15. (a) Each sum is 15. In any magic square, the sum of the numbers in each row, column, and diagonal is the same.

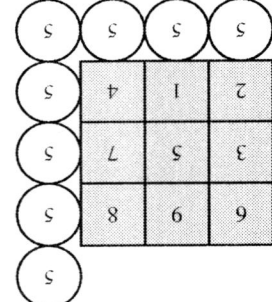

(b) Each result is 5. For example, in the top row, $6 + 8 - 9 = 5$.

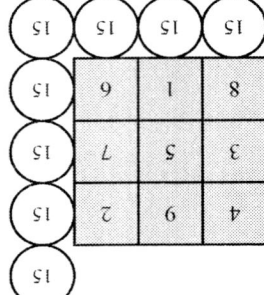

17. (a) Since $2 + 3 = 5$, $3 + 5 = 8$, $5 + 8 = 13$, and $8 + 13 = 21$, the sequence is
1, 2, 3, 5, 8, 13, 21.

(b) Since $8 - 2 = 6$, $6 + 8 = 14$,
$8 + 14 = 22$, $14 + 22 = 36$, and
$22 + 36 = 58$, the sequence is
2, 6, 8, 14, 22, 36, 58.

(c) This can be solved using the guess and check strategy. A more formal solution is as follows: Suppose the second number is n. Then the third number is $3 + n$, and the fourth number is 13. Therefore, $n + (3 + n) = 13$, so $n = 5$. The sequence is 3, 5, 8, 13, 21, 34, 55.

(d) Suppose the second number is n. Then the sequence is $2, n, 2 + n, 2 + 2n, 4 + 3n, 26$. Therefore,
$(2 + 2n) + (4 + 3n) = 26$ or
$6 + 5n = 26$ or $5n = 20$ or $n = 4$.
The sequence is 2, 4, 6, 10, 16, 26.

(e) Suppose the second number is n. Then the sequence is $2, n, 2 + n, 2 + 2n, 4 + 3n, 11$. Therefore $(2 + 2n) + (4 + 3n) = 11$, so $6 + 5n = 11$ and $n = 1$. The sequence is 2, 1, 3, 4, 7, 11.

Section 1.2
Pólya's Problem-Solving Principles and the Standards for Mathematical Practice of the Common Core State State Standards for Mathematics

Problem Set 1.2

1. (a) No, because $5 \times 10 + 13 = 63$, not 48.

(b) We could use Guess and Check, Make a Table, or algebra to find Nancy's number.

(c) Using algebra, if x is Nancy's number, then $5x + 13 = 48$ gives $5x = 35$ or $x = 7$.

3. Construct the following table:

Guess	Twice the guess plus 1	Three times the guess minus 5
4	9	7
5	11	10
6	13	13

Vicky's number must be 6.

5. The four-coin possibilities are:

QQQQ	$1.00	QDNN	0.45
QQQD	0.85	QNNN	0.40
QQQN	0.80	DDDD	0.40
QQDD	0.70	DDDN	0.35
QQDN	0.65	DDNN	0.30
QQNN	0.60	DNNN	0.25
QDDD	0.55	NNNN	0.20
QDDN	0.50		

Since $0.40 appears twice, there are 14 different amounts of money.

7. 258, 285, 528, 582, 825, 852

9. Use the Make a Table strategy.

Number of dimes	Number of nickels	Number of pennies
2	0	1
1	2	1
1	1	6
1	0	11
0	4	1
0	3	6
0	2	11
0	1	16
0	0	21

11. (a) Look for two whole numbers whose product is 120.

1, 120   2, 60   3, 40   4, 30
5, 24   6, 20   8, 15   10, 12

(b) The perimeter is twice the sum of the length and width. Therefore, the $10 \text{ cm} \times 12 \text{ cm}$ rectangle has the smallest perimeter, 44 cm.

13. Find common multiples of 20 and 30 and determine the amount for which Peter worked 5 more days.

Common multiples of 20 and 30	Days worked Jill	Peter
60	2	3
120	4	6
180	6	9
240	8	12
300	10	15

Jill worked 10 days, and Peter worked 15 days.

15. A diagram of the situation will show that Bob has to make 9 cuts to get 10 2-foot sections. Since each cut takes one minute, it will take Bob 9 minutes to do this.

17. Try solving a simpler problem and make a diagram.

Number of sections on each side	Number of posts	Diagram
1	4	
2	8	
3	12	

For 10 sections on each side, the number of posts is 40.

19. (a) The possible perimeters are 12, 14, 16, 18, and 20. See the diagrams. Note that the perimeter cannot be odd because the edges are adjoined in pairs. Of course, other diagrams are possible but none yields a perimeter other than those shown on the next page.

(continued on next page)

Copyright © 2015 Pearson Education, Inc.

4 *Chapter 1 Thinking Critically*

(*continued*)

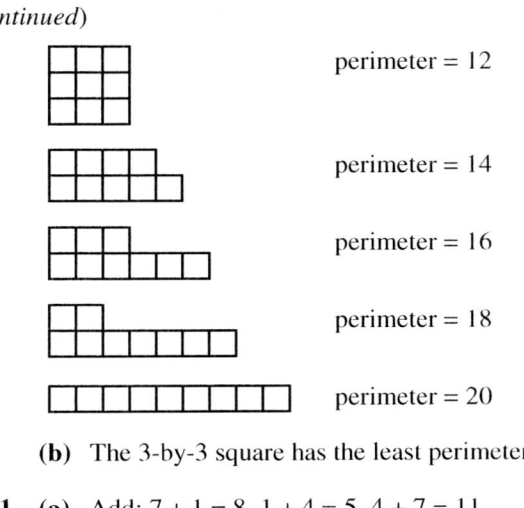

perimeter = 12

perimeter = 14

perimeter = 16

perimeter = 18

perimeter = 20

   (b) The 3-by-3 square has the least perimeter.

21. (a) Add: $7 + 1 = 8$, $1 + 4 = 5$, $4 + 7 = 11$.

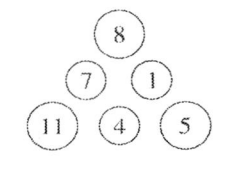

   (b) Note that $11 + 1 = 12$, $19 - 11 = 8$, and $8 + 1 = 9$.

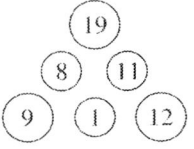

   (c) The sum of the three new numbers must
   be $\dfrac{7 + 11 + 13}{2} = 15.5$. Note that
   $15.5 - 7 = 8.5$, $15.5 - 11 = 4.5$, and
   $15.5 - 13 = 2.5$.

   (d) The sum of the three new numbers must
   be $\dfrac{-2 + 7 + 11}{2} = 8$. Note that
   $8 - (-2) = 10$, $8 - 7 = 1$, and
   $8 - 11 = -3$.

23. (a) The combinations are (skirt, red tee),
   (skirt, blue tee), (skirt, lime green tee),
   (shorts, red tee), (shorts, blue tee), (shorts,
   lime green tee). There are $2 \times 3 = 6$
   combinations.

   (b) There are $2 \times 3 \times 2 = 12$ combinations.

25. Use the Make a Table Strategy.

White Chairs	Black Chairs
1	0
1	4
5	4
5	8
9	8
9	12
13	12
13	16
17	16
17	20

The answer is B. There are 20 black chairs.

Section 1.3
More Problem-Solving Strategies

Problem Set 1.3

1. (a) 2, 5, 8, 11, <u>14, 17, 20</u>.
   Each succeeding term is 3 more than the
   preceding term.

   (b) −5, −3, −1, 1, <u>3, 5, 7</u>.
   Each succeeding term is 2 more than the
   preceding term.

   (c) 1, 1, 3, 3, 6, 6, 10, <u>10, 15, 15</u>.
   Notice that the third, fifth, and seventh
   terms in the sequence are determined by
   adding 2, 3, and 4 respectively to the
   second, fourth, and sixth terms.

   The even numbered terms are the same as
   the preceding term, so the sixth term is
   10. The seventh term is $10 + 5 = 15$, and
   the eighth term is also 15.

3. (a) Each term is 2 more than its predecessor.
   Since $35 = 5 + 30$ and $30 \div 2 = 15$, we
   conclude that 15 2s must be added to 5 to
   get 35. Thus, there are 15 terms after the
   first one, or 16 terms in all.

Copyright © 2015 Pearson Education, Inc.

(b) Each term is 5 more than its predecessor. Since $46 = -4 + 50$ and $50 \div 5 = 10$, we conclude that 10 5s must be added to get -4 to get 46. There are 11 terms.

(c) Each term is 4 more than its predecessor. Since $67 = 3 + 64$ and $64 \div 4 = 16$, we conclude that 16 4s must be added to 3 to get 67. There are 17 terms.

5. (a) The middle term on the left side of the nth equation is n. The next two lines are:
$1 + 2 + 3 + 4 + 5 + 4 + 3 + 2 + 1 = 25$
$1 + 2 + 3 + 4 + 5 + 6 + 5 + 4 + 3 + 2 + 1$
$\qquad\qquad\qquad\qquad = 36.$

(b) Using the pattern observed, the sum is $100^2 = 10{,}000$.

(c) The sum is n^2.

7. (a) Each pattern has one more column of three dots than the previous one. The next two patterns are shown.

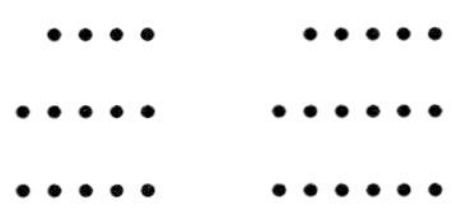

```
. . . .      . . . . .
. . . . .    . . . . . .
. . . . .    . . . . . .
```

(b) 2, 5, 8, 11, 14, 17

(c) The pattern is that each term is increased by 3. Thus, the sequence of numbers can be expressed as:
$2, 2 + 3(1), 2 + 3(2), 2 + 3(3), \ldots$
The 10th term is $2 + 3(9) = 29$.
The 100th term is $2 + 3(99) = 299$.

(d) Since 33 3s must be added to 2 to get 101, the number 101 must be the 34th term.

9. Notice the pattern. Except for the last column of the table, when you move right one square, the units digit increases by 1, and when you move down one square, the tens digit increases by 1. Hence, we fill in the arrays and determine the desired last number as shown.

(a)
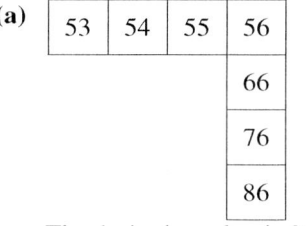

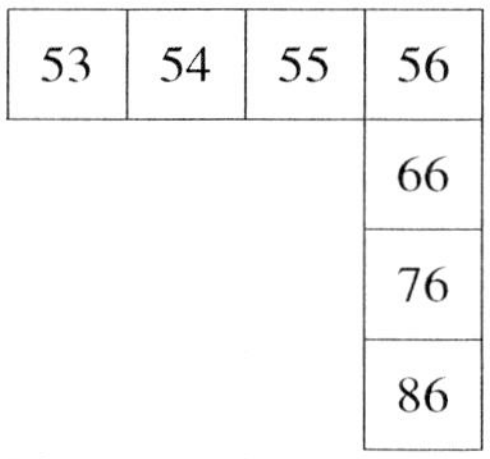

The desired number is 86.

(b)
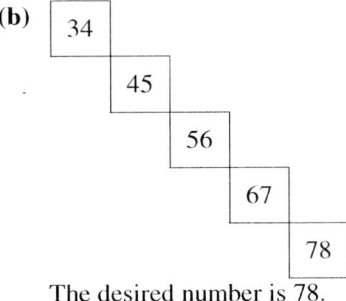

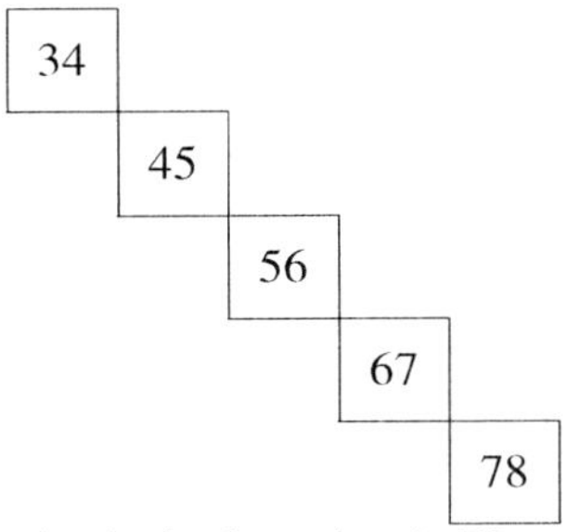

The desired number is 78.

(c)
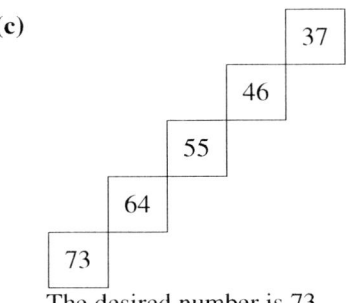

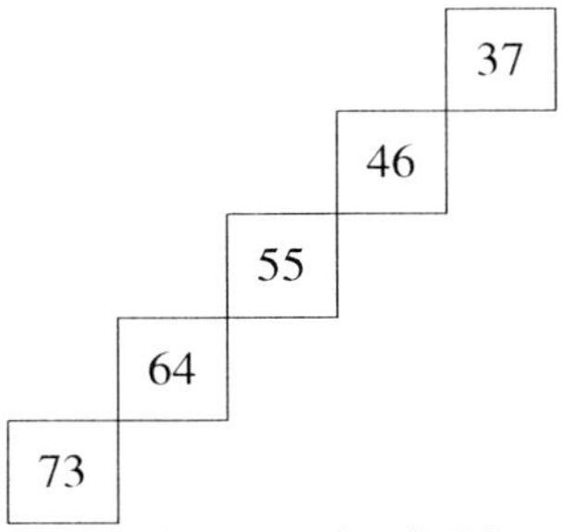

The desired number is 73.

11. The least number of moves is 10. There are many ways to accomplish this in ten moves; one way is shown.

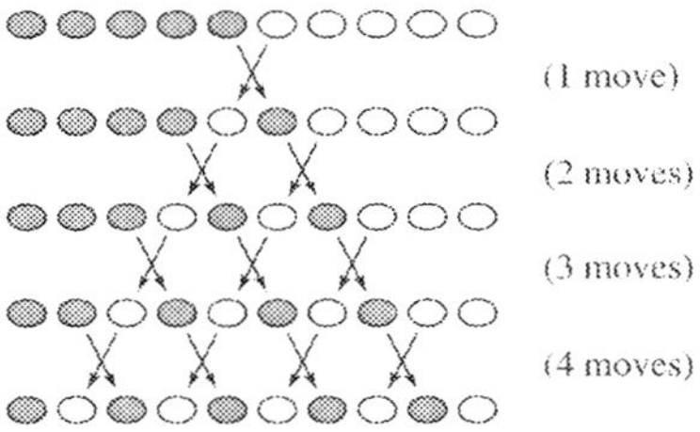

```
◎◎◎◎◎○○○○○
        ╳              (1 move)
◎◎◎◎○○○◎○○
      ╳   ╳            (2 moves)
◎◎○◎○○◎○○○
```

Note that $1 + 2 + 3 + 4 = 10$.

13. (a) In the 1 by 3 figure, there is one 1 by 3 rectangle, two 1 by 2 rectangles, and three 1 by 1 rectangles. The number of rectangles is $1 + 2 + 3 = 6$.
In the 1 by 4 figure, there is one 1 by 4 rectangle, two 1 by 3 rectangles, three 1 by 2 rectangles, and four 1 by 1 rectangles. The number of rectangles is $1 + 2 + 3 + 4 = 10$.

(b) The figure measures 1 by 16, so the number of rectangles is
$$1 + 2 + 3 + \cdots + 16 = t_{16} = \frac{16 \cdot 17}{2} = 136.$$

(c) The number of rectangles is $1 + 2 + 3 + \cdots + n$. This is the nth triangular number, $t_n = \dfrac{n(n+1)}{2}$.
(See section 1.4 in the text.)

6 *Chapter 1 Thinking Critically*

(d) Answers will vary. For example, a 1 by n strip has one 1 by n rectangle starting in square number one; two 1 by $n - 1$ rectangles starting in squares number 1 and 2; three 1 by $n - 2$ rectangles starting in squares number 1, 2, and 3; ...; and n 1 by 1 rectangles starting in squares number 1, 2, ..., n. Thus, the total number of rectangles is

$$1 + 2 + 3 + \cdots + n = \frac{n(n+1)}{2}.$$

15. There are 35 paths from A to B

17. The possible paths are
1. $C \rightarrow A \rightarrow B \rightarrow E$
2. $C \rightarrow B \rightarrow E$
3. $C \rightarrow F \rightarrow D \rightarrow E$
4. $C \rightarrow F \rightarrow E$
5. $C \rightarrow F \rightarrow B \rightarrow E$
There are 5 paths from C to E.

19. (a) $\dfrac{10 \cdot 9}{2} = 45$ games

(b) $\dfrac{11 \cdot 10}{2} = 55$ games

(c) Yes. Each team can be represented as a dot on the circle and each segment represents a game played between the two teams.

21. (a) Answers will vary widely.

(b) Answers will vary widely.

23. None. It's a hole!

25. The coins are a penny and a quarter. The problem states that *one* of the coins is not a quarter. It does not say that neither coin was a quarter.

27. One possibility is simply to try each of the given rules to see which one generates Duane's sequence. Checking this way, we see that rule C generates the given pattern.
$4 \times 0 + 2 = 2$, $4 \times 2 + 2 = 10$,
$4 \times 10 + 2 = 42$, and $4 \times 42 + 2 = 170$.

Section 1.4
Algebra as a Problem-Solving Strategy

Problem Set 1.4

1. (a) Each pattern has one more column of dots than the previous one. The next three patterns are shown:

(b) 2, 4, 6, 8, 10, 12

(c) Each term is twice the number of the term. Therefore, the 10th term is $2(10) = 20$ and the 100th term is $2(100) = 200$.

(d) The nth even number.

(e) Divide 2402 by 2 and obtain 1201.

3. (a) Let m be Jackson's number. To triple m and subtract 13 means to compute $3m - 13$. Thus, $3m - 13 = 2$ or $3m = 15$, so $m = 5$ is Jackson's number.

(b) There are no solutions since the solution of the equation $3n - 13 = 4$ is $n = 17/3$, which is not an integer.

5. Each additional train requires 4 more matchsticks, so the formula has the form $t = k + 4c$ for some constant k. Evaluating this expression at $c = 1$, then $k + 4 = 5$, since the first pattern has $t = 5$ matchsticks. Therefore $k = 1$, and the desired formula is $t = 1 + 4c$.

7. (a) Person A shakes with B and C. Since B and C have already shaken hands with A, only one shake remains, B with C. The total is $2 + 1 = 3$.

(b) If the people are A, B, C, D, E, and F, then A shakes hands with the other five, and B has only four people left with whom to shake hands (C, D, E, and F). Similarly, C has only three handshakes, D has 2, E has 1, and there is nobody new left for F. Total is $5 + 4 + 3 + 2 + 1 = 15$.

Copyright © 2015 Pearson Education, Inc.

(c) The logic is the same as in part (b). The first person shakes hands with the other 199, the second with 198, the third with 197, etc. until we get to the penultimate person who has one new hand to shake. The total is

$$199 + 198 + 197 + 196 + \ldots + 2 + 1$$
$$= \frac{(200)(199)}{2} = 19{,}990,$$ where we have

used Gauss's Insight.

(d) The logic is the same as in part (c) The first person shakes hands with the other $n - 1$, the second with $n - 2$, the third with $n - 3$, etc., until we get to the penultimate (second to last) person who had only one new hand to shake. Thus, the total is

$$(n - 1) + (n - 2) + (n - 3) + \ldots + 2 + 1$$
$$= \frac{n(n - 1)}{2},$$

using Gauss's Insight.

9. There are 100 people in the room. If all of them shake hands exactly once, then (using the result in problem 7), there is a total of $100(99)/2 = 4950$ handshakes. However, since no husband and wife shake each other's hand, there are 50 less according to the condition of the problem. The answer is $4950 - 50 = 4900$.

11. Let c and g denote the number of chickens and goats, respectively. Then $c + g = 100$ and $2c + 4g = 286$, since there are 100 chickens and goats, and chickens have 2 feet each whereas goats have 4 feet each. From the first equation, $c = 100 - g$, which can be substituted into the second equation to obtain $2(100 - g) + 4g = 286$. This is equivalent to $200 - 2g + 4g = 286$, which simplifies to $2g = 286 - 200 = 86$. Therefore, there are

$$g = \frac{86}{2} = 43 \text{ goats and}$$
$$c = 100 - g = 100 - 43 = 57 \text{ chickens.}$$

13. (a) In each figure, dots are added to the upper left, upper right, and lower right sides to complete the next larger pentagon.

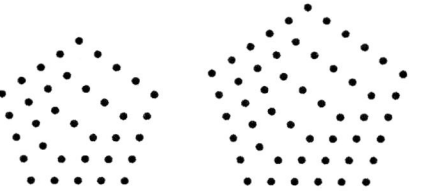

(b) 1, 5, 12, 22, 35, 51, ...

(c) $1 + 4 + 7 + 10 + 13 = 35$
$1 + 4 + 7 + 10 + 13 + 16 = 51$

(d) 10th term $= 1 + 3(9) = 28$

(e) Use Gauss's Insight:
$$\begin{aligned} s &= 1 + 4 + 7 + \cdots + 28 \\ s &= 28 + 25 + 22 + \cdots + 1 \\ \hline 2s &= 29 + 29 + 29 + \cdots + 29 \end{aligned}$$
$$\text{Sum} = \frac{(10)(29)}{2} = 145$$

(f) nth term $= 1 + 3(n - 1) = 3n - 2$

(g) Using Gauss's Insight and the result from part (f), there are n terms of $(3n - 1)$. The sum is $\frac{n(3n - 1)}{2}$. Therefore,

$$p_n = \frac{n(3n - 1)}{2}.$$

15. Using the notation in Example 1.3, we have $a + b = 16, a + c = 11$, and $b + c = 15$. Subtracting the second equation from the first equation gives $b - c = 5$. Adding this result to the third equation yields $2b = 20$, so $b = 10$. Substituting this value into the first equation gives $a = 6$. Substituting $a = 6$ into the second equation gives $c = 5$.

17. Suppose that x denotes the value in the lower small circle. Then the entries in the other small circles are $17 - x$ and $26 - x$, giving the equation $(17 - x) + (26 - x) = 11$. This simplifies to $43 - 2x = 11$, or $2x = 43 - 11 = 32$. Therefore, $x = 16$, and the entries in the other two circles are $17 - 16 = 1$ and $26 - 16 = 10$. Alternatively, one can work clockwise to see that the upper left small circle is $17 - x$ and therefore the remaining small circle value is $11 - (17 - x) = x - 6$. Then $(x - 6) + x = 26$, or $2x - 6 = 26$. As before, $x = 16$.

Copyright © 2015 Pearson Education, Inc.

8 *Chapter 1 Thinking Critically*

19. (a) Let x and w be integers which complete the diagram:

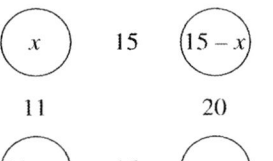

Looking at either $(11 - x) + w = 16$ or $(15 - x) + w = 20$ gives $w = 5 + x$. Thus x can be any integer and $w = x + 5$.

(b) Let x, y, z, w be integers in the circles:

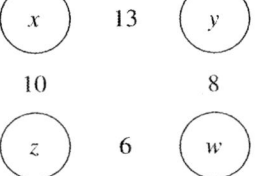

The conditions of the problem are $x + y = 13$, $x + z = 10$, $y + w = 8$, and $w + z = 6$. Subtracting the second equation from the first gives $y - z = 3$. Subtracting the fourth equation from the third gives $y - z = 2$. It is impossible for $y - z = 3$ and $y - z = 2$. Thus, there are no solutions.

21. (a)

n	1	2	3	4	5	6
n^2	1	4	9	16	25	36
$(n+1)^2$	4	9	16	25	36	49
difference	3	5	7	9	11	13

(b) The difference is
$$(n+1)^2 - n^2 = (n^2 + 2n + 1) - n^2 = 2n + 1$$
We can show why this works geometrically.

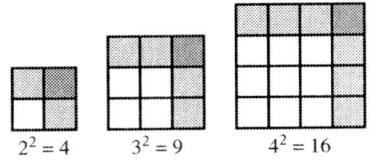

To get from one square to the next, we must add one more row, one more column, and one additional square to fill in the corner. These correspond directly to the $n + n + 1 = 2n + 1$ that we found algebraically.

23. (a) $23 + 32 = 55$
$42 + 24 = 66$
$17 + 71 = 88$
$51 + 15 = 66$
$67 + 76 = 143$
All of the answers are divisible by 11.

(b) The numbers given by the decimal description are $10a + b$ and $10b + a$. Thus, their sum is
$$(10a + b) + (10b + a) = 11a + 11b$$
$$= 11(a + b),$$
which is divisible by 11.

25. Let L and W denote the length and width of the rectangle, respectively, and S the length of the sides of the square. Since the rectangle is 3 times as long as it is wide, $L = 3W$. Therefore, the perimeter of the rectangle is $2L + 2W = 6W + 2W = 8W$, and its area is $LW = 3W^2$. The perimeter of the square is $4S$, and its area is S^2. We know both the rectangle and the square have the same perimeter, so $8W = 4S$, or $2W = S$. Also, the area of the square is 4 square feet larger than the area of the rectangle, so $S^2 = 3W^2 + 4$. By substitution, $(2W)^2 = 3W^2 + 4$, or $4W^2 = 3W^2 + 4$. Therefore, $W^2 = 4$, and the positive width of the rectangle is $W = 2$. Its length is $L = 3W = 6$. The square has sides of length $S = 2W = 4$.

27. If x is the number of students and y is the number of adults, then $x + y = 145$ and $3x + 5y = 601$. Since $x + y = 145$, $y = 145 - x$. Using substitution, we have
$3x + 5(145 - x) = 601 \Rightarrow$
$3x + 725 - 5x = 601 \Rightarrow -2x = -124 \Rightarrow x = 62$
and $y = 145 - 62 = 83$.
There were 62 students and 83 adults.

29. The sum of the two equations gives twice the heart is 20. Thus, the heart is 10 and the star is 4.

31. (a) Daniel is 10 feet away, and Christine is 19 feet away.

(b) Daniel has moved $10 + 5 + 2.5 + 1.25 = 18.75$ feet and is 1.25 feet from the door. Christine has moved four feet and is 16 feet from the door. Daniel is closer.

Copyright © 2015 Pearson Education, Inc.

(c) Daniel is correct. He will get as close to the door as he wishes, but will never actually get there.

Section 1.5
Additional Problem-Solving Strategies

Problem Set 1.5

1. The second player can always add a sufficient number of tallies to reach a multiple of 5 at each step. This will force the first player to go over 30.

3. **(a)** Work backwards. Start with 39 and perform the inverse operations of those indicated.
   $39 \times 2 = 78$
   $78 + 18 = 96$
   $96 \div 6 = 16$
   $16 - 7 = 9$
   The input number is 9.

   (b) Start with 48.
   $48 \times 2 = 96$
   $96 + 18 = 114$
   $114 \div 6 = 19$
   $19 - 7 = 12$
   The input number is 12.

   (c) Answers will vary. The guess and check method is one possibility.

   (d) Let x be the input. After two stages, we have $6(x + 7)$. The output is
   $$(6(x + 7) - 18) \div 2 = \frac{6x + 42 - 18}{2} = 3x + 12.$$
   If the output is 39 as in (a), then $3x + 12 = 39$, so $3x = 27$ or $x = 9$. If the output is 48 as in (b), then the input satisfies $3x + 12 = 48$, so $3x = 36$ or $x = 12$.

5. **(a)**
   $$\frac{6\left(3^2\right) - 18}{2} = \frac{6(9) - 18}{2} = \frac{54 - 18}{2}$$
   $$= \frac{36}{2} = 18$$

   (b) If the input is x and output is y, then the machine starts with x and each step follows as shown below.
   $$x \to x^2 \to 6x^2 \to 6x^2 - 18 \to$$
   $$\frac{6x^2 - 18}{2} \to 3x^2 - 9 = y$$

7. **(a)** Work backward. Before the last jump, Josh had $16, since $16 \times 2 = 32$. Before the second jump, Josh had $\frac{1}{2}(16 + 32) = \$24$. Before the first jump, Josh had $\frac{1}{2}(24 + 32) = \$28$. He started with $28.

   (b) Work backward two more steps.
   $\frac{1}{2}(28 + 32) = 30$ and $\frac{1}{2}(30 + 32) = 31$, so he had $31.

9. Since the number is greater than 20, less than 35, and divisible by 5, it must be either 25 or 30. Since the sum of the digits is 7, it must be 25. Not all information was needed—for example, we did not use the first and second clues.

11. Since the number is an even number between $8^2 = 64$ and $9^2 = 81$, it must be 64, 66, 68, 70, 72, 74, 76, 78, or 80. Since it is not divisible by 4, it must be 66, 70, 74, or 78. Since it is not divisible by 3, it is either 70 or 74. Either answer, 70 or 74, is possible.

13. If the cards start out in the order $a, b, c, d, e, f, g, h, i, j$ then they end up in the order $i, g, e, c, a, b, d, f, h, j$. These need to correspond to 0, 1, 2, 3, 4, 5, 6, 7, 8, 9. Therefore, $a = 4$, $b = 5$, $c = 3$, and so on. The cards must start in the order 4, 5, 3, 6, 2, 7, 1, 8, 0, 9. The work backward strategy is also effective on this problem.

15. Since Anne's husband is an only child and Will is Josie's brother, Taneisha is not married to Will. Combining this fact with the other clues, we immediately have:

	Kitty	Sarah	Josie	Taneisha
David	X	X	O	X
Will	X	O	X	X
Floyd	X	X	X	O
Gus	O	X	X	X

Completing the chart, we see that the following couples are married: Josie and David, Sarah and Will, Taneisha and Floyd, Kitty and Gus. (Note: The fact that Taneisha has two brothers is irrelevant.)

10   *Chapter 1 Thinking Critically*

17. To number pages 1 through 9 takes $9 \cdot 1 = 9$ digits. To number pages 10 through 99 takes $90 \cdot 2 = 180$ digits. This leaves $867 - 180 - 9 = 678$ to number 3- digit pages. Thus, there are $678 \div 3 = 226$ 3-digit pages and $226 + 90 + 9 = 325$ pages in the book.

19. **(a)** 366. Use the Pigeonhole Principle—in this case, the people are the "pigeons" and the 365 birthdays of the year are the "holes."

    (b) 731. With 730 in a room, it is possible that there are exactly 2 people with each of the birthdays: January 1, January 2, ..., December 31.

21. If the difference $a - b$ is divisible by 10, that means that $a - b$ is a multiple of 10. In any set of 11 natural numbers, at least two of the numbers must have the same units digit which implies that their difference is a multiple of 10.

23.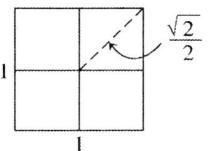

    If five points are chosen in a square with diagonal of length $\sqrt{2}$, then, by the Pigeonhole Principle, at least two of the points must be in or on the boundary of one of the four smaller squares shown. The farthest these two points can be from each other is $\sqrt{2}/2$ units, if they are on opposite corners of the small square.

25. If the cups of marbles are arranged as described, each cup will be part of three different groups of three adjacent cups. The sum of all marbles in all groups of three adjacent cups is $3 \cdot (10 \cdot 11/2) = 165$, since each cup of marbles is counted three times. With the marble count of 165 and 10 possible groups of three adjacent cups, by the Pigeonhole Principle, at least one group of three adjacent cups must have 17 or more marbles, since $165 \div 10 = 16.5 > 16$.

27. The number of people at the party with no friends is none, exactly one, or 2 or more.
    Case (i): If everyone has at least one friend, then each of the 20 people at the party has between 1 and 19 friends, inclusive. By the Pigeonhole Principle, at least two of them have

the same number of friends.
Case (ii): If exactly one person has no friends, then each of the other 19 people has 1 to 18 friends at the party. By the Pigeonhole Principle, at least two of them have the same number of friends.
Case (iii): If 2 or more people have no friends, then they have the same number of friends at the party.

29. Working backward, we have the following sequence of numbers: 3, 6, 12, 24. Dan baked 24 cookies.

31. The answer is D.

Section 1.6
Reasoning Mathematically

Problem Set 1.6

1. **(a)** $1 \times 8 + 1 = 9$
    $12 \times 8 + 2 = 98$
    $123 \times 8 + 3 = 987$

    (b) It looks as if the value of each expression can be determined from the value of the previous expression. The digits for each value start at 9 and decrease by one. To find the new value, just insert another digit to the end of the previous value that is one less than the last digit in the previous value. Note that the number of digits in the value is the number that is added in the expression.

    (c) $1234 \times 8 + 4 = 9876$
    $12,345 \times 8 + 5 = 98,765$
    $123,456 \times 8 + 6 = 987,654$
    $1,234,567 \times 8 + 7 = 9,876,543$
    $12,345,678 \times 8 + 8 = 98,765,432$
    $123,456,789 \times 8 + 9 = 987,654,321$

3. **(a)** $1 \times 1089 = 1089$ $6 \times 1089 = 6534$
    $2 \times 1089 = 2178$ $7 \times 1089 = 7623$
    $3 \times 1089 = 3267$ $8 \times 1089 = 8712$
    $4 \times 1089 = 4356$ $9 \times 1089 = 9801$
    $5 \times 1089 = 5445$

    (b) No. Patterns emerge. For example, the first two digits increase by 1 and the last two digits decrease by 1.

Copyright © 2015 Pearson Education, Inc.

(c) The first and last products are reversals, so are the second and eighth, etc. The product 5445 is a palindrome—that is, it is its own reversal.

5. The three points P, Q, and R are on a line.

7. (a) There are $F_5 = 5$ arrangements of five logs, supporting the generalization.

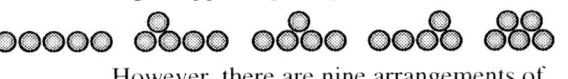

However, there are nine arrangements of six logs, instead of 8 as suggested by the Fibonacci pattern.

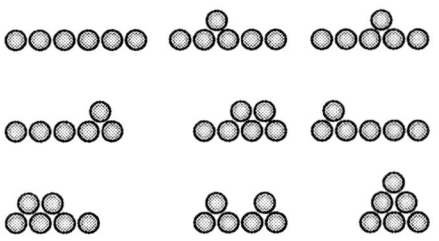

(b) The number of ways to stack n logs in two layers is given by the nth Fibonacci number F_n.

9. (a) Two pennies must be moved.

(b) Three pennies must be moved.

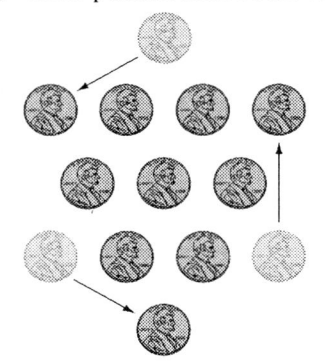

(c) Five pennies must be moved to invert a 15-penny triangle. In general, it can be shown that a triangle with n pennies requires $n/3$ pennies to be moved, where any remainder of the division is dropped. For example, $\dfrac{10}{3} = 3\dfrac{1}{3}$ so the triangle of 10 pennies requires 3 moves, and a 15 penny triangle requires $\dfrac{15}{3} = 5$ pennies to be moved.

11. (a)

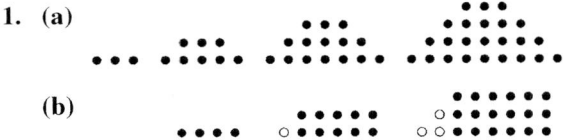

(b)

(c) The nth trapezoidal number is $n(n + 2)$ by inductive reasoning.

13. If n is a multiple of three, then $n = 3s$ for some whole number s and $n^2 = (3s)^2 = 9s^2$, which shows that n^2 is a multiple of 9.

15. Assume that n satisfies $2n + 16 = 35$. Since one side of the equation is $2(n + 8)$ is an even number (it is divisible by 2) and the other side, 35, is odd, there can be no solution to this equation. This is a proof by contradiction.

Chapter 1 Review Exercises

1. *Strategy 1*: Use trial and error and make a chart similar to the one below:

Number of 8′ boards	Number of 10′ boards	Total number of feet
45	45	810 (too small)
40	50	820 (too small)
35	55	830 (too small)
30	60	840 (too small)
28	62	844 (o.k.)

Therefore, there were 28 eight-foot boards.

Strategy 2: Use reasoning. If all boards were 8 feet long, the total length would be $90 \cdot 8 = 720$. He has $844 - 720 = 124$ "extra" feet. Since each ten-foot board has 2 "extra" feet, the number of ten-foot boards is $124 \div 2 = 62$. Therefore, there were 28 eight-foot boards.

3. There are 9 ways to produce 21¢ in change. They are listed in order of number of pennies used:

4N + P	3N + 6P	D + 11P
D + 2N + P	D + N + 6P	N + 16P
2D + P	2N + 11P	21P

5. The flower bed plus the walkway has a total area of $12 \times 14 = 168$ square feet, and the flower bed alone has an area of $8 \times 10 = 80$ square feet. The area of the walkway is $168 - 80 = 88$ square feet.

Copyright © 2015 Pearson Education, Inc.

12 *Chapter 1 Thinking Critically*

7. **(a)** Multiply by 5, then subtract 2.

   (b) Answers will vary. A good strategy is to give Chanty consecutive whole numbers starting with 0.

9. We make a table to show all possibilities and use the clues to delete those that are impossible and, hence, those that are certain. The steps in the argument are numbered and the numbers in the table indicate the corresponding conclusions at each step.

	Doctor	Engineer	Teacher	Lawyer	Writer	Painter
Kimberly	no (5)	no (7)	no (1)	yes (8)	no (1)	yes (3)
Terry	yes (5)	yes (7)	no (6)	no (8)	no (2)	no (3)
Otis	no (4)	no (6)	yes (6)	no (6)	yes (2)	no (3)

(1) By (b), Kimberly is neither the teacher nor the writer.

(2) By (e), Terry is not the writer. Therefore, Otis is the writer.

(3) By (f), neither Otis nor Terry is the painter. So, Kimberly is the painter.

(4) By (g), Otis is not the doctor.

(5) By (d), since the doctor hired the painter (Kimberly) and the doctor is not Otis, Terry is the doctor and Kimberly is not the doctor.

(6) Since Kimberly is not the teacher and, by (a), the doctor (Terry) had lunch with the teacher, Otis is the teacher and Terry is not. Also, since Otis has just two jobs, it follows that he is neither the engineer nor the lawyer.

(7) By (c), the painter (Kimberly) is related to the engineer. Therefore, Kimberly is not the engineer and so Terry is.

(8) Since Terry is the doctor and engineer he is not the lawyer. Thus, finally, Kimberly is the lawyer and the table now shows who holds what jobs.

11. Complete the chart below and then generalize from the results.

Number of chords	Number of regions	Number of intersections	Number of segments
0	1 = 0 + 1	0	0
1	2 = 1 + 1	0	1
2	4 = 3 + 1	1	4
3	7 = 6 + 1	3	9
4	11 = 10 + 1	6	16
5	16 = 15 + 1	10	25
6	22 = 21 + 1	15	36

$\vdots$	$\vdots$	$\vdots$	$\vdots$
n	$\dfrac{n(n+1)}{2}+1$	$\dfrac{n(n-1)}{2}$	n^2

(a) $\dfrac{n(n+1)}{2}+1$

(b) $\dfrac{n(n-1)}{2}$

(c) Each chord is divided into n segments, for a total of n^2 small segments.

13. **(a)** A one-car train uses 6 toothpicks to form the hexagon. Adding a square + hexagon combination requires an additional 8 toothpicks, so the trains with 1, 3, 5, 7, … cars use 6, 6 + 8, 6 + 8 + 8, 6 + 8 + 8 + 8, … toothpicks. In general, a train with $2m + 1$ cars will require $6 + 8m$ toothpicks, where $m = 0, 1, 2, 3, …$. A two-car train uses 9 toothpicks, so trains with 2, 4, 6, 8, … cars use 9, 9 + 8, 9 + 8 + 8, 9 + 8 + 8 + 8, … toothpicks. In general, a train with $2m + 2$ cars uses $9 + 8m$ toothpicks for $m = 0, 1, 2, 3, …$.

   (b) Since $9 + 8m$ is always an odd number, a train with 102 toothpicks has an odd number of cars, say $2m + 1$. Then $6 + 8m = 102$, or $8m = 96$. Therefore, $m = 12$, and there are $2(12) + 1 = 25$ cars in the train.

15. **(a)** There are four "pigeonholes" (suits), so draw 5 cards.

   (b) If only 8 cards are drawn, there could be 2 of each suit. Therefore, draw 9 cards.

Copyright © 2015 Pearson Education, Inc.

(c) If one drew 48 cards, one might get everything except the aces. Therefore, to be absolutely sure of getting two aces, one must draw 50 cards.

17. (a)
$$67 \times 67 = 4489$$
$$667 \times 667 = 444,889$$
$$6667 \times 6667 = 44,448,889$$

(b) $6,666,667 \times 6,666,667$
$$= 44,444,448,888,889.$$
The patterns observed suggest that:
• the number of 4s is one more than the number of 6s in one of the factors.
• the number of 8s is the same as the number of 6s.
• the eights are followed by a single 9.
However, without knowing *why* the pattern holds, or doing the actual calculation, one cannot be completely sure that the guess is correct.

19. If n is odd then $n = 2s + 1$ for some whole number s.
$$n^2 = (2s+1)^2 = 4s^2 + 4s + 1$$
$$= 4(s^2 + s) + 1 = 4\left[2 \cdot \frac{s(s+1)}{2}\right] + 1$$
$$= 8\left[\frac{s(s+1)}{2}\right] + 1$$

Since one of any two consecutive whole numbers must be even, $\dfrac{s(s+1)}{2}$ must be a whole number, say q. Thus $n^2 = 8q + 1$.

Chapter 2   Sets and Whole Numbers

Section 2.1
Sets and Operations on Sets

Problem Set 2.1

1. (a) {Arizona, California, Idaho, Oregon, Utah}

   (b) {Maine, Maryland, Massachusetts, Michigan, Minnesota, Mississippi, Missouri, Montana}

   (c) {Arizona}

3. (a) {7, 8, 9, 10, 11, 12,13}

   (b) {9, 11, 13}

   (c) Since $2 = 2 \cdot 1$, $4 = 2 \cdot 2$, $6 = 2 \cdot 3$, $8 = 2 \cdot 4$, $10 = 2 \cdot 5$, etc., the set is {2, 4, 6, 8, 10, 12, 14, 16, 18, 20}.

5. Answers will vary.

   (a) $\{x \in U \mid 11 \le x \le 14\}$ or $\{x \in U \mid 10 < x < 15\}$

   (b) $\{x \in U \mid x \text{ is even and } 6 \le x \le 16 \text{ and } x \ne 14\}$

   (c) $\{x \in U \mid x = 4n \text{ and } 1 \le n \le 5\}$

   (d) $\{x \in U \mid x = n^2 + 1 \text{ and } 1 \le n \le 4\}$

7. (a) True. The sets contain the same elements.

   (b) True. Every element in {6}—namely, 6—is in {6, 7, 8}.

   (c) True. The sets contain the same elements. Order doesn't matter.

9.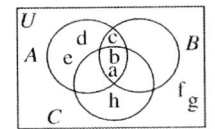

   (a) $B \cup C = \{a, b, c, h\}$

   (b) $A \cap B = \{a, b, c\}$

   (c) $B \cap C = \{a, b\}$

(d) $A \cup B = \{a, b, c, d, e\}$

(e) $\overline{A} = \{f, g, h\}$

(f) $A \cap C = \{a, b\}$

(g) $A \cup (B \cap C) = \{a, b, c, d, e\}$

11. (a) $D = \{1, 2, 3, 4, 6, 8, 9, 12, 16, 18, 24, 36, 48, 72, 144\}$

   (b) $G \cap D = \{1, 2, 3, 6, 9, 18\}$

   (c) 18

13. (a) A and B must be disjoint sets contained in C.

   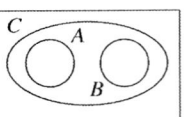

   (b) Answers may vary.

   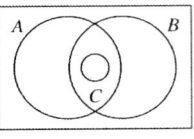

   (c) Answers may vary.

   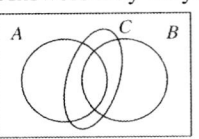

15. No, it is not necessarily true that $F = G$. For example, if D is any subset of the whole numbers, then $F = \{0, 1\}$ and $G = \{7, 9\}$ is a counterexample. In other words, suppose that $D = \{2, 4, 6, 8\}$ and F and G are defined as before. $D \cap F = \varnothing = D \cap G$, but $F \ne G$.

17. (a) $R \cap C$ is the set of red circles. Draw two red circles—one large and one small.

Copyright © 2015 Pearson Education, Inc.

(b) $L \cap H$ is the set of large hexagons. Draw two large hexagons—one red and one blue.

(c) $T \cup H$ is the set of shapes that are either triangles or hexagons. Draw two large triangles, two large hexagons, two small triangles, and two small hexagons—a red and a blue of each.

(d) $L \cap T$ is the set of large triangles. Draw two large triangles—one red and one blue.

(e) $B \cap \overline{C}$ is the set of blue shapes that are not circles. That is, the set of four elements consisting of the large and small blue hexagons and the large and small blue triangles.

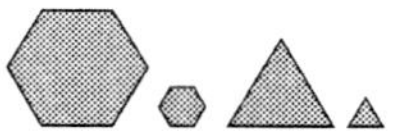

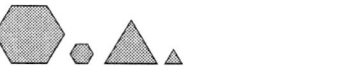

(f) $H \cap S \cap R$ is the set of hexagons that are both small and red. Draw one small red hexagon.

○

19. (a) Answers will vary. One possibility is B = set of students taking piano lessons, C = set of students learning a musical instrument.

(b) Answers will vary.

(c) Answers will vary.

21. Answers will vary.

23. Answers will vary. Sample answer:

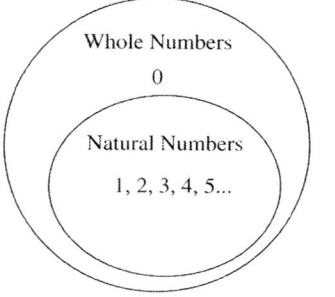

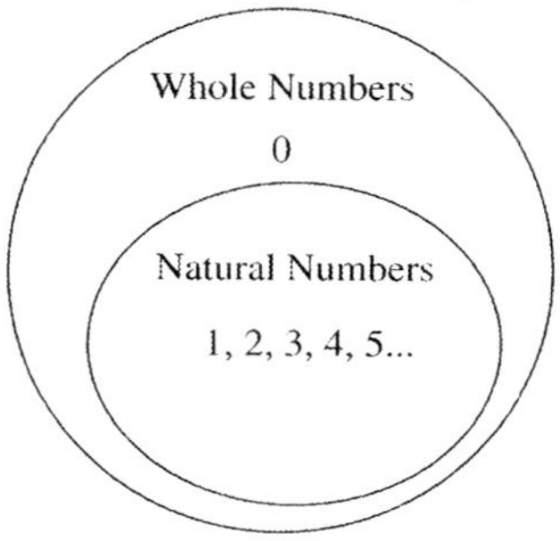

The natural numbers (also called the *counting numbers*) are a subset of the whole numbers. Zero is a whole number but not a natural number, but all natural numbers are a subset of the whole numbers.

25. (a) 8 regions

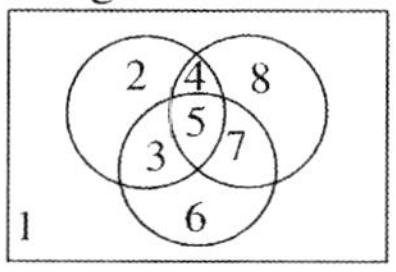

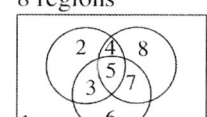

(b) By counting, there are 14 regions.

(c) Verify that $A \cap \overline{B} \cap \overline{C} \cap D$ has no region by observing that the region for $A \cap D$ is entirely contained in $B \cup C$. Likewise, the region for $B \cap C$ is entirely contained in $A \cup D$. The other missing region is $\overline{A} \cap B \cap C \cap \overline{D}$.

(d) Yes. Each loop contains 8 different regions, and there are 16 regions all together.

27. These Venn diagrams show that $\overline{A \cup B} = \overline{A} \cap \overline{B}$:

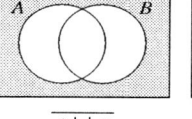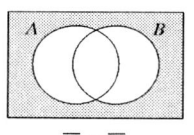

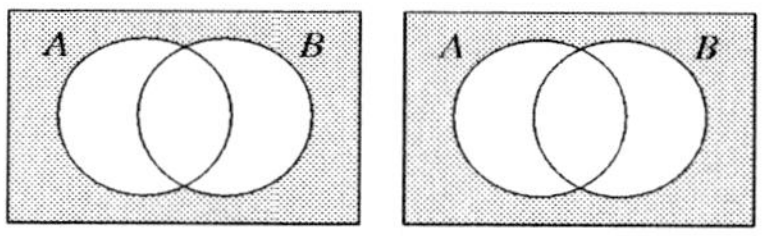

$\overline{A \cup B}$ $\overline{A} \cap \overline{B}$

29. (a) There are eight subsets: ∅, {P}, {N}, {D}, {P, N}, {P, D}, {N, D}, {P, N, D}.

(b) There are 16 subsets: ∅, {P}, {N}, {D}, {P, N}, {P, D}, {N, D}, {P, N, D}, {Q}, {P, Q}, {N, Q}, {D, Q}, {P, N, Q}, {P, D, Q}, {N, D, Q}, {P, N, D, Q}.

(c) Half of the subsets of {P, N, D, Q} contain Q.

(d) The number of subsets doubles with each additional element, so a set with n elements has 2^n subsets.

Copyright © 2015 Pearson Education, Inc.

16 **Chapter 2** Sets and Whole Numbers

31. Answers will vary. (The loop shown on the bottom could have represented negative instead of positive, reversing all signs.) Note that in this diagram, O– refers to the region outside all circles.

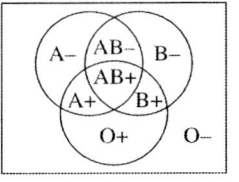

33.

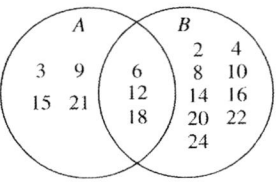

$A \cap B = \{6, 12, 18\}$

These are the elements that are divisible by both 2 and 3.

35. The number must be a multiple of both 10 and 12. The question asks for the least number with that property, so the answer is $LCM(10, 12) = 3 \cdot 2 \cdot 2 \cdot 5 = 60$, choice D.

Section 2.2
Sets, Counting, and the Whole Numbers

Problem Set 2.2

1. **(a)** 13(th): ordinal
first: ordinal

(b) fourth: ordinal; second: ordinal
93: cardinal. In common usage, it represents the number, 93, correct out of 100.

3. **(a)** Equivalent, since there are five letters in the set {A, B, M, N, P}

(b) Not equivalent since the sets have different numbers of elements.

(c) Equivalent, say by the correspondence
o ↔ t,n ↔ w,e ↔ o

(d) Not equivalent, since the set {0} has one element and $\varnothing$ has no elements.

5. **(a)** $n(A)$ is 0, 1, 2, 3, or 4, since it has strictly fewer elements than of set B.

(b) $n(C)$ is 5, 6, 7, ..., any whole number greater than or equal to 5.

7. **(a)** $n(A) = 7$ because
$A = \{21, 22, 23, 24, 25, 26, 27\}$.

(b) $n(B) = 0$ because $B = \varnothing$. There is no natural number x such that $x + 1 = x$.

9. **(a)** The correspondence $0 \leftrightarrow 1, 1 \leftrightarrow 2,$
$2 \leftrightarrow 3, ..., w \leftrightarrow w+1, ...$ shows $W \sim N$.

(b) The correspondence $1 \leftrightarrow 2, 3 \leftrightarrow 4, ...,$
$n \leftrightarrow n+1, ...$ shows $D \sim E$.

(c) The correspondence
$1 \leftrightarrow 10, 2 \leftrightarrow 100 = 10^2, ..., n \leftrightarrow 10^n, ...$
shows that the sets are equivalent.

11. **(a)** Answers will vary. For example,
$Q_1 \leftrightarrow Q_2$, $Q_3 \leftrightarrow Q_4$, and so on. (Note that P need not be the center—it can be any fixed point inside the small circle.)

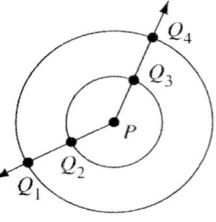

(b) Answers will vary. For example,
$Q_1 \leftrightarrow Q_2$, $Q_3 \leftrightarrow Q_4$, and so on.

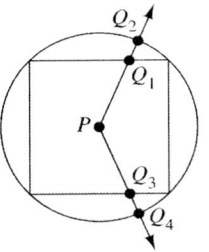

13. **(a)** True. A set B cannot have fewer elements than its subset A.

(b) False. Possible counterexample:
$A = \{1\}, B = \{2, 3\}$. Then $n(A) = 1$ and $n(B) = 2$, so $n(A) < n(B)$, but $A \not\subset B$.

(c) True. The union $A \cup B$ does not include any more elements than just A, so the elements of B must already be elements of A.

(d) True. If an element of A was not also an element of B, then this element would be missing in $A \cap B$ but included in A.

Copyright © 2015 Pearson Education, Inc.

15. (a) $n(A \cap B) \leq n(A)$

The set $A \cap B$ contains only the elements of A that are also elements of B. That is, $A \cap B \subseteq A$. Thus, $A \cap B$ cannot have more elements than A.

(b) $n(A) \leq n(A \cup B)$

The set $A \cup B$ contains all of the elements of the set A *and* any additional elements of B that are not already included. Then $A \subseteq A \cup B$. Therefore, $A \cup B$ must have at least as many elements as A.

(c) Since $n(A \cap B) = n(A \cup B)$ and $A \cap B \subseteq A \subseteq A \cup B$, we can conclude that $A \cap B = A \cup B$. Thus $A \cup B$ has no additional elements besides those in $A \cap B$, and so neither A nor B has any additional elements. Since $A = A \cap B$ and $B = A \cap B$, we conclude that $A = B$. (*Caution*: This reasoning would not be valid if infinite sets were allowed.)

17. Three of the four regions in the 2-loop diagram account for 500 of the households. Thus there are 200 households in the overlapped region representing the number of households having both a TV and a computer.

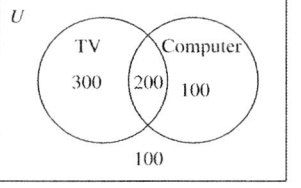

19. The 3-loop Venn diagram can be filled in the manner of problem 18. We now see there are $20 + 5 + 25 = 50$ percent of the students who like just one sport and 5 percent do not like any of the three sports.

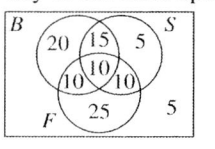

21. (a)

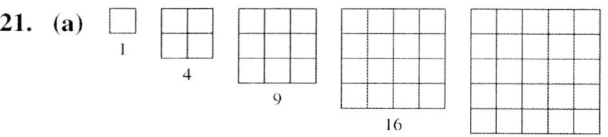

(b)

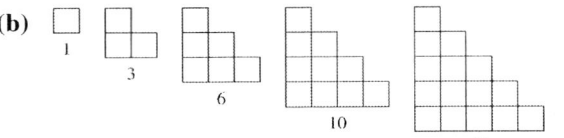

23. Answers will vary. One possible answer uses a cup with 5 marbles. Ask "how many marbles are in the cup?" to get the response "five." Now remove a marble, and again ask how many marbles are in the cup, eliciting the response "four." Continue to remove one marble at a time, until no marbles remain in the cup. Explain that the number of marbles in the empty cup is "zero."

25. 400 is a square number since 400 is the product of two squares, 4 and 100. Also, $400 = 20^2$, as illustrated below.

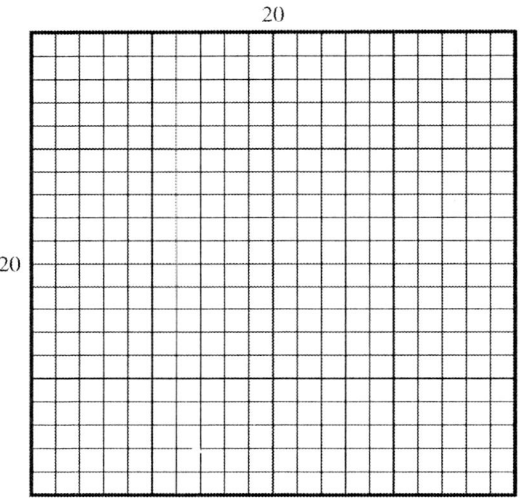

27. (a) The numbers that have been inserted are bold. Explanations to Zack will vary.

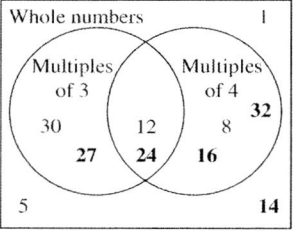

(b) Answers will vary

29. (a) Jeff is multiplying the base by the exponent, which is the wrong procedure. In this case, he is multiplying by three, not taking numbers to the third power.

18 Chapter 2 Sets and Whole Numbers

   (b) Use cubes whose sides are of length 1 cm, 2 cm, 3 cm, 4 cm, and 5 cm.

31. (a)

Row 0	1				
Row 1	1	1			
Row 2	1	2	1		
Row 3	1	3	3	1	
Row 4	1	4	6	4	1

   (b) The table is the same as Pascal's triangle each entry in the table is the sum of the numbers in the row just above that are directly above and one column to the left. Using this pattern, we get
Row 5: 1 5 10 10 5 1 and
Row 6: 1 6 15 20 15 6 1.

33. Consider the 35 students taking Arabic: 20 take only Arabic, 7 take both Arabic and Bulgarian (and possibly Chinese), telling us that 8 students are taking Arabic and Chinese and not Bulgarian. Similarly, there are 5 students taking Bulgarian and Chinese but not Arabic. The rest of the values in the Venn diagram are now easy to fill in. In particular, 3 students take all three languages, and 26 are not taking any of the three languages.

35. Construct a Venn diagram like the one shown, where S, I, and E are the sets of voters willing to raise sales, income, and excise taxes, respectively. Use a guess and check strategy in which you guess how many voters are willing to raise all three taxes and complete the diagram accordingly. The only way to account for all 60 voters is to start with 3 voters approving all three taxes.

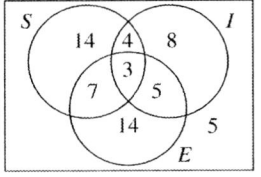

37. (a) All three friends will meet at the mall every $3 \times 4 \times 5 = 60$th day, so there will be 6 days when all three get together. ($365 \div 60 = 6$ with a remainder of 5.)

   (b) Since Letitia and Brianne are both at the mall 30 days, there are 24 of these days they will not be joined by Jake. Letitia and Jake will both be at the mall every $3 \times 5 = 15$th day, and $365 \div 15$ is 24 with a remainder of 5. This means Letitia and

Jake are at the mall 18 days by themselves and the other 6 days are also joined by Brianne. Similar reasoning fills in the remaining regions of the Venn diagram shown.

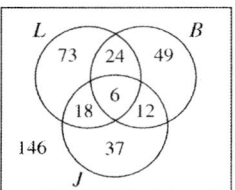

   (c) Jake will be at the mall by himself 37 days.

   (d) None of the three is at the mall on 146 days of the year.

39. Use the Venn diagram below to solve the problem.

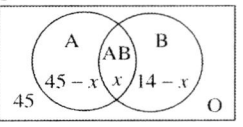

$104 - x = 100$, so $x = 4$
Four people are of type AB. The completed Venn diagram is shown below.

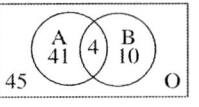

41. A. Seth read a total of $5 + 9 = 14$ books and Anna read 16 books.

43. B. $12 = 4 + 8$. The students taking French and Spanish are the four students taking all three courses and the eight students taking only French and Mandarin.

45. A.

Section 2.3
Addition and Subtraction of Whole Numbers

Problem Set 2.3

1. (a) (i) $A \cup B = \{$apple, berry, peach, lemon, lime$\}$
so $n(A \cup B) = 5$.

    (ii) $A \cup C = \{$apple, berry, peach, lemon, prune$\}$
so $n(A \cup C) = 5$.

Copyright © 2015 Pearson Education, Inc.

(iii) $B \cup C = \{\text{lemon, lime, berry, prune}\}$
so $n(B \cup C) = 4$ since lemon is a member of both sets.

(b) (ii) and (iii) because the two sets in each case are not disjoint, that is, they have at least one member in common.

3. (a)

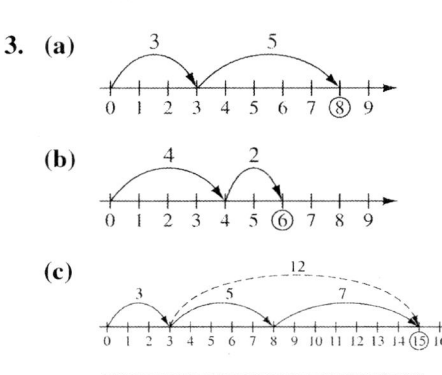

(b)

(c)

5. (a)

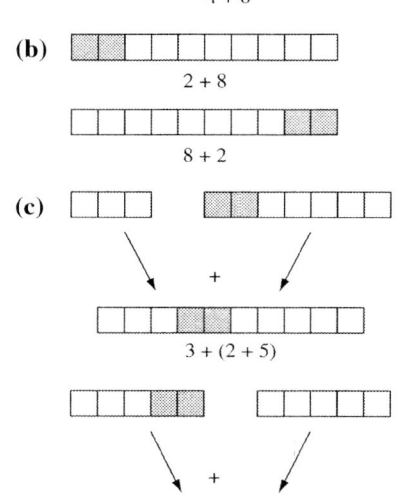

$4 + 6$

(b)

$2 + 8$

$8 + 2$

(c)

$+$

$3 + (2 + 5)$

$+$

$(3 + 2) + 5$

7. Answers will vary. One possibility is given. Town B is 18 miles due east of Town A. Town C is 25 miles due east of Town B. How far is Town C from Town A?

9. (a) Commutative property of addition

(b) Closure property

(c) Additive-identity property of zero

(d) Associative and commutative properties

(e) Associative and commutative properties

11. (a)

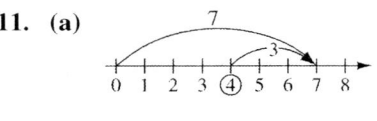

(b)

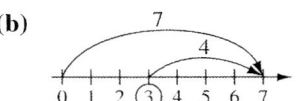

13. (a) $5 + 7 = 12$ $12 - 7 = 5$
$7 + 5 = 12$ $12 - 5 = 7$

(b) $4 + 8 = 12$ $12 - 8 = 4$
$8 + 4 = 12$ $12 - 4 = 8$

15. (a) comparison

(b) measurement (number-line)

17. Answers will vary. For example,

(a) *Take-away*: Maritza bought a booklet of 20 tickets for the amusement park rides. She used 6 tickets for the roller coaster. How many tickets does she have left?

(b) *Missing Addend*: Nancy's school has 23 proof-of-purchase coupons for graphing calculators. Her school will be entitled to a free overhead projection calculator with 40 proof-of-purchase coupons. How many more coupons do they need to get the free calculator?

(c) *Comparison:* Oak Ridge Elementary School has an enrollment of 482 students and Crest Hill School has an enrollment of 393 students. How many more students does Oak Ridge have than Crest Hill?

(d) *Measurement*: A fireman climbed up 11 rungs on a ladder, but the smoke was too thick and he came down 3 rungs. How many rungs up the ladder is the fireman?

19. Use the guess and check method. Some answers will vary.

(a) $(8 - 5) - (2 - 1) = 2$

(b) $8 - (5 - 2) - 1 = 4$

21. (a) First fill in the squares by noting that $3 - 1 = 2$, $4 - 2 = 2$, and $7 - 2 = 5$. Then complete the circles.

5	2	⑦
1	2	③

⑥   ④   ⑦

20 *Chapter 2 Sets and Whole Numbers*

(b) Use the guess and check method.

4	2	⑥
5	3	⑧

⑨ ⑤ ⑦

23. Blake has one more marble than before, and Andrea has one less than before, so now Blake has two more marbles than Andrea. Use a small number of marbles, say five each, to demonstrate what happened. After giving Blake one marble, he has 6 and Andrea has 4, and thus clearly Blake has two more marbles than Andrea.

25. Answers will vary.

27. Answers will vary but should include that given two of the minuend, subtrahend, and answer, the third can be found. Furthermore, students often struggle when problems are presented in "non-conventional" ways. (This is worth the struggle as it is the beginning of algebra.) Many elementary school students have difficulty when there is a blank in the front of the number sentence.

29. First find the top row and the left column. For example, the fourth entry in the left column must be $6 - 2 = 4$, and then the third entry in the top row is $5 - 4 = 1$, and so on. The completed table is shown below.

+	5	4	1	6	9	2	0	8	7	3
3	8	7	4	9	12	5	3	11	10	6
9	14	13	10	15	18	11	9	17	16	12
6	11	10	7	12	15	8	6	14	13	9
4	9	8	5	10	13	6	4	12	11	7
0	5	4	1	6	9	2	0	8	7	3
7	12	11	8	13	16	9	7	15	14	10
5	10	9	6	11	14	7	5	13	12	8
2	7	6	3	8	11	4	2	10	9	5
1	6	5	2	7	10	3	1	9	8	4
8	13	12	9	14	17	10	8	16	15	11

31. (a) This student is "subtracting up", which is a common mistake.

(b) Answers will vary. For example: After reminding her about place value, ask her to do a two-digit subtraction such as $73 - 35$ and reflect on the similarities between how to subtract two-digit numbers and how to subtract three-digit numbers. That could be followed with $736 - 327 = ?$ as a next level problem that is easier than the one in the problem.

33. (a) Yes, Carmen's method is correct using the associativity property and a good understanding of position value.

(b)
$$
\begin{array}{rcr}
60 + 11 &=& 71 \\
-(30 + 8) &=& -38 \\
\hline
30 + 3 &=& 33
\end{array}
$$

35. Let
$$T = \{n \in N \,|\, n \text{ is a multiple of 12 and } n \le 200\}$$
and
$$F = \{n \in N \,|\, n \text{ is a multiple of 5 and } n \le 200\}.$$
The number we need to find is
$$n(T \cup F) = n(T) + n(F) - n(T \cap F)$$
$$= 16 + 40 - 3 = 53$$

Copyright © 2015 Pearson Education, Inc.

37.

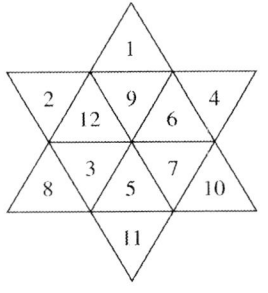

39. (a) $\{2, 3, 4, 5, 6, 7, 8, 9, \ldots\}$ since $2 + 2 = 4$, $2 + 3 = 5$, $2 + 4 = 6$, etc.

(b) $\{0, 1\}$

(c) No, because the set of all whole numbers, $\{0, 1, 2, \ldots\}$, fits the given description.

(d) Whole numbers that must be in C are all even numbers ≥ 2. Zero and odd numbers may or may not be in C.

(c) For example, choose 73 and 74. $73 = 45 + 28$, and $74 = 45 + 28 + 1$. Three is the maximum number of triangular numbers needed.

(d) Any whole number may be written as the sum of at most three triangular numbers.

41. It can be shown that Sameer is correct. The easiest way to find a sum is to use the largest Fibonacci number possible as the next summand, a procedure sometimes called a "greedy algorithm." For example, to express 100 as a Fibonacci sum, first write down 89 since it is the largest Fibonacci number less than 100. This leaves 11, so next write down 8. The 3 that still remains to be accounted for is a Fibonacci number, so $100 = 89 + 8 + 3$.

43. Statement B does not correspond to the number sentence $15 - 8 = \square$. The other problems illustrate take-away (A), comparison (C), and missing addend (D).

45. D. The number must be more than 500.

47. C. $1975 + 16 + 4 = 1995$

Section 2.4
Multiplication and Division of Whole Numbers

Problem Set 2.4

1. (a) $3 \times 5 = 15$, set model (repeated addition)

(b) $6 \times 3 = 18$, number-line model

(c) $5 \times 3 = 15$, set model

(d) $3 \times 6 = 18$, number-line model

(e) $8 \times 4 = 32$, rectangular area model

(f) $3 \times 2 = 6$, multiplication tree model

3. (a) Answers will vary.

(b) (i) $4 \cdot 9 = 9 + 9 + 9 + 9 = 36$

(ii) 7×536
$= 536 + 536 + 536 + 536 + 536$
$+ 536 + 536 = 3752$

(iii) $6 \times 47{,}819$
$= 47{,}819 + 47{,}819 + 47{,}819$
$+ 47{,}819 + 47{,}819$
$+ 47{,}819 = 286{,}914$

(iv) Using the commutative property of multiplication,
$56{,}108 \times 6 = 6 \times 56{,}108$
$= 56{,}108 + 56{,}108 + 56{,}108$
$+ 56{,}108 + 56{,}108$
$+ 56{,}108 = 336{,}648$

5. (a) Not closed. For example, $2 \times 2 = 4$, which is not in the set.

(b) Closed. $0 \times 0 = 0, 0 \times 1 = 0, 1 \times 1 = 1,$ $1 \times 0 = 0$. All products are in the set.

(c) Not closed. For example, $2 \times 4 = 8$, which is not in the set.

(d) Closed. The product of any two even whole numbers is always another even whole number.

(e) Closed. The product of any two odd whole numbers is always another odd whole number.

(f) Not closed. For example, $2 \times 2^3 = 2^4$, which is not in the set.

22 Chapter 2 Sets and Whole Numbers

(g) Closed. $2^m \times 2^n = 2^{m+n}$ for any whole numbers m and n.

(h) Closed. $7^a \times 7^b = 7^{a+b}$ for any whole numbers a and b.

7. (a) Commutative property of multiplication

(b) Distributive property of multiplication over addition

(c) Multiplication-by-zero property

9. Commutative property: $5 \times 3 = 3 \times 5$

11. (a)

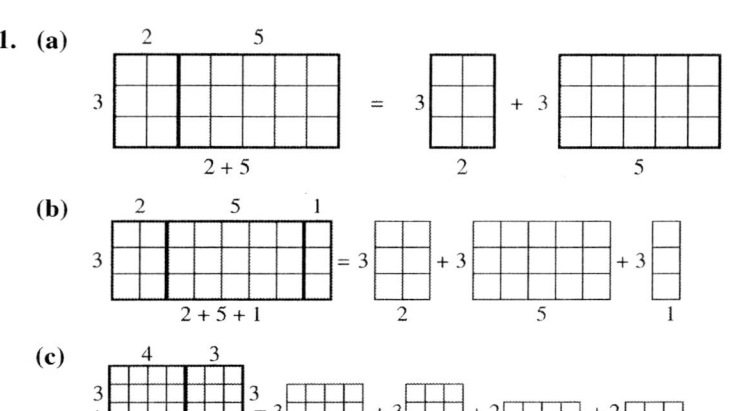

(b)

(c)

13. The rectangle is $a + b$ by $c + d$, so its area is $(a + b) \times (c + d)$. The rectangle labeled F is a by c so its area is ac. Similarly, the areas of the rectangles O, I, and L are given by the respective products ad, bc, and bd. Summing the areas of the four rectangles labeled F, O, I, and L gives the area of the large rectangle, so $(a + b) \times (c + d) = ac + ad + bc + bd$.

15. (a) Distributive property of multiplication over addition:
$$7 \cdot 19 + 3 \cdot 19 = (7 + 3) \cdot 19$$
$$= 10 \cdot 19 = 190$$

(b) Distributive property of multiplication over addition:
$$24 \cdot 17 + 24 \cdot 3 = 24 \cdot (17 + 3) = 24 \cdot 20$$
$$= 480$$

(c) Distributive property, associative property, and/or multiplication property of zero:
$$36 \cdot 15 - 12 \cdot 45 = (12 \cdot 3) \cdot 15 - 12 \cdot (3 \cdot 15) = 12 \cdot (3 \cdot 15) - 12 \cdot (3 \cdot 15)$$
$$= (12 - 12) \cdot (3 \cdot 15) = 0 \cdot (3 \cdot 15) = 0$$

17.

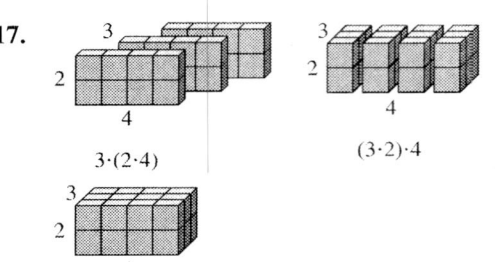

$$3 \cdot (2 \cdot 4)$$

$$(3 \cdot 2) \cdot 4$$

$$3 \cdot 2 \cdot 4$$

19. (a) $4 \times 8 = 32$, $8 \times 4 = 32$, $32 \div 8 = 4$, $32 \div 4 = 8$

(b) $6 \times 5 = 30$, $5 \times 6 = 30$, $30 \div 5 = 6$, $30 \div 6 = 5$

21. (a) Repeated subtraction

(b) Partition

(c) Missing factor or repeated subtraction

23. (a) $14 - 7 = 7$. Since $7 \le 7$, subtract 7 again to get $7 - 7 = 0$. The remainder is 0. We subtracted twice, so the quotient is 2.
$$a = 14 \xrightarrow{-7} 14 - 7 = 7 \xrightarrow{-7} 7 - 7 = 0 \text{ (done)}$$

Copyright © 2015 Pearson Education, Inc.

(b) 7 is already less than 14, so there is no subtraction. The remainder is 7 and, since there were no subtractions, the quotient is 0.

(c) Answered in each part separately.

25. (a) $y = (5 \cdot 5) + 4 = 25 + 4 = 29$

(b) $3x + 2 = 20$, so $x = 6$.

27. (a) $3^{20} \cdot 3^{15} = 3^{20+15} = 3^{35}$

(b) $(3^2)^5 = 3^{2 \cdot 5} = 3^{10}$

(c) $y^3 \cdot z^3 = (y \cdot z)^3$ or $(yz)^3$

29. (a) $8 = 2 \cdot 2 \cdot 2 = 2^3$

(b) $4 \cdot 8 = (2 \cdot 2) \cdot (2 \cdot 2 \cdot 2) = 2^5$

(c) $1024 = 2^{10}$

(d) $8^4 = (2^3)^4 = 2^{3 \cdot 4} = 2^{12}$

31. Answers will vary, but should include the fact that, if we are dividing a by b, then if $a < b$, the process is finished. If $a = b$ or $a > b$, the either $a - b = 0$ or $a - b > 0$. The remainder cannot be negative in either case because the process stops before that can happen.

33. Answers will vary. These are sample problems: Peter has a board 14 feet long, and each box he makes requires 3 feet of board. How many completed boxes can be made? Answer = 4. Tina has a collection of 14 antique dolls. A display box can hold at most three dolls. How many boxes does Tina require to display her entire collection? Answer = 5. Andrea has 14 one-by-one foot paving stones to place on her 3-foot-wide walkway. How many feet of walk can she pave, and how many stones will she have left over for a future project? Answer = 4 feet of walk paved, with 2 stones remaining.

35. As listed in the theorem, starting with the second one,
- It doesn't matter in which order you multiply two numbers.
- It doesn't matter which way you group the terms when multiplying three numbers.

- Multiplying by 1 never changes the number (and 1 is the only number for which that happens.)
- Zero times any number is zero.

37. This student struggles to see the relationship between multiplication and division exercises. In addition, students typically have more difficulty when the blank or box is at the beginning of the equation.

39. The student subtracted the number of cupcakes Nelson baked in each pan instead of dividing to find the number of pans needed to bake all of the cupcakes. This is common because students understand they need to do something with the numbers, but they aren't sure what. In this case, the student is not thinking of grouping the cupcakes into pans, which should be a signal that he should divide rather than subtract, which suggests removal.

41. The equation is
$x^2 - 3 = 33 \Rightarrow x^2 = 36 \Rightarrow x = 6$ or $x = -6$.
Since we are dealing only with whole numbers, Jing's answer is $x = 6$.

43. The operation is closed, commutative, and associative. The circle is the identity, since if it is either of the "factors" the outcome of the operation is the other factor.

45. (a) Since $2 \times 185 = 370, 500 - 370 = 130$, and $130 \div 2 = 65$, the question is "How many tickets must still be sold?'

(b) $67 \div 12 = 5 \, \text{R} \, 7$.
Thus 5 answers "How many cartons will be filled?" and 7 answers "How many eggs are in the partially filled carton?"

47. When divided by 5 the remainder is 4, so the number is in the list 4, 9, 14, 19, ..., 94, 99. When divided by 4 the remainder is 3, the list of remainders are 0, 1, 2, 3, 0, 1, 2, 3, ..., 0, 1, 2, 3, so the number we seek is one of 19, 39, 59, 79, or 99. When divided by 3 the remainders in this list of 5 numbers is 1, 0, 2, 1, 0. Thus the number we seek is 59, which we note has a remainder of 1 when divided by 2.

49. B   **51.** A   **53.** C

Copyright © 2015 Pearson Education, Inc.

24 *Chapter 2 Sets and Whole Numbers*

Chapter 2 Review Exercises

1. **(a)** $S = \{4, 9, 16, 25\}$
     $P = \{2, 3, 5, 7, 11, 13, 17, 19, 23\}$
     $T = \{2, 4, 8, 16\}$

    (b) $\overline{P} = \{4, 6, 8, 9, 10, 12, 14, 15, 16, 18,$
      $20, 21, 22, 24, 25\}$
     $S \cap T = \{4, 16\}$
     $S \cup T = \{2, 4, 8, 9, 16, 25\}$
     $S \cap \overline{T} = \{9, 25\}$

3. **(a)** $A \boxed{\subseteq} A \cup B$

    (b) If $A \subseteq B$ and A is not equal to B, then
      $A \boxed{\subset} B$.

    (c) $A \boxed{\cap} (B \cup C) = (A \cap B) \cup (A \cap C)$

    (d) $A \boxed{\cup} \varnothing = A$

5. 1 $\leftrightarrow$ a $\;\big|\;$ 36 $\leftrightarrow$ f
    4   $\leftrightarrow$ b $\;\big|\;$ 49 $\leftrightarrow$ g
    9   $\leftrightarrow$ c $\;\big|\;$ 64 $\leftrightarrow$ h
    16 $\leftrightarrow$ d $\;\big|\;$ 81 $\leftrightarrow$ i
    25 $\leftrightarrow$ e $\;\big|\;$ 100 $\leftrightarrow$ j

7.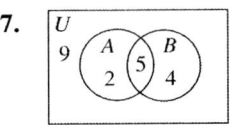

9. **(a)** Commutative property of addition:
      $7 + 3 = 3 + 7$

    (b) Additive-identity property of zero:
      $7 + 0 = 7$

11. **(a)**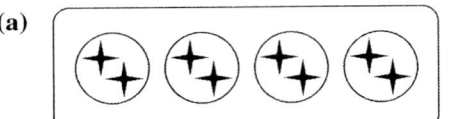

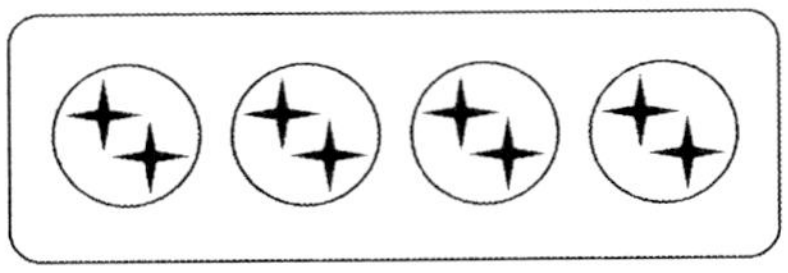

    (b)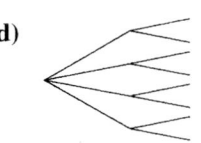

(c)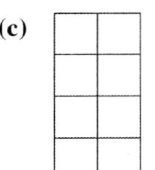

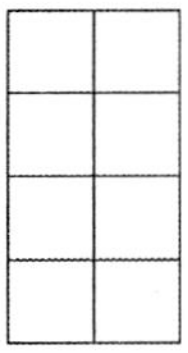

(d)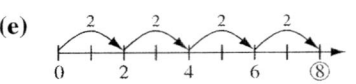

(e)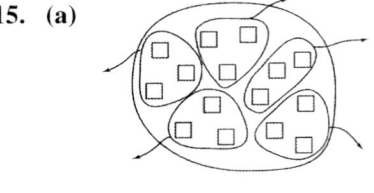

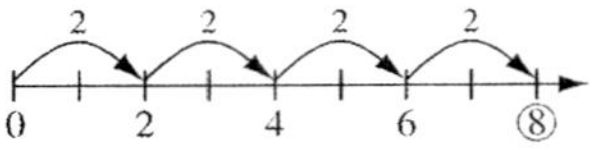

13. Answers will vary. Since $36 = 3 \times 3 \times 4$, one
     possibility is $6'' \times 6'' \times 8''$.

15. **(a)**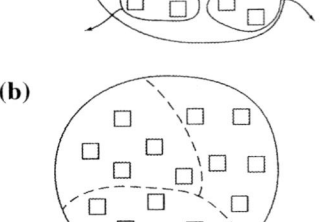

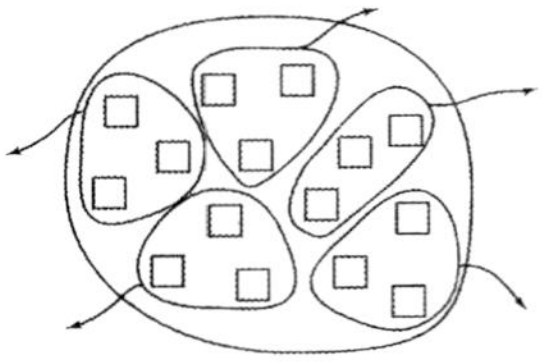

    (b)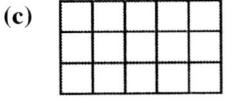

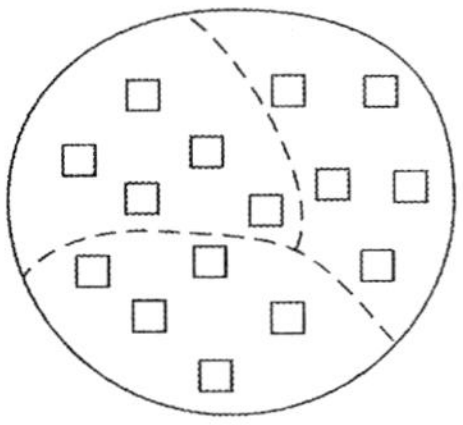

    (c)

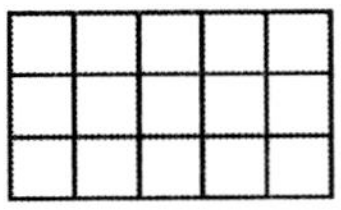

Copyright © 2015 Pearson Education, Inc.

Chapter 3 Numeration and Computation

Section 3.1
Numeration Systems Past and Present

Problem Set 3.1

1. (a) $2 \cdot 1000 + 1 \cdot 100 + 3 \cdot 10 + 7 = 2137$

(b) $1000 + 500 + 200 + 20 + 9 = 1729$

(c) $500 + 100 + 90 + 5 + 2 = 697$

(d) $1 \cdot 60^1 = 60$

(e) $1 \cdot 60^2 = 3600$

(f) $2 \cdot 18 \cdot 20^2 + 6 \cdot 18 \cdot 20 + 18 \cdot 20 + 0$
$= 2 \cdot 7200 + 6 \cdot 360 + 360 + 0$
$= 16{,}920$

3. (a) $11 = 1 \cdot 10 + 1 \cdot 1 = \cap \mid$

(b) $597 = 5 \cdot 100 + 9 \cdot 10 + 7 \cdot 1$

$$9\,9\,9 \quad \cap\cap\cap\cap\cap \quad ||||$$
$$9\,9 \quad\quad \cap\cap\cap\cap \quad |||$$

(c) $1949 = 1 \cdot 1000 + 9 \cdot 100 + 4 \cdot 10 + 9 \cdot 1$

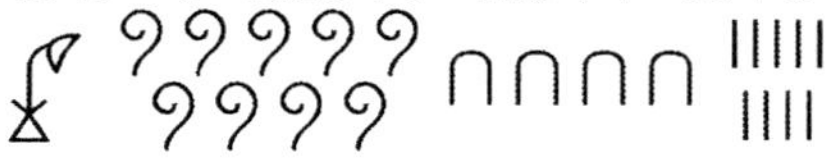

5. (a) $974 = 900 + 70 + 4$
$= (1000 - 100) + 70 + 4$
$= (1000 - 100) + 50 + 20 + (5 - 1)$
$= \text{CMLXXIV}$

(b) $2009 = 2000 + 9 = 2000 + (10 - 1)$
$= \text{MMIX}$

7. (a) $12 = 12 \cdot 1$

(b)

$$360\overline{)584} \qquad 20\overline{)224}$$

with quotients 1 and 11 respectively,

$$\frac{360}{224} \qquad \frac{220}{4}$$

$584 = 1 \cdot 360 + 224$
$= 1 \cdot 18 \cdot 20 + 11 \cdot 20 + 4 \cdot 1$

(c)

$$7200\overline{)12{,}473} \qquad 360\overline{)5273} \qquad 20\overline{)233}$$

with quotients 1, 14, and 11 respectively,

$$\frac{7\,200}{5\,273} \qquad \frac{360}{1673} \qquad \frac{220}{13}$$
$$\qquad\qquad \frac{1440}{233}$$

$12{,}473 = 1 \cdot 18 \cdot 20^2 + 14 \cdot 18 \cdot 20$
$\qquad\qquad + 11 \cdot 20 + 13 \cdot 1$

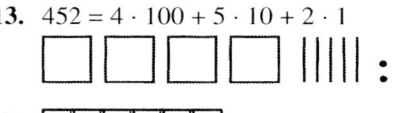

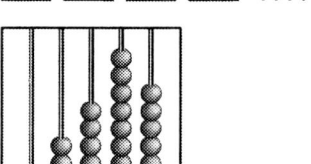

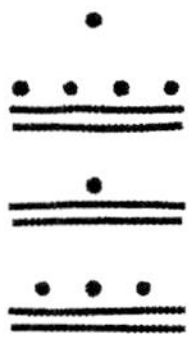

9.

M	D	CCC		X	VI
+ M			CCC	LX	IV
MM	D	CCCCCC		LXX	X

$= \text{MMDDCLXXX} = \text{MMMCLXXX}$

11. In Roman numerals 2002, 2003, and 2004 would be represented respectively by MMII, MMIII, and MMIV.

13. $452 = 4 \cdot 100 + 5 \cdot 10 + 2 \cdot 1$

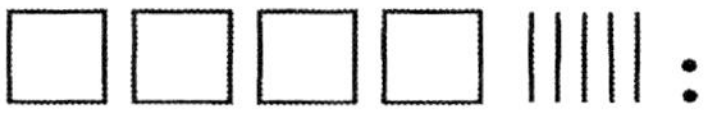

15.

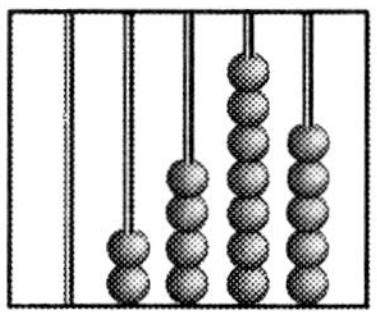

17. First trade 10 of your units for a strip to get 3 mats, 25 strips, and 3 units. Then trade 20 strips for 2 mats to get 5 mats, 5 strips, and 3 units.

19. (a) 1 mat must be exchanged for 10 strips, and then 1 strip must be exchanged for 10 units.

(b) 2 mats, 6 strips, and 8 units

(c) This represents the borrowing from the 100s column and the tens columns in this subtraction:

$$\begin{array}{r} 2\;\;\;9 \\ \not{3}\;\not{0}\,{}^1 6 \\ -\;\;\;3\;\;8 \\ \hline 2\;\;6\;\;8 \end{array}$$

Copyright © 2015 Pearson Education, Inc.

26 Chapter 3 Numeration and Computation

21. (a) Counting on, the girls determine that they have 21 dollars and 44 dimes. But since 10 dimes can be exchanged for 1 dollar, the 44 dimes are the same amount of money as 4 dollars and 4 dimes. Thus, all told they have 25 dollars and 4 dimes worth $25.40.

(b) After buying the perfume, they have 4 dimes worth $0.40.

(c) Answers may vary. See part (a) above.

23. Answers will vary.

(b) Yes. In words, the problem is to find the sum of 2 and 3 sevenths and 2 and 5 sevenths. Counting on gives 4 and 8 sevenths. But since 7 sevenths is equivalent to 1 unit, the final result is 5 and 1 seventh. That is

$$2\frac{3}{7} + 2\frac{5}{7} = 4\frac{8}{7} = 5\frac{1}{7}.$$

25. The problem involves exchanges. Every 60 minutes is one hour and every 8 hours is one workday.

27. Sara wanted to see how the units and the rod were related to each other. She though that the rod would be made out of 10 units of the same length when she cut it apart.

29. $2 \cdot 100 + 3 \cdot 10 + 6 \cdot 5 = 235$. The answer is A.

31. We must select a number between 59,343 mi^2 and 82,234 mi^2. The only choice that meets this criterion is choice D, 67,345 mi^2.

Section 3.2
Algorithms for Adding and Subtracting Whole Numbers

Problem Set 3.2

1. (a) $36 + 75 = 111$

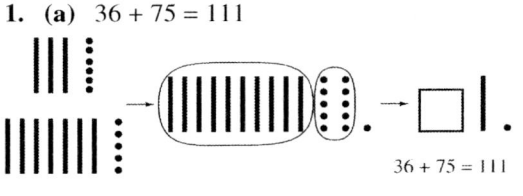

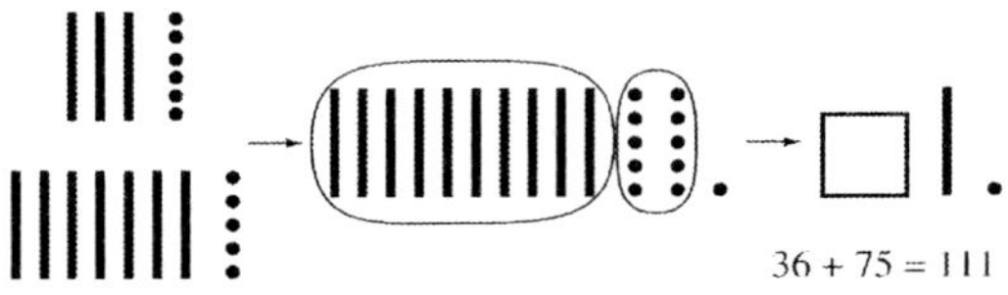

36 + 75 = 111

(b)

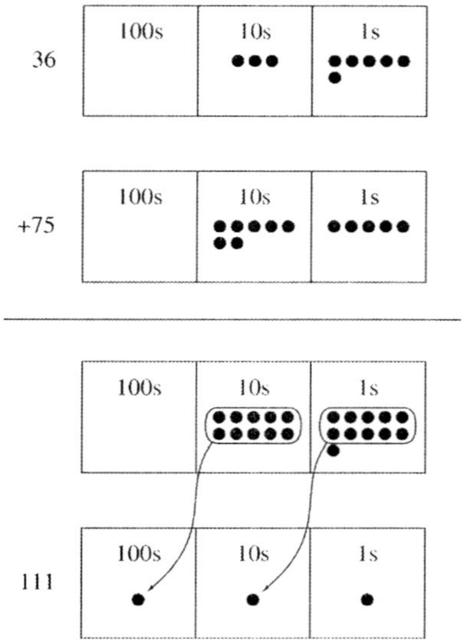

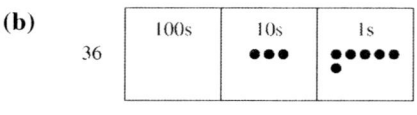

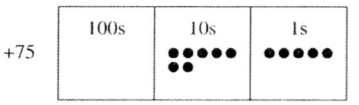

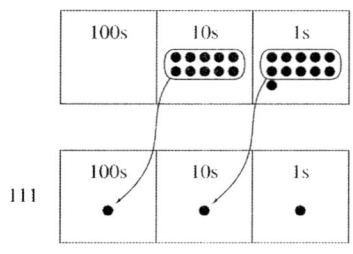

3. (a)
$$\begin{array}{r} 23 \\ +44 \\ \hline 7 \\ 60 \\ \hline 67 \end{array}$$

(b)
$$\begin{array}{r} 57 \\ +84 \\ \hline 11 \\ 130 \\ \hline 141 \end{array}$$

5. (a)
$$\begin{array}{r} 78 \\ -35 \\ \hline 43 \end{array}$$

(b)

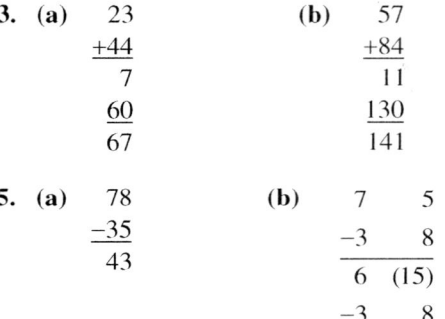

7. In these problems, we must exchange 60 seconds for one minute and 60 minutes for one hour or vice versa.

(a)   3 hours, 24 minutes, 54 seconds
     + 2 hours, 47 minutes, 38 seconds
     ‾‾‾‾‾‾‾‾‾‾‾‾‾‾‾‾‾‾‾‾‾‾‾‾‾‾‾‾‾‾‾‾‾
     5 hours, 71 minutes, 92 seconds
     = 5 hours, 72 minutes, 32 seconds
     = 6 hours, 12 minutes, 32 seconds

(b)   7 hours, 56 minutes, 29 seconds
     + 3 hours, 27 minutes, 52 seconds
     ‾‾‾‾‾‾‾‾‾‾‾‾‾‾‾‾‾‾‾‾‾‾‾‾‾‾‾‾‾‾‾‾‾
     10 hours, 83 minutes, 81 seconds
     = 10 hours, 84 minutes, 21 seconds
     = 11 hours, 24 minutes, 21 seconds

(c)   5 hours, 24 minutes, 54 seconds
     − 2 hours, 47 minutes, 38 seconds
     ‾‾‾‾‾‾‾‾‾‾‾‾‾‾‾‾‾‾‾‾‾‾‾‾‾‾‾‾‾‾‾‾‾
     4 hours, 84 minutes, 54 seconds
     − 2 hours, 47 minutes, 38 seconds
     ‾‾‾‾‾‾‾‾‾‾‾‾‾‾‾‾‾‾‾‾‾‾‾‾‾‾‾‾‾‾‾‾‾
     2 hours, 37 minutes, 16 seconds

Copyright © 2015 Pearson Education, Inc.

(d)   7 hours, 46 minutes, 29 seconds
    – 3 hours, 27 minutes, 52 seconds
    7 hours, 45 minutes, 89 seconds
    – 3 hours, 27 minutes, 52 seconds
    4 hours, 18 minutes, 37 seconds

9. Using numerals, we have 9 + 4 = 13 ones. Ten of the 1s are exchanged for one 10.
    1 + 7 + 8 = 16 tens.
    Ten of the 10s are exchanged for one 100.

11. (a) Yes, the addition is correct.

    (b) The sum of the digits in the ones column is 107. Sylvia wrote down the 7 in the units position and exchanged 100 ones for 1 hundred The sum of the digits in the hundreds column is then 19 and Sylvia wrote 9 and exchanged 10 hundreds for 1 thousand.

    (c) Answers will vary, but Sylvia's method is not the standard one.

    (d) Sylvia might have carried 10, actually 100, to the tens column.

    (e) Answers will vary. A good response to the student would be, "That's good thinking and it's certainly correct. Tell me what you would have done if the sum in the ones column had been 127 instead of 107." (You might have carried the 2, actually 20, to the tens column and the 1 to the hundreds column as before.)

13. Answers will vary.

15. (a) One possibility is 8642 and 7531. See parts (c) and (d).

    (b) One possibility is 1357 and 2468. See parts (c) and (d).

    (c) Of the eight digits, two will become "thousands," two will be "hundreds," two will be "tens", and two will be "ones". To make the sum as large as possible, the 8 and 7 should be thousands, since they are largest. Then the 6 and 5 should be hundreds, 4 and 3 should be tens, and 1 and 2 should be ones. A similar strategy can be applied to part (b).

    (d) No. For example, in part (a) 8 and 7 are the thousands digits and 6 and 5 are the hundreds digits, but it doesn't matter if the 8 is paired with the 6 or the 5. One other solution for (a) could be 8542 and 7631.

17. (a) Thomas made a common addition error in the ones place, yet was also able to "carry in his head" from the tens to the hundreds place. Perhaps he wasn't paying enough attention in the ones place.

    (b) Annabelle did not carry in either the tens or the hundreds place.

    (c) Xiao made a common addition error of forgetting to add two tens; he just carried the 1 ten and added that to 9 tens. He then wrote down 0 in the tens place and correctly carried 10 tens to become an addition 1 hundreds unit.

19. One of the 2 tens in 523 was exchanged for 10 ones, and then one of the 5 hundreds was exchanged for 10 tens.

21. Choice (a) is correct because
    $$273 - (152 + 1) = (273 - 152) - 1$$
    $$= 121 - 1 = 120$$

23. (a)   Work from right to   1 1 1
          left.   2437
                                   281
                                 + 3476
                                  6194

    (b)   Work from right to   4721
          left.   9012
                                   7193
                                 20,926

    (c) Work from right to left. Notice that the only way to get a 5 in the hundreds place of the sum is to "carry a 2," so the missing digits in the tens place of the addends must be 9s.
        3891
        2493
       +5125
       11,509

28   *Chapter 3 Numeration and Computation*

(d) Work from right to left. Notice that the only way to get a 6 in the hundreds place of the sum is to "carry a 2," so the missing digits in the tens place of the addends must be 9s.

$$\begin{array}{r} 594 \\ 6121 \\ +891 \\ \hline 7606 \end{array}$$

25. Every time the sum in a column becomes ten or more, a digit is scratched out in this column. This scratch becomes a 1 in the next column to the left, in effect exchanging 10 ones for 1 ten, or 10 tens for 1 hundred, or 10 hundreds for 1 thousand, and so on.

27. (a)
$$\begin{array}{r} 3'8'' \\ 4'2'' \\ 6'10'' \\ +5'11'' \\ \hline 18'31'' \\ =20'7'' \end{array}$$

(b)
$$\begin{array}{r} 20'7'' \\ -9'10'' \\ \hline 19'19'' \\ -9''10'' \\ \hline 10'9'' \end{array}$$

29. H. Select one start and end time, then subtract.
11 : 35 = 11 hours, 35 minutes
9 : 05 = 9 hours, 5 minutes

$$\begin{array}{r} 11 \text{ hours, } 35 \text{ minutes} \\ -9 \text{ hours, } 5 \text{ minutes} \\ \hline 2 \text{ hours, } 30 \text{ minutes} \end{array}$$

31. C, since $46 + 56 + 39 = 141$

33. $619{,}581 - 23{,}183 = 596{,}398$
$6747 + 321{,}105 = 327{,}852$

Section 3.3
Algorithms for Multiplication and Division of Whole Numbers

Problem Set 3.3

1. (a)

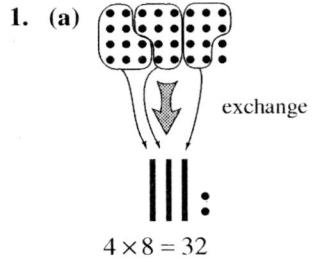

$4 \times 8 = 32$

(b)

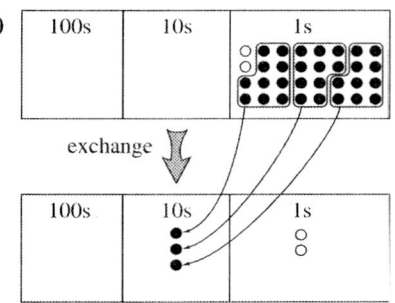

$4 \times 8 = 32$

3. (a) The red 2 represents the number of hundreds in $30 \times 70 + 100$. You can see this by thinking of the calculation 30×274. The 100 comes from $30 \times 4 = 120 = 100 + 20$.

(b) Twenty 10s are being exchanged for two 100s.

5. (a) Distributive property of multiplication over addition

(b) Commutative property of multiplication

(c) Associative property of addition

(d) Distributive property of multiplication over addition

7. $437 \div 3 = 145$ R 2
The first sketch shows 437 as 4 mats, 3 strips, and 7 units. Since 4 mats cannot be divided evenly into 3 groups, one mat is exchanged for 10 strips. Likewise, one strip is exchanged for 10 units. Then the mats, strips, and units are split into 3 groups, but there are 2 units left over. Thus, the remainder of 2.

9.
$$\begin{array}{r} 241 \\ \times\ 35 \\ \hline 5 \\ 200 \\ 1000 \\ 30 \\ 1200 \\ \underline{6000} \\ 8435 \end{array}$$

Copyright © 2015 Pearson Education, Inc.

11. (a)

$$
\begin{array}{r}
51 \ \text{R } 4 \\
17\overline{)871} \\
\underline{85} \\
21 \\
\underline{17} \\
4
\end{array}
$$

$871 = 17 \cdot 51 + 4$

(b)

$$
\begin{array}{r}
34 \ \text{R } 9 \\
21\overline{)723} \\
\underline{63} \\
93 \\
\underline{84} \\
9
\end{array}
$$

$723 = 21 \cdot 34 + 9$

13. (a)

$$
\begin{array}{r}
14 \\
4 \\
10 \\
213\overline{)3175} \\
\underline{2130} \\
1045 \\
\underline{852} \\
193
\end{array}
$$

The division checks because
$3175 = 213 \cdot 14 + 193$.

(b)

$$
\begin{array}{r}
191 \\
1 \\
90 \\
100 \\
43\overline{)8250} \\
\underline{4300} \\
3950 \\
\underline{3870} \\
80 \\
\underline{43} \\
37
\end{array}
$$

The division checks because
$8250 = 43 \cdot 191 + 37$.

15. (a) Follow these steps:

$10 \div 8 = 1 \ \text{R } 2$
$20 \div 8 = 2 \ \text{R } 4$
$49 \div 8 = 6 \ \text{R } 1$
$15 \div 8 = 1 \ \text{R } 7$

$$
\begin{array}{r}
1\ 2\ 6\ 1 \ \text{R } 7 \\
8\overline{)10^2 0^4 9^1 5}
\end{array}
$$

Check: $10{,}095 = 8 \cdot 1261 + 7$

(b) Follow these steps:

$8 \div 6 = 1 \ \text{R } 2$
$24 \div 6 = 4$
$3 \div 6 = 0 \ \text{R } 3$
$32 \div 6 = 5 \ \text{R } 2$

$$
\begin{array}{r}
1\ 4\ 0\ 5 \ \text{R } 2 \\
6\overline{)8^2 4\ 3\ 2}
\end{array}
$$

Check: $8432 = 6 \cdot 1405 + 2$

17. (a) The product is 80,410. The small numerals show the exchanges from the previous diagonal.

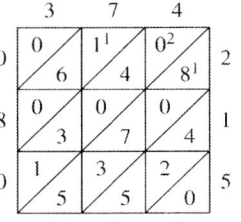

(b) Answers will vary. If this multiplication were performed using the Instructional Algorithm, it would appear as follows.

$$
\begin{array}{r}
374 \\
\times\ 215 \\
\hline
20 \\
1 \\
35 \\
2 \\
15 \\
4 \\
7 \\
3 \\
8 \\
1 \\
14 \\
6 \\
\hline
80{,}410
\end{array}
$$

The lattice algorithm is simply another form of this algorithm with the 2-digit products of the digits in the entire product placed in the appropriate box in the lattice with the units digit of the product below the diagonal and the tens digit above the diagonal. Making the appropriate "carriers" to the next diagonal is the same as making the same "carry" to the next column in Instructional Algorithm II. To connect the lattice algorithm directly to the standard algorithm, find the sums along the diagonals of each row shown. The top row gives $600 + 140 + 8 = 748$, the middle row gives $300 + 70 + 4 = 374$, and the bottom row gives $1500 + 350 + 20 = 1870$. These are the same numbers (in reverse order) that appear in the standard algorithm.

(c) Answers will vary. However, since this algorithm also depends on positional notation, studying it and comparing it with the standard algorithm should help students to more fully understand the concepts involved.

30 Chapter 3 Numeration and Computation

19. (a) The digits in the top number need to be written in decreasing order (to maximize the number of thousands, and so on). This eliminates all but five possibilities:

$$\begin{array}{r} 9753 \\ \times\ 1 \\ \hline 9753 \end{array} \qquad \begin{array}{r} 9751 \\ \times\ \ 3 \\ \hline 29,253 \end{array} \qquad \begin{array}{r} 9731 \\ \times\ \ 5 \\ \hline 48,655 \end{array}$$

$$\begin{array}{r} 9531 \\ \times\ \ 7 \\ \hline 66,717 \end{array} \qquad \boxed{\begin{array}{r} 7531 \\ \times\ \ 9 \\ \hline 67,779 \end{array}}$$

The largest product is 7531×9.

(b) Again, the digits in each number need to be written in decreasing order. There are ten possibilities.

$$\begin{array}{r} 531 \\ \times\ 97 \\ \hline 51,507 \end{array} \quad \begin{array}{r} 731 \\ \times\ 95 \\ \hline 69,445 \end{array} \quad \boxed{\begin{array}{r} 751 \\ \times\ 93 \\ \hline 69,843 \end{array}} \quad \begin{array}{r} 753 \\ \times\ 91 \\ \hline 68,523 \end{array}$$

$$\begin{array}{r} 931 \\ \times\ 75 \\ \hline 69,825 \end{array} \quad \begin{array}{r} 951 \\ \times\ 73 \\ \hline 69,423 \end{array} \quad \begin{array}{r} 953 \\ \times\ 71 \\ \hline 67,663 \end{array}$$

$$\begin{array}{r} 971 \\ \times\ 53 \\ \hline 51,463 \end{array} \quad \begin{array}{r} 973 \\ \times\ 51 \\ \hline 49,623 \end{array} \quad \begin{array}{r} 975 \\ \times\ 31 \\ \hline 30,225 \end{array}$$

The largest product is 751×93.

21. Henry did the problem correctly with conceptual understanding.

23. (a) In this division, the student ignored the remainder in each place value (i.e., $8 \div 3$ is 2 rather than 2 R 2 in the hundreds place.)

(b) In this division, the student understood the need to bring down the next digit, but instead she brought both digits down and then couldn't solve the division. There is an awareness of the process, but obviously the student is far away from a conceptual understanding of the steps.

(c) In this division, the student ignored the zero in the dividend. Students will often ignore zeros and assume that no matter where the zeros are, they have no value. You could also ask the student to estimate $806 \div 3$ to show that a number close to 28 (28 R 2) can't be correct.

25. Yes, Marsha's algorithm is correct. She has done the multiplication in the reverse order from the usual order.

27. (a) They are the same because of the associative property of multiplication. $34 \cdot 54 = (17 \cdot 2) \cdot 54 = 17 \cdot (2 \cdot 54) = 17 \cdot 108$ since 2 evenly divides 34.

(b) Explanations may vary. They differ by 108 because
$$\begin{aligned} 17 \cdot 108 &= (8 \cdot 2 + 1) \cdot 108 \\ &= (8 \cdot 2) \cdot 108 + 1 \cdot 108 \\ &= 8 \cdot 216 + 108. \end{aligned}$$
This happens because 17 is odd, and remainders are ignored.

(c) 2 evenly divides 8, 4, and 2, so the associative property can be used as in part (a).

(d) When the number in the left column is even, the product is unchanged by halving the number on the left and doubling the number on the right. When the number on the left is odd, the product of the numbers in the next line will be reduced by the number in the right column, as explained in part (b). By adding these "shortages" to the bottom number (the one opposite 1), we recover the original product.

(e)

29	81	11	243
~~14~~	~~162~~	5	486
7	324	~~2~~	~~972~~
3	648	1	1944
1	1296		2673
	2349		

29. (a) $48 + 60 + 60 + 35 = 203$
She practiced 203 minutes.

(b) $203 - 3 \cdot 60 = 23$
Melody practiced for 3 hours and 23 minutes.

31. Lori's calculator is adding 10 to every answer. If she enters 9×6, her calculator will show 64. Lori must subtract 10 to get the correct answer of 54.

33. B, since $32 \div 4 = 8$.

Copyright © 2015 Pearson Education, Inc.

Section 3.4 Mental Arithmetic and Estimation

Problem Set 3.4

1. Thought process may vary.

   (a)
   $$7 + 11 + 5 + 3 + 9 + 16 + 4 + 3$$

   10, 30, 50, 55, 58

   (b)
   $$6 + 9 + 17 + 5 + 8 + 12 + 3 + 6$$

   15, 20, 40, 60, 66

   (c) $27 + 42 + 23$

   50, 90, 92

3. Thought process may vary.

   (a) $78 + 64 = 80 + 62 = 140 + 2 = 142$

   (b) $294 + 177 = 300 + 171 = 400 + 71 = 471$

   (c) $306 - 168 = 308 - 170 = 300 - 170 + 8$
   $= 130 + 8 = 138$

   (d) $294 - 102 = 292 - 100 = 192$

5. Thought process may vary.

   (a) $425 + 362$
   $= (400 + 20 + 5) + (300 + 60 + 2)$
   $= (400 + 300) + (20 + 60) + (5 + 2)$
   $= 700 + 80 + 7 = 787$
   700, 780, 787

   (b) $363 + 274$
   $= (300 + 60 + 3) + (200 + 70 + 4)$
   $= (300 + 200) + (60 + 70) + (3 + 4)$
   $= 500 + 130 + 7 = 637$
   500, 630, 637

   (c) $572 - 251$
   $= (500 + 70 + 2) - (200 + 50 + 1)$
   $= (500 - 200) + (70 - 50) + (2 - 1)$
   $= 300 + 20 + 1 = 321$
   300, 320, 321

7. (a) $425 + 362$
   Ignore the second and third digits and first think of $400 + 300 = 700$. Since there are 2 tens in the first number and 6 tens in the second number, adjust the estimate by adding 8 tens, or about 100. Thus, the estimate is $700 + 100 = 800$.

   (b) $363 + 274$
   Ignore the second and third digits and first think of $300 + 200 = 500$. Since there are 6 tens in the first number and 7 tens in the second number, adjust the estimate by adding 13 tens, or about 100. Thus, the estimate is $500 + 100 = 600$.

   (c) $572 - 251$
   Ignore the second and third digits and first think of $500 - 200 = 300$. Since there are 5 tens in the second number which can be subtracted from 7 tens in the first number, no adjustment is needed. The estimate is 300.

   (d) $764 - 282$
   Ignore the second and third digits and first think of $700 - 200 = 500$. Since there are 8 tens in the second number which must be subtracted from 6 tens in the first number, no adjustment is needed. Thus, the estimate is 500.

   (e) No, because the problems in number 5 are calculated while these problems were estimated.

9. (a) Round down because there is a 4 in the tens place. 900

   (b) Round up because there is a 5 in the tens place. 900

   (c) Round down because there is a 4 in the hundred thousands place. 27,000,000

   (d) Round down because there is a 4 in the hundreds place. 2000

11. (a) 4,340,000   (b) 25,000,000

13. (a) $17,000 + 7000 + 12,000 + 2000 + 14,000$
    $= 52,000$

    (b) $3000 + 4000 + 2000 + 3000 + 7000$
    $= 19,000$

    (c) $4000 - 1000 = 3000$

15. (a) $3000 \cdot 30 = 90,000$

    (b) $5000 \cdot 300 = 1,500,000$

    (c) $20,000 \cdot 400 = 8,000,000$

    (d) Use a calculator.

       (a) $2748 \cdot 31 = 85,188$

       (b) $4781 \cdot 342 = 1,635,102$

Copyright © 2015 Pearson Education, Inc.

32 *Chapter 3 Numeration and Computation*

(c) $23,247 \cdot 357 = 8,299,179$

17. (a) Not a particularly good estimate, since the last three digits of each number name a number just a bit less than 500. Thus, rounding to the left-most digit will give an estimate that is almost 2000 too small.

(b) Suggest rounding to the nearest hundred to obtain 10,900.

(c) 10,846

19. Answers will vary.

21. Diley is solving her problems by adding first and then rounding, which defeats the purpose of the estimation process. She should be asked to round off each of the addends and then add for estimation by rounding.

23. Students will often round the tens digit, but leave the one's digit unchanged as Raphael did.

25. Ask them which digit is the tens digit. Then, on receiving the answer that it is a 1, say that the two competitors for an answer are 1 or 2 in the tens place. Since the hundreds digit does not change, the diagram should be

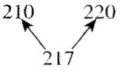

27. Theresa. Using rounding to the left-most digit, we obtain an approximate answer of $400 + 500 = 900$. Thus, Theresa is probably correct.

29. (a) Since $3 \cdot 7 = 21$, the last digit should be 1. 27,451 is correct.

(b) The last digit of a product must be the same as the last digit of the product of the last digits of the original numbers. This is easily seen by thinking about the standard multiplication algorithm.

31. (a) Answers will vary, but one possibility is 120, 235, 340, 420, (rounding to the nearest 5000). The combined area is about 420,000 square miles. Other estimates should be near this.

(b) $82,168 + 103,730 + 121,365 + 113,642 = 420,905$

33. C. Round 599 to 600 and 218 to 200.
$600 - 200$

35. $81 \div 9 = 9$. About 820 divided by 9 must be about 90.

37. B

Section 3.5
Nondecimal Positional Systems

Problem Set 3.5

1.

Base Ten	Base 5
0	0
1	1
2	2
3	3
4	4
$5 = 1 \cdot 5 + 0$	10
$6 = 1 \cdot 5 + 1$	11
$7 = 1 \cdot 5 + 2$	12
$8 = 1 \cdot 5 + 3$	13
$9 = 1 \cdot 5 + 4$	14
$10 = 2 \cdot 5 + 0$	20
$11 = 2 \cdot 5 + 1$	21
$12 = 2 \cdot 5 + 2$	22
$13 = 2 \cdot 5 + 3$	23
$14 = 2 \cdot 5 + 4$	24
$15 = 3 \cdot 5 + 0$	30
$16 = 3 \cdot 5 + 1$	31
$17 = 3 \cdot 5 + 2$	32
$18 = 3 \cdot 5 + 3$	33
$19 = 3 \cdot 5 + 4$	34
$20 = 4 \cdot 5 + 0$	40
$21 = 4 \cdot 5 + 1$	41
$22 = 4 \cdot 5 + 2$	42
$23 = 4 \cdot 5 + 3$	43
$24 = 4 \cdot 5 + 4$	44
$25 = 1 \cdot 5^2 + 0 \cdot 5 + 0$	100

Copyright © 2015 Pearson Education, Inc.

Solutions to Problem Set 3.5 **33**

3. Only the digits 0 through 5 are used. There are 6 columns, and in each column the representations have the same right-most digit and the left-most digit increases from 0 to 5. Also, the left digits are the same in any row.

5. (a) $413_{\text{five}} = 4 \cdot 5^2 + 1 \cdot 5 + 3 = 108$

   (b) $2004_{\text{five}} = 2 \cdot 5^3 + 0 \cdot 5^2 + 0 \cdot 5 + 4 = 254$

   (c) $10_{\text{five}} = 1 \cdot 5 + 0 = 5$

7. (a) $2 \cdot 5 + 4 = 14$　　(b) $1 \cdot 5 = 5$

   (c) $2 \cdot 5 + 3 = 13$

9. (a) $4 \cdot 5^2 + 1 \cdot 5 + 3 = 108$

   (b) 413

   (c)

11. (a) $362 = 2 \cdot 125 + 112$
    $$= 2 \cdot 125 + 4 \cdot 25 + 12$$
    $$= 2 \cdot 125 + 4 \cdot 25 + 2 \cdot 5 + 2$$
    $$= 2422_{\text{five}}$$

    (b) $27 = 1 \cdot 25 + 0 \cdot 5 + 2 = 102_{\text{five}}$

    (c) $5 = 1 \cdot 5 + 0 = 10_{\text{five}}$

    (d) $25 = 1 \cdot 25 + 0 \cdot 5 + 0 = 100_{\text{five}}$

13. (a) $41_{\text{five}} + 14_{\text{five}} = (4 \cdot 5 + 1) + (1 \cdot 5 + 4)$
    $$= 1 \cdot 5^2 + 1 \cdot 5 + 0 = 110_{\text{five}}$$
    $$= 25_{\text{ten}} + 5_{\text{ten}} = 30_{\text{ten}}$$
    Check:
    $$41_{\text{five}} + 14_{\text{five}} = (4 \cdot 5 + 1) + (1 \cdot 5 + 4)$$
    $$= 21_{\text{ten}} + 9_{\text{ten}} = 30_{\text{ten}}$$

    (b) $213_{\text{five}} + 432_{\text{five}}$
    $$= (2 \cdot 5^2 + 1 \cdot 5 + 3) + (4 \cdot 5^2 + 3 \cdot 5 + 2)$$
    $$= 6 \cdot 5^2 + 4 \cdot 5 + 5 = 6 \cdot 5^2 + 5 \cdot 5$$
    $$= 1 \cdot 5^3 + 2 \cdot 5^2 = 1200_{\text{five}}$$
    $$= 125_{\text{ten}} + 50_{\text{ten}} = 175_{\text{ten}}$$

Check:
$$213_{\text{five}} + 432_{\text{five}}$$
$$= (2 \cdot 5^2 + 1 \cdot 5 + 3) + (4 \cdot 5^2 + 3 \cdot 5 + 2)$$
$$= 58_{\text{ten}} + 117_{\text{ten}} = 175_{\text{ten}}$$

15. (a) $23_{\text{six}} + 35_{\text{six}} = (2 \cdot 6 + 3) + (3 \cdot 6 + 5)$
    $$= (5 \cdot 6) + (6 + 2) = (6 \cdot 6) + 2$$
    $$= 6^2 + 2 = 102_{\text{six}}$$
    $$= 36_{\text{ten}} + 2_{\text{ten}} = 38_{\text{ten}}$$
    Check:
    $$23_{\text{six}} + 35_{\text{six}} = (2 \cdot 6 + 3) + (3 \cdot 6 + 5)$$
    $$= 15_{\text{ten}} + 23_{\text{ten}} = 38_{\text{ten}}$$

    (b) $423_{\text{six}} + 43_{\text{six}} = (4 \cdot 6^2 + 2 \cdot 6 + 3) + (4 \cdot 6 + 3)$
    $$= 4 \cdot 6^2 + 6 \cdot 6 + 6$$
    $$= 5 \cdot 6^2 + 1 \cdot 6 + 0 = 510_{\text{six}}$$
    $$= 180_{\text{ten}} + 6_{\text{ten}} = 186_{\text{ten}}$$
    Check:
    $$423_{\text{six}} + 43_{\text{six}} = (4 \cdot 6^2 + 2 \cdot 6 + 3) + (4 \cdot 6 + 3)$$
    $$= 159_{\text{ten}} + 27_{\text{ten}} = 186_{\text{ten}}$$

17. (a)
    $$\begin{array}{r} {}^{3}\cancel{4}\,{}^{1}1_{\text{five}} \\ -\ 1\ \ 4_{\text{five}} \\ \hline 2\ \ 2_{\text{five}} \end{array}$$
    Check:
    $$41_{\text{five}} - 14_{\text{five}} = (4 \cdot 5 + 1) - (1 \cdot 5 + 4)$$
    $$= 12_{\text{ten}}$$
    $$22_{\text{five}} = 2 \cdot 5 + 2 = 12_{\text{ten}}$$

    (b)
    $$\begin{array}{r} {}^{1}\cancel{2}\,{}^{1}1\ 3_{\text{five}} \\ -\ \ \ 3\ \ 2_{\text{five}} \\ \hline 1\ \ 3\ \ 1_{\text{five}} \end{array}$$
    Check:
    $$213_{\text{five}} - 32_{\text{five}}$$
    $$= (2 \cdot 5^2 + 1 \cdot 5 + 3) - (3 \cdot 5 + 2)$$
    $$= 53_{\text{ten}} - 13_{\text{ten}} = 41_{\text{ten}}$$
    $$131_{\text{five}} = 1 \cdot 5^2 + 3 \cdot 5 + 1 = 41_{\text{ten}}$$

Copyright © 2015 Pearson Education, Inc.

34 *Chapter 3 Numeration and Computation*

19. (a)
$$\overset{4}{\cancel{5}}\,\overset{1}{0}5_{six}$$
$$-\quad 35_{six}$$
$$\overline{\quad 4\ 30_{six}}$$

Check:
$$505_{six} - 35_{six}$$
$$= \left(5\cdot 6^2 + 5\right) - (3\cdot 6 + 5) = 162_{ten}$$
$$430_{six} = 4\cdot 6^2 + 3\cdot 6 = 162_{ten}$$

(b)
$$\overset{3}{\cancel{4}}\,\overset{1}{2}3_{six}$$
$$-\quad 43_{six}$$
$$\overline{\quad 3\ 40_{six}}$$

Check:
$$423_{six} - 43_{six}$$
$$= \left(4\cdot 6^2 + 2\cdot 6 + 3\right) - (4\cdot 6 + 3) = 132_{ten}$$
$$340_{six} = 3\cdot 6^2 + 4\cdot 6 = 132_{ten}$$

21. (a) Reason as follows: $3\times 3 = 9_{ten} = 14_{five}$, so write 4 and exchange 5 ones for 1 five. Then $3\times 2 + 1 = 7_{ten} = 12_{five}$.

$$\begin{array}{r} \overset{1}{} \\ 23_{five} \\ \times 3_{five} \\ \hline 124_{five} \end{array}$$
Check:
$$\begin{array}{r} 13_{ten} \\ \times 3_{ten} \\ \hline 39_{ten} \end{array}$$

(b)
$$\begin{array}{r} {}_{21} \\ 432_{five} \\ \times 41_{five} \\ \hline 432 \\ 33\ 33 \\ \hline 34,312_{five} \end{array}$$
Check:
$$\begin{array}{r} 117_{ten} \\ \times 21_{ten} \\ \hline 117 \\ 234 \\ \hline 2457_{ten} \end{array}$$

(c)
$$\begin{array}{r} {}^{1} \\ 2013_{five} \\ \times 23_{five} \\ \hline 11\ 044 \\ 40\ 31 \\ \hline 101,404_{five} \end{array}$$
Check:
$$\begin{array}{r} 258_{ten} \\ \times 13_{ten} \\ \hline 774 \\ 258 \\ \hline 3354_{ten} \end{array}$$

(d) See "checks" shown above.

23. (a) A positional system of base twenty.

(b) Answers will vary. Suggest that they envision a classroom abacus with twenty beads per wire.

Chapter 3 Review Exercises

1. (a) $2000 + 300 + 50 + 3 = 2353$

(b) $8\cdot 7200 + 2\cdot 360 + 0\cdot 20 + 11 = 58{,}331$

(c) $1000 + (1000 - 100) + (100 - 10) + 5 + 3$
$= 1000 + 900 + 90 + 5 + 3 = 1998$

3. Exchange 30 units for 3 strips, then exchange all 30 strips for 3 mats. The result is 8 mats, 0 strips, and 2 units.

5. (a)
$$\begin{array}{r} 42 \\ + 54 \\ \hline 6 \\ 90 \\ \hline 96 \end{array}$$
(b)
$$\begin{array}{r} 47 \\ + 35 \\ \hline 12 \\ 70 \\ \hline 82 \end{array}$$

(c)
$$\begin{array}{r} 59 \\ + 63 \\ \hline 12 \\ 110 \\ \hline 122 \end{array}$$

7. (a)
$$\begin{array}{r} 357 \\ \times\ 4 \\ \hline 28 \\ 200 \\ 1200 \\ \hline 1428 \end{array}$$
(b)
$$\begin{array}{r} 642 \\ \times\ 27 \\ \hline 14 \\ 280 \\ 4\ 200 \\ 40 \\ 800 \\ 12{,}000 \\ \hline 17{,}334 \end{array}$$

9. (a)
$$5\overline{\smash{)}2\ 7^2 4^4 3^3 6} = 5487\ \text{R1}$$

(b)
$$8\overline{\smash{)}3\ 9^7 5^3 8^6 4} = 4948\ \text{R0}$$

11. 657 rounds to 700, 439 rounds to 400, 1657 rounds to 2000 and 23 rounds to 20. Thus,

(a) $657 + 439$ is approximately $700 + 400 = 1100$. The actual sum is 1096.

(b) $657 - 439$ is approximately $700 - 400 = 300$. The actual answer is 218.

Copyright © 2015 Pearson Education, Inc.

Solutions to Chapter 3 Review Exercises **35**

(c) $657 \cdot 439$ is approximately
$700 \cdot 400 = 280{,}000.$
The actual answer is 288,423.

(d) $1657 \div 23$ is approximately
$2000 \div 20 = 100.$ The actual answer is
approximately 72.04.

13. a.
$$\begin{array}{r} \overset{2}{\not{1}} \\[-2pt] 2\,3_{\text{five}} \\ \times\ 42_{\text{five}} \\ \hline 101 \\ 202 \\ \hline 2121_{\text{five}} \end{array}$$

(b)
$$\begin{array}{r} \overset{2}{\not{1}}\ {}_{1} \\[-2pt] 2\,4\,1\,3_{\text{five}} \\ \times\ \ \ 332_{\text{five}} \\ \hline 10331 \\ 13244 \\ 13244 \\ \hline 2023221_{\text{five}} \end{array}$$

Copyright © 2015 Pearson Education, Inc.

Chapter 4 Number Theory

Problem Set 4.1

1. (a)

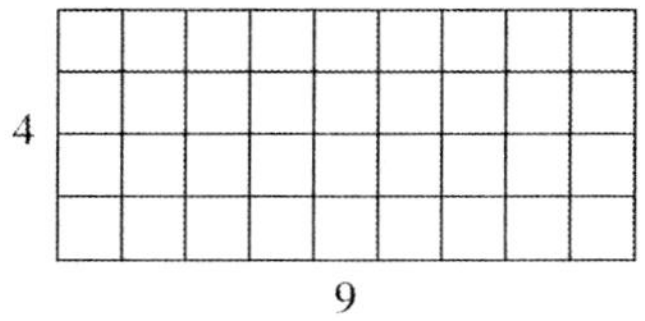

$36 = 4 \cdot 9$

4 divides 36.

(b)

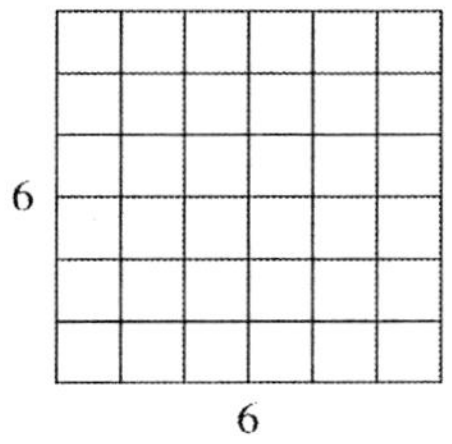

$36 = 6 \cdot 6$

6 divides 36.

3. (a) $1 \cdot 8 = 8, 2 \cdot 8 = 16, 3 \cdot 8 = 24, 4 \cdot 8 = 32,$
$5 \cdot 8 = 40, 6 \cdot 8 = 48, 7 \cdot 8 = 56,$
$8 \cdot 8 = 64, 9 \cdot 8 = 72, 10 \cdot 8 = 80$

(b) $1 \cdot 6 = 6, 2 \cdot 6 = 12, 3 \cdot 6 = 18, 4 \cdot 6 = 24,$
$5 \cdot 6 = 30, 6 \cdot 6 = 36, 7 \cdot 6 = 42,$
$8 \cdot 6 = 48, 9 \cdot 6 = 54, 10 \cdot 6 = 60$

(c) Both 24 and 48 are natural numbers that are multiples of 6 and 8, but 24 is the least natural number that is a multiple of both.

5. Using proof by contradiction: Consider any product ab, where at least one of the factors, say a, is even. There is then a natural number k so that $a = 2k$. But then $ab = (2k)b = 2(kb)$, so 2 is a factor of ab which shows ab is even. Thus, if ab is odd, then both a and b must be odd.

7. Let b be odd, so $a = 2j + 1$ for some natural number j. Then
$$b^2 = (2j+1)^2 = 4j^2 + 4j + 1 = 4(j^2 + j) + 1,$$
so b^2 divided by 4 has a quotient of $j^2 + j$ and a remainder of 1.

9. Divide 18 by each factor to find the corresponding quotient.

Factors of 18	1	2	3	6	9	18
Corresponding Quotients	18	9	6	3	2	1

11. (a)

$$\begin{array}{r} 5 \\ 5\overline{)25} \\ 2\overline{)50} \\ 2\overline{)100} \\ 7\overline{)700} \end{array}$$
$700 = 2 \cdot 2 \cdot 5 \cdot 5 \cdot 7$

(b)

$$\begin{array}{r} 11 \\ 3\overline{)33} \\ 3\overline{)99} \\ 2\overline{)198} \end{array}$$
$198 = 2 \cdot 3 \cdot 3 \cdot 11$

(c)

$$\begin{array}{r} 2 \\ 3\overline{)6} \\ 3\overline{)18} \\ 5\overline{)90} \\ 5\overline{)450} \end{array}$$
$450 = 5 \cdot 5 \cdot 3 \cdot 3 \cdot 2$

(d)

$$\begin{array}{r} 11 \\ 3\overline{)33} \\ 2\overline{)66} \\ 2\overline{)132} \\ 2\overline{)264} \\ 2\overline{)528} \end{array}$$
$528 = 2 \cdot 2 \cdot 2 \cdot 2 \cdot 3 \cdot 11$

13. (a) $136 = 2^3 \cdot 17^1$, $102 = 2^1 \cdot 3^1 \cdot 17^1$

(b) The divisors of 136 are
$2^0 \cdot 17^0 = 1$, $2^1 \cdot 17^0 = 2$, $2^2 \cdot 17^0 = 4$,
$2^3 \cdot 17^0 = 8$, $2^0 \cdot 17^1 = 17$, $2^1 \cdot 17^1 = 34$,
$2^2 \cdot 17^1 = 68$, $2^3 \cdot 17^1 = 136$.

(c) The divisors of 102 are
$2^0 \cdot 3^0 \cdot 17^0 = 1$, $2^0 \cdot 3^1 \cdot 17^0 = 3$,
$2^0 \cdot 3^0 \cdot 17^1 = 17$, $2^0 \cdot 3^1 \cdot 17^1 = 51$,
$2^1 \cdot 3^0 \cdot 17^0 = 2$, $2^1 \cdot 3^1 \cdot 17^0 = 6$,
$2^1 \cdot 3^0 \cdot 17^1 = 34$, $2^1 \cdot 3^1 \cdot 17^1 = 102$.

(d) From parts (b) and (c), we see that the greatest divisor of 136 and 102 is 34.

15. (a) $91 = 7 \cdot 13$

(b) $6125 = 5^3 \cdot 7^2$

(c) $23,000 = 2^3 \cdot 5^3 \cdot 23$

17. (a) Yes. $28 = 2^2 \cdot 7^1$. Since 2^3 and 7^2 appear in the factorization of a, all the factors of 28 appear in a and to at least as high a power.

(b) No. Since 3^2 is a factor of 126, the prime 3 appears to a higher power in 126 than it does in a.

(c) The exponent on 2 is $3 - 2 = 1$, since 2^3 appears in a and 2^2 appears in b. Likewise, the exponent on 3 is $1 - 1 = 0$, and the exponent on 7 is 2 (unchanged from a). Thus

$$\frac{a}{b} = 2^1 \cdot 3^0 \cdot 7^2 = 2^1 \cdot 7^2 = 98.$$

This may be written as

$$\frac{a}{b} = \frac{2^3 \cdot 3^1 \cdot 7^2}{2^2 \cdot 3^1} = 2^{3-2} \cdot 3^{1-1} \cdot 7^2 = 2^1 \cdot 7^2.$$

(d) 24. There are 4 ways to choose the power of 2 (0, 1, 2, or 3), 2 ways to choose the power of 3 (0 or 1), and 3 ways to choose the power of 7 (0, 1, or 2). The number of factors is $4 \cdot 2 \cdot 3 = 24$.

(e) $2^0 \cdot 3^0 \cdot 7^0 = 1$, $2^0 \cdot 3^0 \cdot 7^1 = 7$, $2^0 \cdot 3^0 \cdot 7^2 = 49$, $2^0 \cdot 3^1 \cdot 7^0 = 3$, $2^0 \cdot 3^1 \cdot 7^1 = 21$, $2^0 \cdot 3^1 \cdot 7^2 = 147$, $2^1 \cdot 3^0 \cdot 7^0 = 2$, $2^1 \cdot 3^0 \cdot 7^1 = 14$, $2^1 \cdot 3^0 \cdot 7^2 = 98$, $2^1 \cdot 3^1 \cdot 7^0 = 6$, $2^1 \cdot 3^1 \cdot 7^1 = 42$, $2^1 \cdot 3^1 \cdot 7^2 = 294$, $2^2 \cdot 3^0 \cdot 7^0 = 4$, $2^2 \cdot 3^0 \cdot 7^1 = 28$, $2^2 \cdot 3^0 \cdot 7^2 = 196$, $2^2 \cdot 3^1 \cdot 7^0 = 12$, $2^2 \cdot 3^1 \cdot 7^1 = 84$, $2^2 \cdot 3^1 \cdot 7^2 = 588$, $2^3 \cdot 3^0 \cdot 7^0 = 8$, $2^3 \cdot 3^0 \cdot 7^1 = 56$, $2^3 \cdot 3^0 \cdot 7^2 = 392$, $2^3 \cdot 3^1 \cdot 7^0 = 24$, $2^3 \cdot 3^1 \cdot 7^1 = 168$, $2^3 \cdot 3^1 \cdot 7^2 = 1176$ The factors in order are 1, 2, 3, 4, 6, 7, 8, 12, 14, 21, 24, 28, 42, 49, 56, 84, 98, 147, 168, 196, 294, 392, 588, 1176.

19. (a) $\sqrt{271} = 16.46\ldots$, so check if any of 2, 3, 5, 7, 11, 13 divide 271. None do, so 271 is prime.

(b) $\sqrt{319} = 17.86\ldots$, so check if any of 2, 3, 5, 7, 11, 13, 17 divide 319. Since 11 is a divisor, 319 is composite. Its prime factorization is $11 \cdot 29$.

(c) $\sqrt{731} = 27.03\ldots$, so check if any of 2, 3, 5, 7, 11, 13, 17, 19, 23 divide 731. Since 17 is a divisor, 731 is composite. Its prime factorization is $17 \cdot 43$.

(d) $\sqrt{1801} = 42.43\ldots$, so check if any of 2, 3, 5, 7, 11, 13, 17, 19, 23, 29, 31, 37, 41 divide 1801. None do, so 1801 is prime.

21. Simply show them a counterexample. For example, $34 = 2 \cdot 17$ with 2 and 17 both primes. Yet $17 > \sqrt{34}$.

23. Answers will vary. The prime factorization of ab is the product of the prime factorizations of a and b. This means that if p is a prime factor of ab, the prime p must have also been a factor of at least one of a or b. This means that p is necessarily a factor of a or b, or possible both a and b.

25. Point out that 90, when divided by 3, has the quotient 30, and 30 is also divisible by 3. Thus, $90 = 3 \times 3 \times 10$, and therefore, $90 = 3 \times 3 \times 2 \times 5$ is the correct prime factorization.

27. Point out that zero is "something", namely an especially important member of the whole number system. Since $0 = 0 \times 2$, point out that 0 leaves a 0 remainder when divided by 2, which by definition tells us that 0 is an even whole number.

29. (a) 1, 3, $3^2 = 9$

(b) $5^2 = 25$ with factors 1, 5, and 25; $7^2 = 49$ with factors 1, 7, and 49; $11^2 = 121$ with factors 1, 11, and 121.

(c) $3^3 = 27$ with factors 1, 3, 9, and 27; $5^3 = 125$ with factors 1, 5, 25, and 125; $7^3 = 343$ with factors 1, 7, 49, and 343.

38 *Chapter 4 Number Theory*

31. If p divides n then p appears in the prime factorization of n and similarly for q. But p and q are different so they both appear in the prime factorization of n, so then pq divides n.

33. (a) The 16 by 13 rectangle has area $16 \times 13 = 208$ square units, so it requires 104 dominoes each of area 2.

(b) Since 208 is not divisible by 3, the rectangle cannot be tiled by triominoes of area 3 square units.

35. (a) F_{999} is even, and F_{1000} is odd.

(b) The Fibonacci number F_n is even if, and only if, n is divisible by 3.

37. (a)
$$P_{13} = P_{10} + P_{11} = 17 + 22 = 39$$
$$P_{14} = P_{11} + P_{12} = 22 + 29 = 51$$
$$P_{15} = P_{12} + P_{13} = 29 + 39 = 68$$
$$P_{16} = P_{13} + P_{14} = 39 + 51 = 90$$
$$P_{17} = P_{14} + P_{15} = 51 + 68 = 119$$
$$P_{18} = P_{15} + P_{16} = 68 + 90 = 158$$
$$P_{19} = P_{16} + P_{17} = 90 + 119 = 209$$
$$P_{20} = P_{17} + P_{18} = 119 + 158 = 277$$
$$P_{21} = P_{18} + P_{19} = 158 + 209 = 367$$
$$P_{22} = P_{19} + P_{20} = 209 + 277 = 486$$
$$P_{23} = P_{20} + P_{21} = 277 + 367 = 644$$
$$P_{24} = P_{21} + P_{22} = 367 + 486 = 853$$
$$P_{25} = P_{22} + P_{23} = 486 + 644 = 1130$$
$$P_{26} = P_{23} + P_{24} = 644 + 853 = 1497$$
$$P_{27} = P_{24} + P_{25} = 853 + 1130 = 1983$$
$$P_{28} = P_{25} + P_{26} = 1130 + 1497 = 2627$$
$$P_{29} = P_{26} + P_{27} = 1497 + 1983 = 3480$$
$$P_{30} = P_{27} + P_{28} = 1983 + 2627 = 4610$$
$$P_{31} = P_{28} + P_{29} = 2627 + 3480 = 6107$$
$$P_{32} = P_{29} + P_{30} = 3480 + 4610 = 8090$$

n	0	1	2	3	4	5
P_n	3	0	2	3	2	5
n	6	7	8	9	10	11
P_n	5	7	10	12	17	22
n	12	13	14	15	16	17
P_n	29	39	51	68	90	119
n	18	19	20	21	22	23
P_n	158	209	277	367	486	644

n	24	25	26	27	28	29
P_n	853	1130	1497	1983	2627	3480
n	30	31	32			
P_n	4610	6107	8090			

(b) It appears from the table that n divides P_n if and only if n is a prime number. For example, 2 divides $P_2 = 2$, 3 divides $P_3 = 3$, 5 divides $P_5 = 5$, 7 divides $P_7 = 7$, 11 divides $P_{11} = 22$, 13 divides $P_{13} = 39$, ..., 31 divides $P_{31} = 6107 = 31 \cdot 197$. Perrin proved that if n is prime then n must divide P_n. However, the converse ("only if") conjecture was proved to be incorrect in 1982, when it was shown that the composite number $n = 271,441 = 521^2$ is a factor of $P_{271,441}$. It is the smallest nonprime n that divides its corresponding Perrin number P_n.

39. (a) The 21 digit string
010001010101010001000

(b) There are 35 digits, so we fill in either a 5×7 or a 7×5 rectangle to get, respectively:

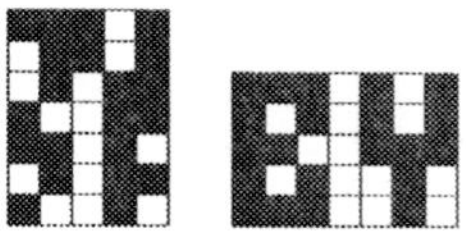

The second rectangle has the desired message, "BY."

(c) Since $96 = 2^5 \cdot 3$, there are 12 rectangles, including those with one row or one column. The product of two prime numbers gives just two rectangles with more than one row and more than one column.

41. B.
$$440 = 10 \cdot 44 = 2 \cdot 5 \cdot 4 \cdot 11 = 2 \cdot 5 \cdot 2 \cdot 2 \cdot 11$$
$$= 2^3 \cdot 5 \cdot 11$$

43. A

Copyright © 2015 Pearson Education, Inc.

45. 1 is neither a prime nor a product of primes.

$2 = 2, 3 = 3, 4 = 2^2, 5 = 5, 6 = 2 \cdot 3, 7 = 7,$

$8 = 2^3, 9 = 3^2, 10 = 2 \cdot 5, 11 = 11,$

$12 = 2^2 \cdot 3, 13 = 13, 14 = 2 \cdot 7, 15 = 3 \cdot 5,$

$16 = 2^4, 17 = 17, 18 = 2 \cdot 3^2, 19 = 19,$

$20 = 2^2 \cdot 5, 21 = 3 \cdot 7, 22 = 2 \cdot 11, 23 = 23,$

$24 = 2^3 \cdot 3, 25 = 5^2, 26 = 2 \cdot 13,$

$27 = 3^3, 28 = 2^2 \cdot 7, 29 = 29, 30 = 2 \cdot 3 \cdot 5,$

$31 = 31, 32 = 2^5, 33 = 3 \cdot 11, 34 = 2 \cdot 17,$

$35 = 5 \cdot 7, 36 = 2^2 \cdot 3^2, 37 = 37, 38 = 2 \cdot 19,$

$39 = 3 \cdot 13, 40 = 2^3 \cdot 5, 41 = 41, 42 = 2 \cdot 3 \cdot 7,$

$43 = 43, 44 = 2^2 \cdot 11, \quad 45 = 3^2 \cdot 5, 46 = 2 \cdot 23,$

$47 = 47, 48 = 2^4 \cdot 3, 49 = 7^2, 50 = 2 \cdot 5^2$

Section 4.2
Tests for Divisibility

Problem Set 4.2

1.

		2	3	4	5	6	8	9	10
(a)	684	✓	✓	✓	✗	✓	✗	✓	✗
(b)	1950	✓	✓	✗	✓	✓	✗	✗	✓
(c)	2014	✓	✗	✗	✗	✗	✗	✗	✗
(d)	2015	✗	✗	✗	✓	✗	✗	✗	✗
(e)	51,120	✓	✓	✓	✓	✓	✓	✓	✓

(a) 684 is divisible by 2 because units digit is even. Divisible by 3 because 6 + 8 + 4 = 18 and 3 divides 18. Not divisible by 5 because units digit is not 0 or 5. Divisible by 6 because it's an even number divisible by 3. Not divisible by 8. Divisible by 9 because 6 + 8 + 4 = 18 and 9 divides 18. Not divisible by 10 because the last digit is not 0.

(b) 1950 is divisible by 2 because units digit is even. Divisible by 3 because 1 + 9 + 5 + 0 = 15 and 3 divides 15. Not divisible by 4 because 4 does not divide 50. Divisible by 5 because last digit is 0. Divisible by 6 because it's an even number divisible by 3. Not divisible by 8 because 8 does not divide 950. Not divisible by 9 because 1 + 9 + 5 + 0 = 15 and 9 does not divide 15. Divisible by 10 because the last digit is 0.

(c) 2014 is divisible by 2 because units digit is even. Not divisible by 3 because 2 + 0 + 1 + 4 = 7 and 3 does not divide 7. Not divisible by 5 because units digit is not 0 or 5. Not divisible by 6 because it's not an even number divisible by 3. Not divisible by 8 because 14 is not divisible by 8. Not divisible by 9 because 2 + 0 + 1 + 4 = 7 and 9 does not divide 7. Not divisible by 10 because the last digit is not 0.

(d) 2015 is not divisible by 2 because units digit is odd. Not divisible by 3 because 2 + 0 + 1 + 5 = 8 and 3 does not divide 8. Divisible by 5 because units digit is 5. Not divisible by 6 because it's not an even number divisible by 3. Not divisible by 8 because 15 is not divisible by 8. Not divisible by 9 because 2 + 0 + 1 + 5 = 8 and 9 does not divide 8. Not divisible by 10 because the last digit is not 0.

(e) 51,120 is divisible by 2 because units digit is even. Divisible by 3 because 5 + 1 + 1 + 2 + 0 = 9 and 3 divides 9. Divisible by 4 because 4 does divides 20. Divisible by 5 because last digit is 0. Divisible by 6 because it's an even number divisible by 3. Divisible by 8 because 8 divides 120. Divisible by 9 because 5 + 1 + 1 + 2 + 0 = 9 and 9 divides 9.. Divisible by 10 because the last digit is 0.

3. **(a)** 123,456,78*d* is divisible by 2 if the last digit *d* is 0, 2, 4, 6, or 8.

(b) 123,456,78*d* is divisible by 3 if 1 + 2 + 3 + 4 + 5 + 6 + 7 + 8 + *d* = 36 + *d* is divisible by 3. 36 + 0 = 36 36 + 3 = 39, 36 + 6 = 42, and 36 + 9 = 45 are divisible by 3, so 123,456,78*d* is divisible by 3 if the last digit *d* is 0, 3, 6, or 9.

(c) 123,456,78*d* is divisible by 4 if the number formed by last two digits 8*d* = 80 + *d* is divisible by 4. 80 + 0 = 80, 80 + 4 = 84, and 80 + 8 = 88 are divisible by 4, so 123,456,78*d* is divisible by 4 if the last digit *d* is 0, 4, or 8.

(d) 123,456,78*d* is divisible by 5 if the last digit *d* is 0 or 5.

40   Chapter 4   Number Theory

(e) Using the results from parts (a) and (b), 123,456,78d is divisible by 6 if the last digit d is 0 or 6.

(f) 123,456,78d is divisible by 8 if the number formed by the last three digits 78d = 780 + d is divisible by 8. 780 + 4 = 784 is divisible by 8, so 123,456,78d is divisible by 8 if the last digit d is 4.

(g) 123,456,78d is divisible by 9 if 1 + 2 + 3 + 4 + 5 + 6 + 7 + 8 + d = 36 + d is divisible by 9. 36 + 0 = 36 and 36 + 9 = 45 are divisible by 9, so 123,456,78d is divisible by 9 if the last digit d is 0 or 9.

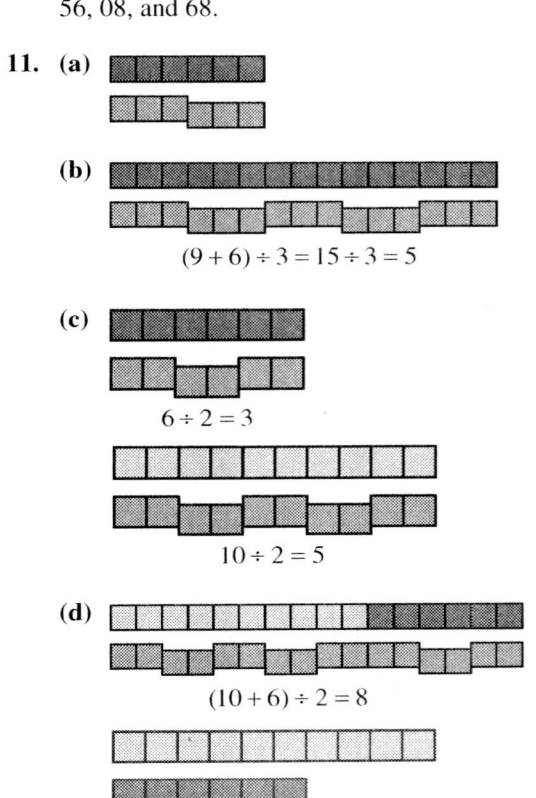

(h) 123,456,78d is divisible by 10 if the last digit d is 0.

5. (a) Make six label boxes (pigeonholes) as follows:

0	1, 9	2, 8	3, 7	4, 6	5

Place the seven numbers into the box according to the last digit of the number. Since there are six boxes, at least one box has two of the numbers. If there are two numbers in the 0 box, or two numbers in the 5 box, both their sum and difference is divisible by 10. Suppose there are two numbers in the 1 and 9 box. If the last digit of the numbers is the same, either a 1 or a 9, then there difference is divisible by 10. If the two units digits are different, a 1 and a 9, then the sum of the two numbers is divisible by 10. The same reasoning applies to the other three boxes.

(b) Answers may vary. Sample answer: 1, 2, 3, 4, 5, 10

7. 34,d40,318 is divisible by 3 if 3 + 4 + d + 4 + 0 + 3 + 1 + 8 = 23 + d is divisible by 3. 23 + 1 = 24, 23 + 4 = 27, and 23 + 7 = 30 are divisible by 3, so 34,d40,318 is divisible by 3 if d = 1, 4, or 7. However, only 23 + 4 = 27 is divisible by 9, so 34,d40,318 is divisible by 3 and not 9 if d = 1 or 7.

9. To be divisible by 12, n = 473,246,8de must be divisible by both 3 and 4. Therefore, the sum of digits, which is 34 + d + e must be a multiple of 3 and de must be divisible by 4. 34 + 2 = 36, 34 + 5 = 39, 34 + 8 = 42,

34 + 11 = 45, 34 + 14 = 48, and 34 + 17 = 51 are all divisible by 3, so d + e must be 2, 5, 8, 11, 14, or 17. To be divisible by 4, e must be even—0, 2, 4, 6, or 8. If e = 0, then d is 2 or 8; if e = 2, then d is 3 or 9; if e = 4, then d is 4; if e = 6, then d is 5; if e = 8, then d is 0 or 6. This gives eight choices of de: 20, 80, 32, 92, 44, 56, 08, and 68.

11. (a)

(b)

$$(9 + 6) \div 3 = 15 \div 3 = 5$$

(c)

$$6 \div 2 = 3$$

$$10 \div 2 = 5$$

(d)

$$(10 + 6) \div 2 = 8$$

$$10 - 6 = 4$$

13. (a) It is true when the sum of the first and last digits is less than 10.

(b) It is not true when the sum of the first and last digits is 10 or more.

(c) Yes.

(d) Answers will vary. One might illustrate parts (a) and (b) with specific multiplication examples such as

(a)	23	(b)	47
	× 11		× 11
	23		47
	23		47
	253		517

(continued on next page)

Copyright © 2015 Pearson Education, Inc.

Solutions to Problem Set 4.3 **41**

(continued)

In (a) $2 + 3 = 5$ which is less than 10 and in (b) $4 + 7 = 11$ and we record the 1 (for one 10) and exchange ten 10s for one 100. One might illustrate part (c) with specific division examples such as

$$
\begin{array}{r}
52 \\
11)\overline{572} \\
\underline{55} \\
22 \\
\underline{22} \\
0
\end{array}
$$

15. (a) Max is not correct.

    (b) Note that $177 + 48 = 225$ and 15 *does* divide 225. Indeed, $225 = 15^2$.

17. (a) $686 \leftrightarrow 68 - 12 = 56 = 7 \cdot 8$, so 686 is divisible by 7.

    (b) $2951 \leftrightarrow 295 - 2 = 293 \leftrightarrow 29 - 6 = 23$, so 2951 is not divisible by 7.

    (c) $18,487 \leftrightarrow 1848 - 14 = 1834 \leftrightarrow 183 - 8 = 175 \leftrightarrow 17 - 10 = 7$, so 18,487 is divisible by 7.

19. (a) 2,262,942
$(2 + 6 + 9 + 2) - (2 + 2 + 4) = 19 - 8 = 11$, so divisible by 11

    (b) 48,360,219
$(4 + 3 + 0 + 1) - (8 + 6 + 2 + 9) = 8 - 25 = -17$, so not divisible by 11

    (c) 7,654,592
$(7 + 5 + 5 + 2) - (6 + 4 + 9) = 19 - 19 = 0$, so divisible by 11

    (d) 8,352,607
$(8 + 5 + 6 + 7) - (3 + 2 + 0) = 26 - 5 = 21$, so not divisible by 11

    (e) 718,161,906
$(7 + 8 + 6 + 9 + 6) - (1 + 1 + 1 + 0)$
$= 36 - 3 = 33$, so divisible by 11

21. (a) For any palindrome with an even number of digits, the digits in the odd positions are the same as the digits in the even positions, but with the order reversed. Thus, the difference of the sums of the digits in the even and odd positions is zero, which is divisible by 11.

    (b) Yes. For example, $121 = 11 \cdot 11$.

23. (a) The Fibonacci numbers F_{5k} are divisible by $F_5 = 5$ for every $k = 1, 2, 3, \dots$.

    (b) The Fibonacci numbers F_{3k} are divisible by $F_3 = 2$ and the Fibonacci numbers F_{4k} are divisible by $F_4 = 3$ .for every $k = 1, 2, 3, \dots$

    (c) If n is a prime number, then F_n is only divisible by itself and $F_1 = 1$. If n is a composite number and $n = ij$, then F_n is divisible by F_i. It can be shown that F_i divides F_n if, and only if, i divides n.

25. Each calculation is in error.

    (a) $35,874 + 7531 + 69,450 \leftrightarrow$
$0 + 7 + 6 = 13 \leftrightarrow 4$ and $113,855 \leftrightarrow 5$.
The sum should be 112,855.

    (b) $8514 + 6854 + 2578 + 6014 \leftrightarrow$
$0 + 5 + 4 + 2 = 11 \leftrightarrow 2$ and $23,860 \leftrightarrow 1$.
The sum should be 23,960.

    (c) $78,962 - 3621 \leftrightarrow 5 - 3 = 2$ and
$75,331 \leftrightarrow 1$.
The difference should be 75,341.

    (d) $358 \times 592 \leftrightarrow 7 \times 7 = 49 \leftrightarrow 4$ and
$221,936 \leftrightarrow 5$.
The product should be 211,936.

27. C. All of the choices are divisible by 4 except for choice C, a class of 46 students.

Section 4.3
Greatest Common Divisors and Least Common Multiples

Problem Set 4.3

1. (a) Let D_{24} and D_{27} represent the sets of divisors of 24 and 27.
$D_{24} = \{1, 2, 3, 4, 6, 8, 12, 24\}$
$D_{27} = \{1, 3, 9, 27\}$
$D_{24} \cap D_{27} = \{1, 3\}$
GCF(24, 27) = 3

Copyright © 2015 Pearson Education, Inc.

42 Chapter 4 Number Theory

(b) Let D_{14} and D_{22} represent the sets of divisors of 14 and 22.

$D_{14} = \{1, 2, 7, 14\}$

$D_{22} = \{1, 2, 11, 22\}$

$D_{14} \cap D_{22} = \{1, 2\}$

GCF(14, 22) = 2

(c) Let D_{48} and D_{72} represent the sets of divisors of 48 and 72.

$D_{48} = \{1, 2, 3, 4, 6, 8, 12, 16, 24, 48\}$

$D_{72} = \{1, 2, 3, 4, 6, 8, 9, 12, 18, 24, 36, 72\}$

$D_{48} \cap D_{72} = \{1, 2, 3, 4, 6, 8, 12, 24\}$

GCF(48, 72) = 24

3. (a) Let M_{24} and M_{27} represent the set of multiples of 24 and 27.

$M_{24} = \{24, 48, 72, 96, 120, 144, 168,$
$\qquad 192, 216, 240, \cdots\}$

$M_{27} = \{27, 54, 81, 108, 135, 162, 189,$
$\qquad 216, 243, \cdots\}$

$M_{24} \cap M_{27} = \{216, \cdots\}$

LCM(24, 27) = 216

(b) Let M_{14} and M_{22} represent the sets of multiples of 14 and 22.

$M_{14} = \{14, 28, 42, 56, 70, 84, 98, 112,$
$\qquad 126, 140, 154, 168, 182, \cdots\}$

$M_{22} = \{22, 44, 66, 88, 110, 132, 154,$
$\qquad 176, 198, \cdots\}$

$M_{14} \cap M_{22} = \{154, \cdots\}$

LCM(14, 22) = 154

(c) Let M_{48} and M_{72} represent the sets of multiples of 48 and 72.

$M_{48} = \{48, 96, 144, 192, 240, 288, 336,$
$\qquad 384, 432, \cdots\}$

$M_{72} = \{72, 144, 216, 288, 360, 432,$
$\qquad 504, \cdots\}$

$M_{48} \cap M_{72} = \{144, 288, 432, \cdots\}$

LCM(48, 72) = 144

5. (a) For the GCD we choose the smaller of the two exponents with which each prime appears in r and s:

$r = 2^2 \cdot 3^1 \cdot 5^3$

$s = 2^1 \cdot 3^3 \cdot 5^2$

$GCD(r, s) = 2^1 \cdot 3^1 \cdot 5^2 = 150$

For the LCM we choose the larger of the two exponents with which each prime appears in r and s:

$LCM(r, s) = 2^2 \cdot 3^3 \cdot 5^3 = 13{,}500$.

(b) $u = 2^0 \cdot 5^1 \cdot 7^2 \cdot 11^1$

$v = 2^2 \cdot 5^3 \cdot 7^1 \cdot 11^0$

$GCD(u, v) = 2^0 \cdot 5^1 \cdot 7^1 \cdot 11^0 = 35$

$LCM(u, v) = 2^2 \cdot 5^3 \cdot 7^2 \cdot 11^1 = 269{,}500$

(c) $w = 2^2 \cdot 3^3 \cdot 5^2 \cdot 7^0$

$x = 2^1 \cdot 3^0 \cdot 5^3 \cdot 7^2$

$GCD(w, x) = 2^1 \cdot 3^0 \cdot 5^2 \cdot 7^0 = 50$

$LCM(w, x) = 2^2 \cdot 3^3 \cdot 5^3 \cdot 7^2 = 661{,}500$

7. (a)

$550 \overline{)3500}$ 6 R 200 $200 \overline{)550}$ 2 R 150

$150 \overline{)200}$ 1 R 50 $50 \overline{)150}$ 3 R 0

GCD(550, 3500) = 50

$LCM(550, 3500) = \dfrac{550 \cdot 3500}{GCD(550, 3500)}$

$\qquad = \dfrac{1{,}925{,}000}{50} = 38{,}500$

(b)

$825 \overline{)3915}$ 4 R 615 $615 \overline{)825}$ 1 R 210

$210 \overline{)615}$ 2 R 195

$195 \overline{)210}$ 1 R 15 $15 \overline{)195}$ 13 R 0

GCD(825, 3915) = 15

$LCM(825, 3915) = \dfrac{825 \cdot 3915}{GCD(825, 3915)}$

$\qquad = \dfrac{3{,}229{,}875}{15} = 215{,}325$

(c)

$624 \overline{)1044}$ 1 R 420 $420 \overline{)624}$ 1 R 204

$204 \overline{)420}$ 2 R 12 $12 \overline{)204}$ 17 R 0

GCD(624, 1044) = 12

$LCM(624, 1044) = \dfrac{624 \cdot 1044}{GCD(624, 1044)}$

$\qquad = \dfrac{651{,}456}{12} = 54{,}288$

9. (a) GCD(40, 40) = 40

(b) GCD(19, 190) = 19

(c) GCD(59, 0) = 59

11. (a) For all natural numbers m,
GCD(m, m) = m.

(b) For all natural numbers m and n,
GCD(m, mn) = m.

(c) For all natural numbers n, GCD($n, 0$) = n.

(d) For all natural numbers n, LCM(n, n) = n.

(e) For all natural numbers m and n,
LCM(m, mn) = mn.

(f) For all natural numbers n, LCM($n, 1$) = n.

13. (a) The 4-rods can measure trains of length 4, 8, 12, 16, ⋯. The 6-rods can measure trains of length 6, 12, 18, ⋯. The shortest train that can be measured by 4-rods and by 6-rods has length 12.

(b) LCM(4, 6) = 12

(c) The 6-rods can measure trains of length 6, 12, 18, 24, ⋯. The 9-rods can measure trains of length 9, 18, 27, ⋯. The shortest train that can be measured by 6-rods and by 9-rods has length 18.

(d) LCM(6, 9) = 18

(e) The least common multiple of two numbers a and b is the length of the shortest train that can be measured by both the a train and the b train.

15. The GCD(0, n) is the greatest natural number d that divides both 0 and n. If GCD(0, n) = n, then n is the greatest number that divides both 0 and n. If n divides 0, then $0 = nq$ for a unique whole number q. However, if $n = 0$, since there is no unique whole number q for which $0 = nq$, n cannot divide 0 and n cannot be the GCD (0, n). Note, the definition of GCD(m, n) specifies that m and n are natural numbers.

17. Note that mn is certainly a common multiple of m and n, but it is often not the least common multiple. For example, $4 \cdot 6 = 24$ is a common multiple of 4 and 6, but their least common multiple is 12. But use Tawana's comment as a springboard for class discussion by asking, "What must be true if mn is the least common multiple of m and n?" The answer, of course, is that m and n must have no common divisor other than 1. For, in this case,

$$\text{LCM}(m, n) = \frac{mn}{\text{GCD}(m, n)} = \frac{mn}{1} = mn$$

19. (a) Since $60 = 2^2 \times 3 \times 5$ and $105 = 3 \times 5 \times 7$, we see that GCD(60, 105) = $3 \times 5 = 15$. The tile sizes that can be used to tile the floor are 1″ by 1″, 3″ by 3″, 5″ by 5″ and 15″ by 15″. Thus, the largest square tile possible is 15″ by 15″.

(b) Since $60 \div 15 = 4$ and $105 \div 15 = 7$, there will be 4 rows of 7 tiles each. Thus, $4 \times 7 = 28$ tiles are required.

21. (a) The side length of any tiled square must be a multiple of both dimensions of the brick. Since

$$\text{LCM}(8, 12) = \frac{8 \cdot 12}{\text{GCD}(8, 12)} = \frac{96}{4} = 24 \text{ in.},$$

the smallest square tiled by $8'' \times 12''$ bricks is $24'' \times 24''$. This is tiled with three rows of two bricks each.

(b) Reasoning as in part (a), the smallest square that can be tiled with $9'' \times 12''$ bricks has sides of length

$$\text{LCM}(9, 12) = \frac{9 \cdot 12}{\text{GCD}(9, 12)} = \frac{108}{3} = 36 \text{ in.}$$

The smallest square is tiled with four rows of three bricks each.

23. (a) $a = 2^2 \cdot 3^1 \cdot 5^2 \cdot 7^0$
$b = 2^1 \cdot 3^3 \cdot 5^1 \cdot 7^0$
$c = 2^0 \cdot 3^2 \cdot 5^3 \cdot 7^1$
For the GCD we choose the smallest of the three exponents with which each prime appears in a, b, and c:
GCD(a, b, c) = $2^0 \cdot 3^1 \cdot 5^1 \cdot 7^0 = 15$.
For the LCM we choose the largest of the three exponents with which each prime appears in a, b, and c:
LCM(a, b, c) = $2^2 \cdot 3^3 \cdot 5^3 \cdot 7^1 = 94,500$.

(b) No. For example, the numbers given in part (a) give GCD(a, b, c) · LCM(a, b, c) = 1,417,500, but $abc = 637,875,000$.

Copyright © 2015 Pearson Education, Inc.

44 *Chapter 4 Number Theory*

(c) Answers will vary. One possibility is
GCD(3, 5, 7) = 1, LCM(3, 5, 7) = 105,
GCD(3, 5, 7) · LCM(3, 5, 7) = 105
$= 3 \cdot 5 \cdot 7$.

25. (a) $6 = 2^1 \cdot 3^1 = 2^1 \cdot 3^1 \cdot 5^0 \cdot 7^0 \cdot 11^0 \cdot 13^0$

$35 = 5^1 \cdot 7^1 = 2^0 \cdot 3^0 \cdot 5^1 \cdot 7^1 \cdot 11^0 \cdot 13^0$

$143 = 11^1 \cdot 13^1 = 2^0 \cdot 3^0 \cdot 5^0 \cdot 7^0 \cdot 11^1 \cdot 13^1$

$GCD(6, 35, 143) = 2^0 \cdot 3^0 \cdot 5^0 \cdot 7^0 \cdot 11^0 \cdot 13^0$
$= 1$

$LCM(6, 35, 143) = 2^1 \cdot 3^1 \cdot 5^1 \cdot 7^1 \cdot 11^1 \cdot 13^1$
$= 30,030$

(b) Yes. GCD(6, 35, 143) · LCM(6, 35, 143)
= 1 · 30,030 = 6 · 35 · 143

(c) GCD(a, b, c) · LCM(a, b, c) = abc
if and only if,
GCD(a, b) = GCD(a, c)
$= $ GCD(b, c) = 1.
That is, if no pair of numbers from a, b,
and c have any common factors other than
1. *Note*: Students may guess that the
condition is GCD(a, b, c) = 1. However,
this is not correct. For example,
GCD(15, 35, 21) = 1 but
GCD(15, 35, 21) · LCM(15, 35, 21)
$= 1 \cdot 105 \neq 15 \cdot 35 \cdot 21$.

27. (a)

	48	25	35
2	24	25	35
2	12	25	35
2	6	25	35
2	3	25	35
3	1	25	35
5	1	5	7

$LCM(48, 25, 35) = 2^4 \cdot 3^1 \cdot 5^2 \cdot 7^1$

(b)

	40	28	12	63
2	20	14	6	63
2	10	7	3	63
2	5	7	3	63
3	5	7	1	21
3	5	7	1	7

$LCM(40, 28, 12, 63) = 2^3 \cdot 3^2 \cdot 5^1 \cdot 7^1$

(c)

	250	28	44	110
2	125	14	22	55
2	125	7	11	55
5	25	7	11	11
5	5	7	11	11
11	5	7	1	1

$LCM(250, 28, 44, 110) = 2^2 \cdot 5^3 \cdot 7^1 \cdot 11^1$

29. (a) $6 = 2^1 \cdot 3^1$

$8 = 2^1 \cdot 4^1$

$GCD(6, 8) = 2^1 = 2$

(b) $18 = 2^1 \cdot 3^2$

$24 = 2^3 \cdot 3^1$

$GCD(18, 24) = 2^1 \cdot 3^1 = 6$

(c) $132 = 2^2 \cdot 3^1 \cdot 11^1$

$209 = 11^1 \cdot 19^1$

$GCD(132, 209) = 11^1 = 11$

(d) $315 = 3^2 \cdot 5^1 \cdot 7^1$

$375 = 3^1 \cdot 5^3$

$GCD(315, 375) = 3^1 \cdot 5^1 = 15$

(e) $\dfrac{6}{8} = \dfrac{6 \div GCD(6, 8)}{8 \div GCD(6, 8)} = \dfrac{6 \div 2}{8 \div 2} = \dfrac{3}{4}$

(f) $\dfrac{18}{24} = \dfrac{18 \div GCD(18, 24)}{24 \div GCD(18, 24)} = \dfrac{18 \div 6}{24 \div 6} = \dfrac{3}{4}$

(g) $\dfrac{132}{209} = \dfrac{132 \div GCD(132, 209)}{209 \div GCD(132, 209)}$
$= \dfrac{132 \div 11}{209 \div 11} = \dfrac{12}{19}$

(h) $\dfrac{315}{375} = \dfrac{315 \div GCD(315, 375)}{375 \div GCD(315, 375)}$
$= \dfrac{315 \div 15}{375 \div 15} = \dfrac{21}{25}$

31. (a) $\dfrac{3}{4} + \dfrac{1}{6} = \dfrac{9}{12} + \dfrac{2}{12} = \dfrac{11}{12}$

(b) $\dfrac{7}{10} + \dfrac{2}{15} = \dfrac{21}{30} + \dfrac{4}{30} = \dfrac{25}{30} = \dfrac{5 \cdot 5}{5 \cdot 6} = \dfrac{5}{6}$

Copyright © 2015 Pearson Education, Inc.

(c) $\dfrac{8}{9} - \dfrac{4}{15} = \dfrac{40}{45} - \dfrac{12}{45} = \dfrac{28}{45}$

(d) $\dfrac{11}{25} - \dfrac{3}{10} = \dfrac{22}{50} - \dfrac{15}{50} = \dfrac{7}{50}$

33. 32 revolutions. Since the speed at which the gears rotate does not affect the answer, we may assume that one tooth-width passes the contact point every second. Then tooth A returns to this point every 45 seconds and tooth B returns every 96 seconds. Since
$45 = 3^2 \cdot 5^1$ and $96 = 2^5 \cdot 3^1$,
LCM(45, 96) $= 2^5 \cdot 3^2 \cdot 5^1 = 1440$, so the first time both teeth return to this position is after 1440 seconds. At this time, the small gear has made $\dfrac{1440}{45} = 32$ revolutions.

35. (a) 221 years will pass, since GCD(13, 17) = 1 and LCM(13, 17) = 13 × 17 = 221.

(b) In just 36 years, the two species will emerge together, since GCD(12, 18) = 6 and therefore
$$\text{LCM}(12,18) = \frac{12 \cdot 18}{\text{GCD}(12,18)} = \frac{12 \cdot 18}{6} = 36.$$

(c) In 45 years, the species will emerge together again, since GCD (9, 15) = 3 and
$$\text{LCM}(9,15) = \frac{9 \cdot 15}{3} = 45.$$

37. B.
$12 = 2^2 \cdot 3$
$16 = 2^4$
$24 = 2^3 \cdot 3$
$40 = 2^3 \cdot 5$
GCF(12, 16, 24, 40) $= 2^3 = 8$

39. (a) The 30th customer received a muffin only, since 6 divides 30 but 4 does not divide 30.

(b) Casey is the 12th customer. Since LCM(4, 6) = 12, he was the first to receive both a cookie and a muffin.

(c) Only cookies are given to customers 4, 8, 16, 20, 28, 32, 40, 44, 52, 56, … These are the multiples of 4 that are not multiples of 6. Since Tom came after

Casey, he may be customer 16, 20, 28, 32, 40, 44, 52, 56, …

(d) Since $4 \cdot 29 = 116$, there were 116, 117, 118, or 119 customers. None of these are multiple of 6, since $6 \cdot 19 = 114$. Thus, 19 muffins were given away.

41. B.

Chapter 4 Review Exercises

1.

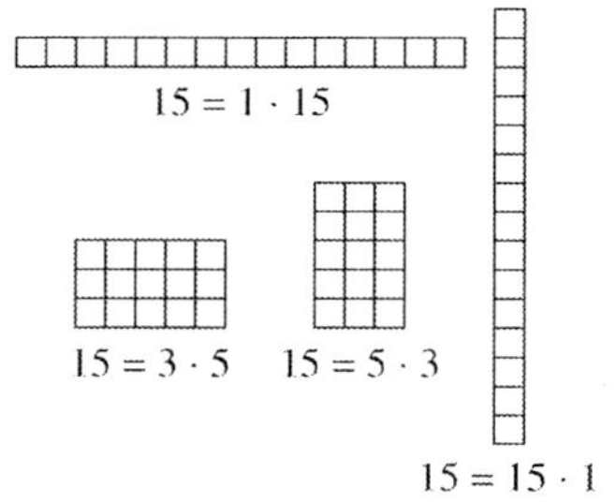

3. (a) $D_{60} = \{1, 2, 3, 4, 5, 6, 10, 12, 15, 20, 30, 60\}$

(b) $D_{72} = \{1, 2, 3, 4, 6, 8, 9, 12, 18, 24, 36, 72\}$

5. (a) $N = (1 + 1)(1 + 1)(1 + 1)(1 + 1)(1 + 1)$
$= 2^5 = 32$

(b) $N = (2 + 1)(2 + 1)(2 + 1)(2 + 1)(2 + 1)$
$= 3^5 = 243$

7. (a) Answers will vary. For example,
$15 = 3 \cdot 5;\ 5 > \sqrt{15}$.

(b) Yes. $3 \le \sqrt{15}$.

9. (a) The exponents in the prime power representation are 4 and 2, and
$(4 + 1)(2 + 1) = 15$.
There are 15 divisors.

(b) Make an orderly list:
$3^0 \cdot 7^0 = 1, \ 3^0 \cdot 7^1 = 7,$
$3^0 \cdot 7^2 = 49, \ 3^1 \cdot 7^0 = 3,$
$3^1 \cdot 7^1 = 21, \ 3^1 \cdot 7^2 = 147,$
$3^2 \cdot 7^0 = 9, \ 3^2 \cdot 7^1 = 63,$
$3^2 \cdot 7^2 = 441, \ 3^3 \cdot 7^0 = 27,$
$3^3 \cdot 7^1 = 189, \ 3^3 \cdot 7^2 = 1323,$
$3^4 \cdot 7^0 = 81, \ 3^4 \cdot 7^1 = 567,$
$3^4 \cdot 7^2 = 3969$
In order, the divisors are 1, 3, 7, 9, 21, 27, 49, 63, 81, 147, 189, 441, 567, 1323, and 3969.

46 Chapter 4 Number Theory

11. **(a)** True. 67,275 is divisible by 5 because the last digit is 5. 67,275 is divisible by 3 because $6 + 7 + 2 + 7 + 5 = 27$ and 3 divides 27. Therefore, 67,275 is divisible by $3 \cdot 5 = 15$.

(b) True. 578,940 is divisible by 3 because $5 + 7 + 8 + 9 + 4 + 0 = 33$ and 3 divides 33. 578,940 is divisible by 4 because 4 divides 40. Therefore, 578,940 is divisible by $3 \cdot 4 = 12$.

(c) True. 42,720 is divisible by 3 because $4 + 2 + 7 + 2 + 0 = 15$ and 3 divides 15. 42,720 is divisible by 8 because 8 divides 720. Therefore, 42,720 is divisible by $3 \cdot 8 = 24$, and 24 is a factor of 42,720

(d) False. 62,730 does not have 4 as a factor because 30 is not divisible by 4. Therefore, 62,730 is not divisible by $4 \cdot 9 = 36$, and 36 is not a factor of 62,730.

13.

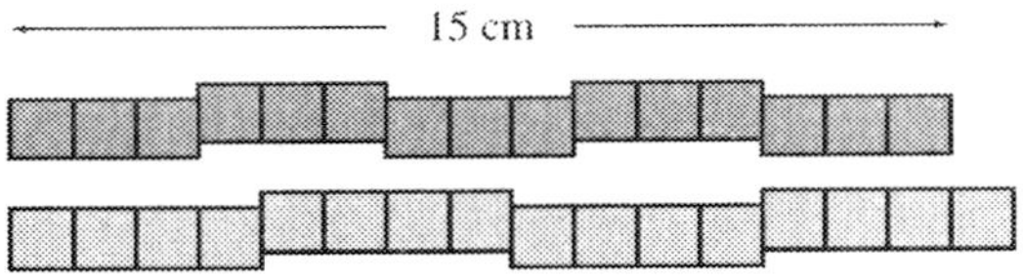

15. **(a)** $D_{63} = \{1, 3, 7, 9, 21, 63\}$
$D_{91} = \{1, 7, 13, 91\}$
$D_{63} \cap D_{91} = \{1, 7\}$ so GCD(91, 63) = 7.

(b) $M_{63} = \{63, 126, 189, 252, 315, 378, 441,$ $504, 567, 630, 693, 756, 819,$ $882, 945, 1008, ...\}$
$M_{91} = \{91, 182, 273, 364, 455, 546, 637,$ $728, 819, 910, ...\}$
$M_{63} \cap M_{91} = \{819, 1638, \cdots\}$, so
LCM (63, 91) = 819.

(c) $7 \cdot 819 = 5733 = 63 \cdot 91$

17. **(a)**
$$12,100 \overline{)119,790} \quad \begin{matrix} 9 & R\ 10,890 \end{matrix}$$

$$10,890 \overline{)12,100} \quad \begin{matrix} 1 & R\ 1210 \end{matrix}$$

$$1210 \overline{)10,890} \quad \begin{matrix} 9 & R\ 0 \end{matrix}$$

Thus, GCD(119,790, 12,100) = 1210

(b) $\text{LCM}(119,790, 12,100) = \dfrac{119,790 \cdot 12,100}{1210}$
$= 1,197,900$

Copyright © 2015 Pearson Education, Inc.

Chapter 5 Integers

Section 5.1
Representations of Integers

Problem Set 5.1

1. (a) The loop must have 5 more black counters than red counters. Two possibilities are

and

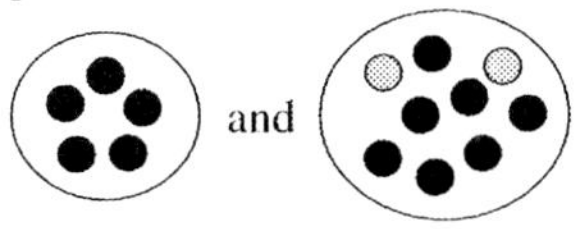

(b) The loop must have 2 more red counters than black counters. Two possibilities are

and

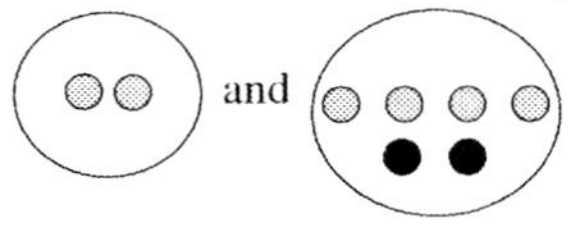

3. (a) The loop must contain 3 more black counters than red counters. To use the least number of counters, use 3 black counters only:

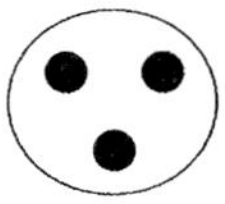

(b) The loop must contain 4 more red counters than black counters. To use the least number of counters, use 4 red counters only:

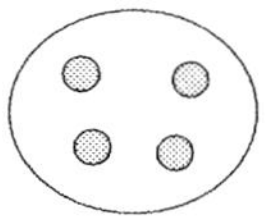

5. (a) At mail time, you are delivered a check for $14.

(b) At mail time, you receive a bill for $27.

7. (a) Since the check is $17 more than the bill you are richer by $17.

(b) $48 + (-31) = 17$

9. (a) The arrow points 4 units to the right, so it represents 4.

(b) The arrow begins and ends at the same point, so it represents 0.

(c) The arrows point 6 units to the right, so each arrow represents 6.

(d) The arrows point 5 units to the left, so each arrow represents –5.

11. (a) The arrow must point 7 units to the right. One possibility is shown.

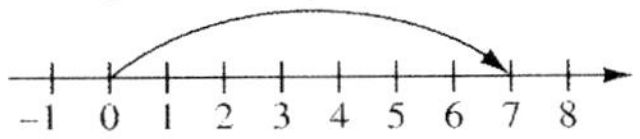

(b) The arrow must begin and end at the same point. One possibility is shown.

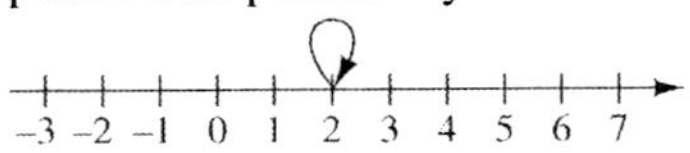

(c) The arrow must point 9 units to the left. One possibility is shown.

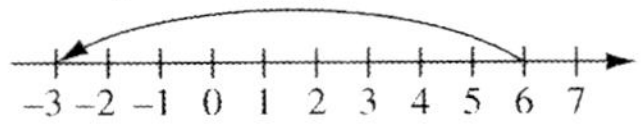

(d) The arrow must point 9 units to the right. One possibility is shown.

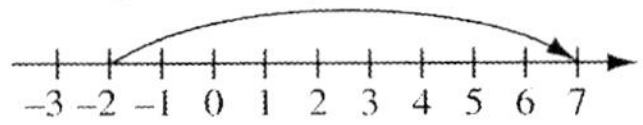

13. (a) The equation $|x| = 13$ means that x is 13 units from 0 on a number line. Therefore, $x = 13$ or $x = -13$.

(b) The equation $|x| + 1 = 2$ is equivalent to $|x| = 1$, so $x = 1$ or $x = -1$.

(c) The equation $|x| + 5 = 0$ is equivalent to $|x| = -5$, which has no solution because an absolute value is always positive or zero. There is no value of x that will make the equation true.

15. (a) Form a loop for 4, turn all the markers over to show –4, and turn them all again to show –(–4). Since the number of markers has not changed, we finish with the original loop for 4. Thus, –(–4) = 4.

(b) Refer to part (a). Since there is nothing special about 4, we could do the same thing with any integer. Therefore, it follows that –(–n) = n for any integer n.

17. Congratulate her since her claim is true. If she is not really sure, lay out a loop for 0 and turn all the markers over to obtain a loop for –0. But the last loop will still be a loop for zero. Therefore, –0 = 0 as she thought.

Copyright © 2015 Pearson Education, Inc.

48 *Chapter 5 Integers*

19. (a) To represent –3, the loop needs to have 3 more red counters than black counters. Since there are 3 black counters, we could add red counters until there are 6 red counters—that is, add 4 red counters.

(b) Yes, for example we could have added 5 red counters and 1 black counter.

21. (a) 12 red: –12
11 red + 1 black: –10
10 red + 2 black: –8
9 red + 3 black: –6
8 red + 4 black: –4
7 red + 5 black: –2
6 red + 6 black: 0
5 red + 7 black: 2
4 red + 8 black: 4
3 red + 9 black: 6
2 red + 10 black: 8
1 red + 11 black: 10
12 black: 12

(b) 11 red: –11
10 red + 1 black: –9
9 red + 2 black: –7
8 red + 3 black: –5
7 red + 4 black: –3
6 red + 5 black: –1
5 red + 6 black: 1
4 red + 7 black: 3
3 red + 8 black: 5
2 red + 9 black: 7
1 red + 10 black: 9
11 black: 11

23. Any integer from –10, –9, … , 9, 10

25. (a) 1, –2, 3, –4, 5

(b) n if n is odd, $-n$ if n is even

27. (a) 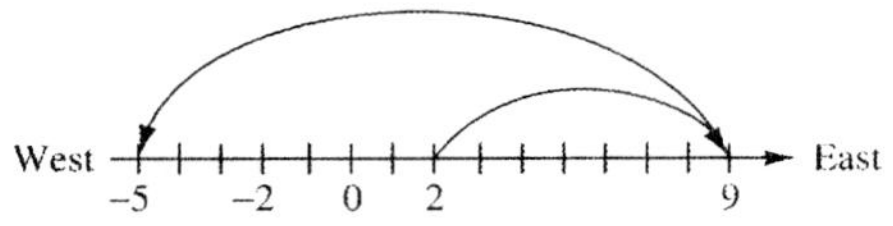 West ⟵├┼┼┼┼┼┼┼┼┼┼┼┼┼├→ East
  –5   –2   0   2   9

(b) The total distance traveled from the warehouse to East 9th Street is
$\left|9-2\right|+\left|(-5)-9\right|=\left|7\right|+\left|-14\right|=7+14=21$
blocks.

(c) The gas station is at West 2nd Street. It is
$\left|2-(-2)\right|=\left|4\right|=4$ blocks from the warehouse.

29. (a) If we consider the ground floor of a European building as floor zero, the numbering of the floors matches a number line. Starting at –5 and going up 27 floors, we arrive at the 22nd floor.

(b) The 22nd floor in Europe is what an American would call the 23rd floor. In New York, the elevator would stop on the 23rd floor.

31. This can be shown by counting on a number line. Start at 33, count backward (to the left) 9 units, then count backwards 6 more steps, and finally count forward (to the right) 29 units to end at 47. The line of scrimmage is at the 47-yard line.

33. (a) The rock will fall
$$\left|29,028\right|+\left|-35,840\right|=29,028+35,840$$
$$=64,868 \text{ ft}$$

(b) 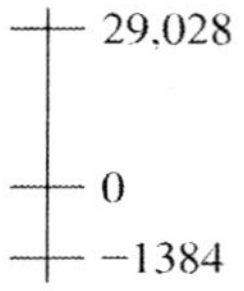 ┼─ 29,028

   ┼─ 0
   ┼─ –1384

The diagram shows that the difference in elevation is given by the sum of the elevation of Mount Everest above sea level and the depth of the Dead Sea below sea level.
$$\left|29,028-\left(-1384\right)\right|=29,028+1384$$
$$=30,412 \text{ ft}$$

Section 5.2
Addition and Subtraction of Integers

Problem Set 5.2

1. (a)

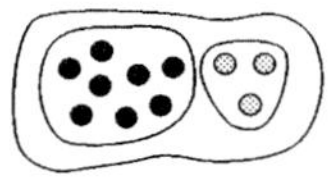

$8 + (-3) = 5$

(b)

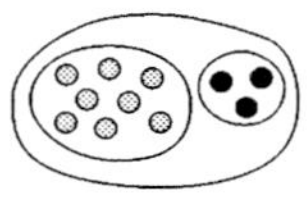

$(-8) + 3 = -5$

(c)

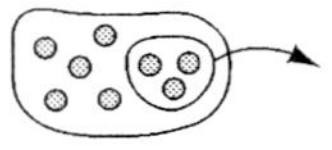

$-8 - (-3) = -5$

Copyright © 2015 Pearson Education, Inc.

(d)

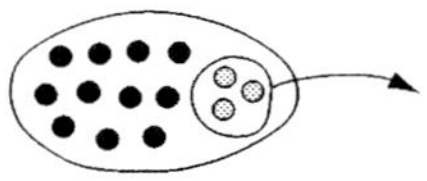

$$8 - (-3) = 11$$

3. (a) At mail time you receive a bill for $27 and a bill for $13.
$$(-27) + (-13) = -40$$

(b) At mail time you receive a bill for $27 and the letter carrier takes away a check for $13.
$$(-27) - 13 = -40$$

(c) The mail carrier brings you a check for $27 and a check for $13.
$$27 + 13 = 40$$

(d) The mail carrier brings you a check for $27 and takes away a check for $13.
$$27 - 13 = 14$$

5. (a)

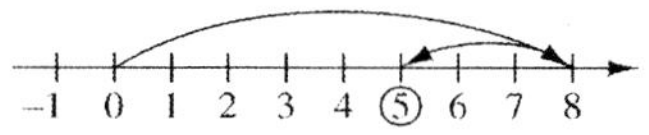

$$8 + (-3) = 5$$

(b)

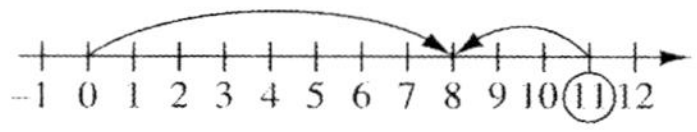

$$8 - (-3) = 11$$

(c)

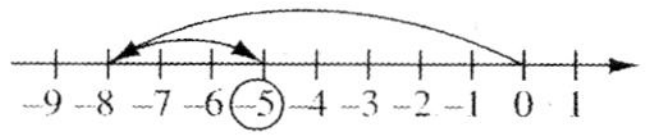

$$(-8) + 3 = -5$$

(d)

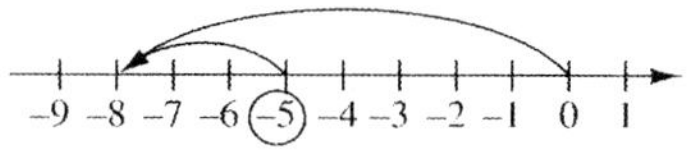

$$(-8) - (-3) = -5$$

7. (a) $13 - 7 = 13 + (-7)$

(b) $13 - (-7) = 13 + 7$

(c) $(-13) - 7 = (-13) + (-7)$

(d) $(-13) - (-7) = -13 + 7$

9. (a) $27 - (-13) = 27 + 13 = 40$

(b) $12 + (-24) = 12 - 24 = -(24 - 12)$
$$= -12$$

(c) $(-13) - 14 = (-13) + (-14)$
$$= -(13 + 14) = -27$$

(d) $-81 + 54 = -(81 - 54) = -27$

11. (a) $(-356) + 148 \approx -350 + 150 = -200$

(b) $728 + (-273) \approx 725 - 275$
$$= 700 - 250 = 450$$

(c) $298 - (-454) \approx 300 + 450 = 750$

(d) $-827 - 370 \approx -(830 + 370) = -1200$

13. (a) $2 - (-27) = 2 + 27 = 29$
The temperature rose 29°.

(b) $2 - (-27) = 29$

15. (a) $314 - 208 = 106$
Sam's net worth was more, by $106.

(b) $314 - 208 = 106$

17. (a) $-117 < -24$ **(b)** $0 > -4$

(c) $18 > 12$

19. $-17, -5, -2, 0, 3, 5, 27$

21. Answers will vary. The important point is that negation is a unary operation, in which pressing the button $\boxed{(-)}$ takes the opposite, or negative, of the number x currently in the display. The subtraction function $\boxed{-}$ is a binary operation that sets up a pending operation in which the currently displayed number x will be the minuend, the next number entered will become the subtrahend y, and pressing $\boxed{=}$ (or $\boxed{\text{ENTER}}$) calculates the difference $x - y$.

23. Complement Melanie on good thinking. For example, if you add 4 to 8 you obtain 12. If you then add −4 to 12, you again obtain 8. This is what the term "additive inverse" actually implies and it will be helpful to the entire class to discuss Melanie's assertion with the group.

25. This is essentially the same as the question Keyshawn asked in problem 22.

(a) Armand is correct. You can begin by representing 7 with 7 black counters, and −3 with 3 red counters. To subtract the 3 red counters, you flip the 3 red counters over, which gives 10 black counters. Then you have a black score of 10.

(b) $7 - (-3) = 10$

50 Chapter 5 Integers

(c) One way of subtracting m from n is to add the additive inverse of m to n. that is $n - m = n + (-m)$. Since the negative of -3 is 3, $7 - (-3) = 7 + 3 = 10$.

27. (a) Yes. $a \le b$ is true whenever $a < b$ or $a = b$. Thus anytime $a < b$, $a \le b$ is true.

(b) No. If $a \ge b$, then $a > b$ or $a = b$. Thus, $a \ge b$ does not imply that $a > b$ is true.

(c) Yes. $2 \ge 2$ is true if $2 > 2$ or $2 = 2$. Since $2 = 2$, $2 \ge 2$ is true.

29. (a) $5 - 11 = 5 + (-11) = -(11 - 5) = -6$, so $|5 - 11| = |-6| = 6$.

(b) $(-4) - (-10) = (-4) + 10 = 10 - 4 = 6$, so $|(-4) - (-10)| = |6| = 6$.

(c) $8 - (-7) = 8 + 7 = 15$, so $|8 - (-7)| = |15| = 15$.

(d) $(-9) - 2 = (-9) + (-2) = -(9 + 2) = -11$, so $|(-9) - 2| = |-11| = 11$.

31. For any two integers a and b, the distance between a and b on a number line is $|a - b|$ (or, equivalently, $|b - a|$).

33.

-1	-6	8	1
4	5	-5	-2
-7	0	2	7
6	3	-3	-4

The magic sum is 2.

35. (a) (i) $7 - (-3) = 7 + 3 = 10$ and
$(-3) - 7 = (-3) + (-7) = -(3 + 7)$
$= -10$

(ii) $(-2) - (-5) = (-2) + 5 = 5 - 2 = 3$
and $(-5) - (-2) = (-5) + 2 = -(5 - 2) = -3$

(b) No. It is not true that $a - b = b - a$ in general, because, for example, $7 - (-3) \ne (-3) - 7$.

37. Let $a < b$ so there is a positive integer $c > 0$ for which $a + c = b$. Add n to both sides of this equation to see that $(a + n) + c = b + n$.

Therefore, $a + n < b + n$. Each step is reversible, so the if and only if statement has been proved.

39. (a) $101 + 3 = 104$

(b) $101 + 3 - 5 = 99$

41. $4500 - 510 - 87 - 212 - 725 + 600 + 350 = 3916$, so his balance was \$3916.

43. (a)

t	v
0	$(-32) \cdot (0) + 96 = 96$
1	$(-32) \cdot (1) + 96 = 64$
2	$(-32) \cdot (2) + 96 = 32$
3	$(-32) \cdot (3) + 96 = 0$
4	$(-32) \cdot (4) + 96 = -32$
5	$(-32) \cdot (5) + 96 = -64$
6	$(-32) \cdot (6) + 96 = -96$
7	$(-32) \cdot (7) + 96 = -128$

(b) The ball is moving upward at 96 feet per second.

(c) The ball is not moving at that instant. It has stopped traveling upward and is just about to start falling back down.

(d) The ball is moving downward at 32, 64, 96, and 128 feet per second, respectively (assuming it does not hit anything).

(e) They are additive inverses. At height 0 ft (that is, $t = 0$ and $t = 6$) the ball has the same speed (but opposite direction) on the way up as on the way down. Similar results hold for $t = 1$ and $t = 5$, and also for $t = 2$ and $t = 4$.

45. B. $-4 + 6 = 2$

Section 5.3
Multiplication and Division of Integers

Problem Set 5.3

1. (a) $7 \cdot 11 = 77$

(b) $7 \cdot (-11) = -(7 \cdot 11) = -77$

(c) $(-7) \cdot 11 = -(7 \cdot 11) = -77$

(d) $(-7) \cdot (-11) = 7 \cdot 11 = 77$

3. (a) $36 \div 9 = 4$

(b) $(-36) \div 9 = -(36 \div 9) = -4$

(c) $36 \div (-9) = -(36 \div 9) = -4$

Copyright © 2015 Pearson Education, Inc.

(d) $(-36) \div (-9) = 36 \div 9 = 4$

5. Multiplication: $(-25{,}753) \cdot (-11) = 283{,}283$
Division: $283{,}283 \div (-11) = -25{,}753$;
$283{,}283 \div (-25{,}753) = -11$

7. **(a)** Richer by \$78; $6 \cdot 13 = 78$

(b) Poorer by \$92; $4 \cdot (-23) = -92$

9. **(a)** $6 \cdot 3 = 18$

(b) $4 \cdot (-4) = -16$

(c) $(-3) \cdot (-5) = 15$

11. **(a)**

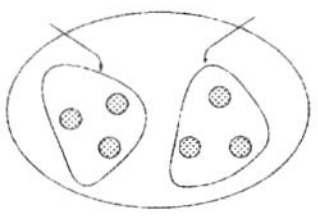

(b)

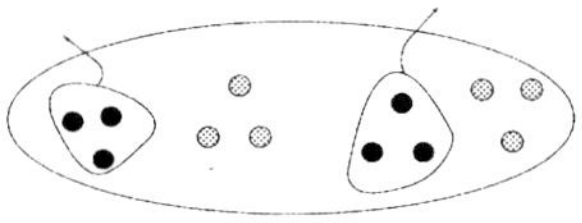

(c)

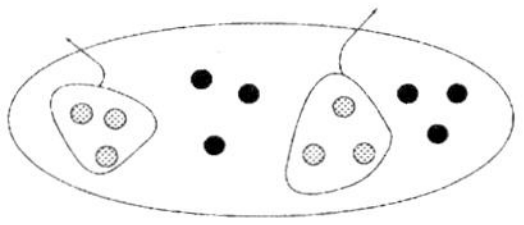

13.

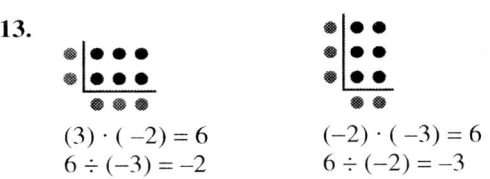

$(3) \cdot (-2) = 6$
$6 \div (-3) = -2$

$(-2) \cdot (-3) = 6$
$6 \div (-2) = -3$

15. Answers will vary.

17. Answers will vary.

19. Kris overlooked the case $n = 0$. When she divided by n, she had assumed that n was not zero.

21. **(a)** Can't tell

(b) Always nonnegative

(c) Always positive

(d) Can't tell

(e) Always nonpositive

23. Yes. $a(b - c) = a[b + (-c)] = ab + a(-c)$
$= ab - ac$. This can be illustrated as shown.

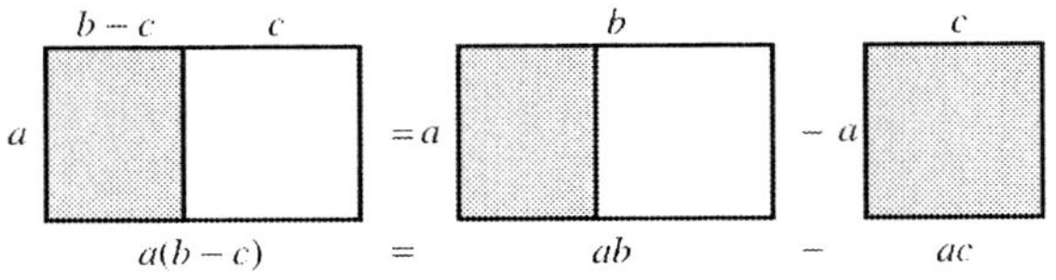

25. **(a)** The products are increasing by positive 3 at each step, so the next terms are
$0 \cdot (-3) = 0$, $(-1) \cdot (-3) = 3$, $(-2) \cdot (-3) = 6$.
That is, negative times negative is positive.

(b) The products are decreasing by positive 3 at each step, so the next terms are
$(-3) \cdot 0 = 0$, $(-3) \cdot 1 = -3$, $(-3) \cdot 2 = -6$.
That is, negative time positive is negative.

27. Since each row (column and diagonal) has the same sum and there are three rows, the sums should be
$$\frac{(-8) + (-6) + (-4) + (-2) + 0 + 2 + 4 + 6 + 8}{3} = 0$$
Also, since there are four sums involving 0 that equal 0 $[(-2) + 0 + 2, (-4) + 0 + 4,$
$(-6) + 0 + 6,$ and $(-8) + 0 + 8]$, and just two sums involving 4 that equal 0 $[(-4) + 0 + 4$
and $(-6) + 2 + 4]$, it follows that 0 must be the middle number and 4 must appear as the middle number on one of the sides. This is enough to complete the array, as shown. Also, except for rotation or reflection of the array, this is the only arrangement possible.

-6	8	-2
4	0	-4
2	-8	6

29. **(a)**

$+_{12}$	0	1	2	3	4	5	6	7	8	9	10	11
0	0	1	2	3	4	5	6	7	8	9	10	11
1	1	2	3	4	5	6	7	8	9	10	11	0
2	2	3	4	5	6	7	8	9	10	11	0	1
3	3	4	5	6	7	8	9	10	11	0	1	2
4	4	5	6	7	8	9	10	11	0	1	2	3
5	5	6	7	8	9	10	11	0	1	2	3	4
6	6	7	8	9	10	11	0	1	2	3	4	5
7	7	8	9	10	11	0	1	2	3	4	5	6
8	8	9	10	11	0	1	2	3	4	5	6	7
9	9	10	11	0	1	2	3	4	5	6	7	8
10	10	11	0	1	2	3	4	5	6	7	8	9
11	11	0	1	2	3	4	5	6	7	8	9	10

(continued on next page)

52 *Chapter 5 Integers*

(*continued*)

$\times_{12}$	0	1	2	3	4	5	6	7	8	9	10	11
0	0	0	0	0	0	0	0	0	0	0	0	0
1	0	1	2	3	4	5	6	7	8	9	10	11
2	0	2	4	6	8	10	0	2	4	6	8	10
3	0	3	6	9	0	3	6	9	0	3	6	9
4	0	4	8	0	4	8	0	4	8	0	4	8
5	0	5	10	3	8	1	6	11	4	9	2	7
6	0	6	0	6	0	6	0	6	0	6	0	6
7	0	7	2	9	4	11	6	1	8	3	10	5
8	0	8	4	0	8	4	0	8	4	0	8	4
9	0	9	6	3	0	9	6	3	0	9	6	3
10	0	10	8	6	4	2	0	10	8	6	4	2
11	0	11	10	9	8	7	6	5	4	3	2	1

(b) Clock addition is closed, commutative, and associative. 0 is the additive identity, and every number n has $12 - n$ as its additive inverse.

(c) I_{12} is a number system with the additive inverse property.

(d) Clock multiplication is closed, commutative, and associative, and distributes over clock addition. The number 1 is the multiplicative identity and $0 \times_{12} n = 0$ for all $n \in I_{12}$.

(e) No. For example, $5 \times_{12} 5 = 1$, $7 \times_{12} 7 = 1$, and $11 \times_{12} 11 = 1$.

(f) No. For example, $2 \times_{12} 6 = 0$, $3 \times_{12} 4 = 0$, $3 \times_{12} 8 = 0$, $4 \times_{12} 6 = 0$, and others.

31. (a) The total cost is the cost of one sweatshirt multiplied by the number of sweatshirts purchased, or $585.

(b) $15 \cdot 39 = 585$. If there are 15 sweatshirts, at $39 each, the total cost is $585.

33. Since the rules for Diffy are to always subtract the smaller number from the larger, the difference will always be positive no matter what signs the two number have. Thus, after the first step, all the numbers will be positive, and the process will then continue as before.

35. $(-2)(-5)(-1) = (10)(-1) = -10$

Chapter 5 Review Exercises

1. (a) $7 - 8 = -1$

(b) 10 are black and 5 are red, so the number is $10 - 5 = 5$.

(c) Since there are 15 counters and 15 is an odd number, the counters can represent any odd integer from -15 to 15. The possible numbers are $-15, -13, -11, \ldots,$ 11, 13, 15.

3. (a) 12

(b) -24

5. Answers will vary.

(a) At mail time you receive a bill for $114 and a check for $29. $(-114) + 29 = -85$

(b) The mail carrier brings you a bill for $19 and a check for $66. $(-19) + 66 = 47$

7. The loop on the left represents $5 - 3 = 2$ and the loop on the right represents $1 - 5 = -4$, so the diagram represents $2 + (-4) = -2$.

9. (a) $45 + (-68) = -(68 - 45) = -23$. You are poorer by $23.

(b) $45 - (-68) = 45 + 68 = 113$. You are richer by $113.

11. (a) $5 + (-7) = -(7 - 5) = -2$

(b) $(-27) - (-5) = (-27) + 5$
$= -(27 - 5) = -22$

(c) $(-27) + (-5) = -(27 + 5) = -32$

(d) $5 - (-7) = 5 + 7 = 12$

(e) $8 - (-12) = 8 + 12 = 20$

(f) $8 - 12 = -(12 - 8) = -4$

13. (a) $(-12) + 37 = 37 - 12 = 25$. Her balance is $25.

(b) $(-12) + 37 = 25$

15. (a) $3 \cdot 4 = 12$ (b) $3 \cdot (-4) = -12$

(c) $(-3) \cdot 4 = -12$ (d) $(-3) \cdot (-4) = 12$

Copyright © 2015 Pearson Education, Inc.

Solutions to Chapter 5 Review Exercises **53**

17. (a) $(-8) \cdot (-7) = 8 \cdot 7 = 56$

(b) $8 \cdot (-7) = -(8 \cdot 7) = -56$

(c) $(-8) \cdot 7 = -(8 \cdot 7) = -56$

(d) $84 \div (-12) = -(84 \div 12) = -7$

(e) $(-84) \div 7 = -(84 \div 7) = -12$

(f) $(-84) \div (-7) = 84 \div 7 = 12$

19. If d divides n, there is an integer c such that
$dc = n$. But then
$d \cdot (-c) = -n, (-d) \cdot (-c) = dc = n$, and
$(-d) \cdot c = -dc = -n$. Thus, d divides $-n$,
$-d$ divides n, and $-d$ divides $-n$.

Copyright © 2015 Pearson Education, Inc.

Chapter 6 Fractions and Rational Numbers

Section 6.1
The Basic Concepts of Fractions and Rational Numbers

Problem Set 6.1

1. **(a)** One of the six portions is shaded, so $\dfrac{1}{6}$

    (b) Two of the six portions are shaded, so $\dfrac{2}{6}$

    (c) None of the figure is shaded, so $\dfrac{0}{1}$

3. **(a)** Shade 1 of 8 equal subregions.
    One way to do this is shown.

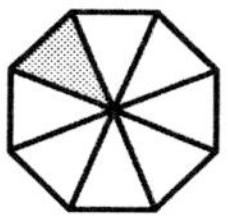

    $\dfrac{1}{8}$

    (b) Shade 6 of 8 equal subregions.
    One way to do this is shown.

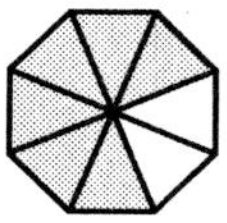

    $\dfrac{6}{8}$

5. **(a)** $\dfrac{4}{5}$ **(b)** $\dfrac{4}{5}$

    (c) $\dfrac{4}{6}$ **(d)** $\dfrac{2}{5}$

    (e) $\dfrac{5}{6}$ **(f)** $\dfrac{7}{10}$

    (g) $\dfrac{9}{14}$

7. **(a)** $\dfrac{4}{10}$ **(b)** $\dfrac{3}{10}$

    (c) $\dfrac{4}{10}$ (squares are also rectangles)

    (d) $\dfrac{6}{10}$

(e) $\dfrac{2}{10}$ (the right triangle and the red parallelogram have no line of symmetry)

9. **(a)** Each tick mark represents $\dfrac{1}{3}$ of one unit.
    From left to right, the arrows represent the fractions $-\dfrac{5}{3}, \dfrac{2}{3}, \dfrac{5}{3}$.

    (b) Each tick mark represents $\dfrac{1}{5}$ of one unit.
    From left to right, the arrow represent the fractions $-\dfrac{5}{5}, \dfrac{3}{5}, \dfrac{7}{5}$.

11. **(a)** One method is shown.

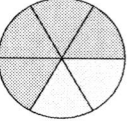

    (b) The set model shows that $\dfrac{4}{6}$ of the coins came up heads.

    (c)

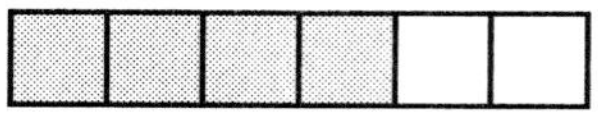

    (d)

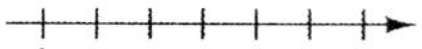

13. **(a)** $\dfrac{3}{6} = \dfrac{1}{2}$ **(b)** $\dfrac{2}{3} = \dfrac{4}{6}$

    (c) $\dfrac{2}{3} = \dfrac{10}{15}$

15. **(a)**

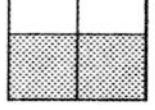

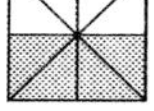

    $\dfrac{2}{4} \quad = \quad \dfrac{4}{8}$

54   Copyright © 2015 Pearson Education, Inc.

(b)

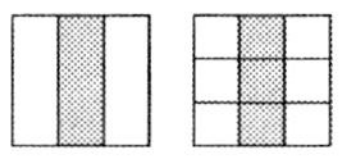

$$\frac{1}{3} = \frac{3}{9}$$

(c)

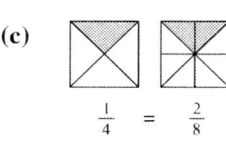

$$\frac{1}{4} = \frac{2}{8}$$

17. (a) A common denominator is LCM(42, 7) = 42. We can write the fractions as $\dfrac{18}{42}$ and $\dfrac{3 \cdot 6}{7 \cdot 6} = \dfrac{18}{42}$. They are equivalent.

(b) A common denominator is LCM(49, 14) = 98. We can write the fractions as $\dfrac{18 \cdot 2}{49 \cdot 2} = \dfrac{36}{98}$ and $\dfrac{5 \cdot 7}{14 \cdot 7} = \dfrac{35}{98}$. Since $\dfrac{36}{98} \neq \dfrac{35}{98}$, they are not equivalent.

(c) A common denominator is LCM(25, 500) = 500. We can write the fractions as $\dfrac{9 \cdot 20}{25 \cdot 20} = \dfrac{180}{500}$ and $\dfrac{140}{500}$. Since $\dfrac{180}{500} \neq \dfrac{140}{500}$, they are not equivalent.

(d) We can write the fractions as $\dfrac{24}{144} = \dfrac{2 \cdot 12}{12 \cdot 12} = \dfrac{2}{12}$ and $\dfrac{32}{96} = \dfrac{8 \cdot 4}{8 \cdot 12} = \dfrac{4}{12}$. Since $\dfrac{2}{12} \neq \dfrac{4}{12}$, the fractions are not equivalent.

19. (a) Yes. $4 \cdot 3 = 12$ and $9 \cdot 3 = 27$.

(b) Yes. This is a consequence of the fundamental law of fractions.

(c) Yes. $4 + 3 = 7$ and $9 + 3 = 12$.

(d) No. $\dfrac{7}{12} \neq \dfrac{4}{9}$ because $7 \cdot 9 \neq 12 \cdot 4$.

21. (a)

$$\begin{array}{r} 3 \\ 2\overline{)6} \\ 2\overline{)12} \\ 2\overline{)24} \end{array} \qquad \begin{array}{r} 3 \\ 3\overline{)9} \\ 2\overline{)18} \\ 2\overline{)36} \\ 2\overline{)72} \end{array}$$

$$\begin{array}{r} 2\overline{)48} \\ 2\overline{)96} \end{array} \qquad \begin{array}{r} 2\overline{)144} \\ 2\overline{)288} \end{array}$$

$$\frac{96}{288} = \frac{2^5 \cdot 3^1}{2^5 \cdot 3^2} = \frac{1}{3^1} = \frac{1}{3}$$

(b)

$$\begin{array}{r} 19 \\ 13\overline{)247} \end{array} \qquad \begin{array}{r} 5 \\ 5\overline{)25} \\ 3\overline{)75} \end{array}$$

$$\frac{247}{-75} = \frac{13^1 \cdot 19^1}{(-1) \cdot 3^1 \cdot 5^2} = \frac{247}{-75} = \frac{-247}{75}$$

(c)

$$\begin{array}{r} 7 \\ 5\overline{)35} \\ 3\overline{)105} \\ 3\overline{)315} \\ 2\overline{)630} \\ 2\overline{)1260} \\ 2\overline{)2520} \end{array} \qquad \begin{array}{r} 7 \\ 3\overline{)21} \\ 3\overline{)63} \\ 3\overline{)189} \\ 2\overline{)378} \end{array}$$

$$\frac{2520}{378} = \frac{2^3 \cdot 3^2 \cdot 5^1 \cdot 7^1}{2^1 \cdot 3^3 \cdot 7^1} = \frac{2^2 \cdot 5^1}{3^1} = \frac{20}{3}$$

23. (a) The least common denominator is LCM(8, 6) = 24. We write $\dfrac{3 \cdot 3}{8 \cdot 3} = \dfrac{9}{24}$ and $\dfrac{5 \cdot 4}{6 \cdot 4} = \dfrac{20}{24}$. The fractions are equivalent to $\dfrac{9}{24}$ and $\dfrac{20}{24}$.

(b) The least common denominator is LCM(7, 5, 3) = 105. We write $\dfrac{1 \cdot 15}{7 \cdot 15} = \dfrac{15}{105}, \dfrac{4 \cdot 21}{5 \cdot 21} = \dfrac{84}{105}$, and $\dfrac{2 \cdot 35}{3 \cdot 35} = \dfrac{70}{105}$. The fractions are equivalent to $\dfrac{15}{105}, \dfrac{84}{105}$, and $\dfrac{70}{105}$.

(c) The least common denominator is LCM(12, 32) = 96. We write $\dfrac{17 \cdot 8}{12 \cdot 8} = \dfrac{136}{96}$ and $\dfrac{7 \cdot 3}{32 \cdot 3} = \dfrac{21}{96}$. The fractions are equivalent to $\dfrac{136}{96}$ and $\dfrac{21}{96}$.

56 *Chapter 6 Fractions and Rational Numbers*

(d) The fractions can be simplified to $\frac{1}{3}$ and $\frac{4}{3}$, which have a least common denominator of 3.

25. (a) Since $2 \cdot 12 > 3 \cdot 7, \frac{2}{3} > \frac{7}{12}$. The rational numbers in order are $\frac{7}{12}, \frac{2}{3}$.

(b) Since $2 \cdot 6 < 3 \cdot 5, \frac{2}{3} < \frac{5}{6}$. The rational numbers in order are $\frac{2}{3}, \frac{5}{6}$.

(c) Since $5 \cdot 36 > 6 \cdot 29, \frac{5}{6} > \frac{29}{36}$. The rational numbers in order are $\frac{29}{36}, \frac{5}{6}$.

(d) The denominators are positive. Since $(-5) \cdot 9 > 6 \cdot (-8), \frac{-5}{6} > \frac{-8}{9}$. The rational numbers in order are $\frac{-8}{9}, \frac{-5}{6}$.

(e) Since $\frac{2}{3} = \frac{24}{36}, \frac{5}{6} = \frac{30}{36}$, and $\frac{8}{9} = \frac{32}{36}$, the rational numbers in order are $\frac{24}{36}, \frac{29}{36}, \frac{30}{36}, \frac{32}{36},$ or $\frac{2}{3}, \frac{29}{36}, \frac{5}{6}, \frac{8}{9}$.

27. Two numbers. $\frac{27}{36} = \frac{21}{28} = \frac{3}{4}$, and $4 = \frac{24}{6} = \frac{-8}{-2}$.

29. (a) trapezoid = $\frac{1}{2}$, triangle = $\frac{1}{6}$, rhombus = $\frac{1}{3}$

(b) hexagon = $\frac{2}{1}$, triangle = $\frac{1}{3}$, rhombus = $\frac{2}{3}$

(c) hexagon = $\frac{3}{1}$, trapezoid = $\frac{3}{2}$, triangle = $\frac{1}{2}$

31. Answers will vary, but typically include a figure of pizzas cut into fourths and sixths to show that $\frac{1}{4}$ of the pizza is larger than $\frac{1}{6}$ of the pizza.

33. Miley has used excellent reasoning to draw a valid conclusion.

35. (a) There are three groups of two, so the unit consists of four groups of two.

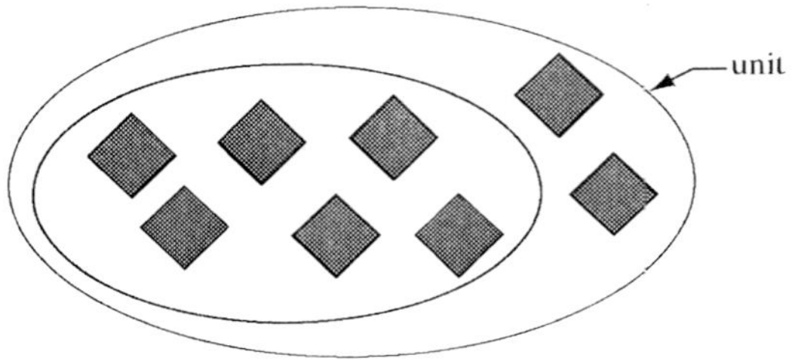

(b) There are five groups of three, so the unit consists of three units of three.

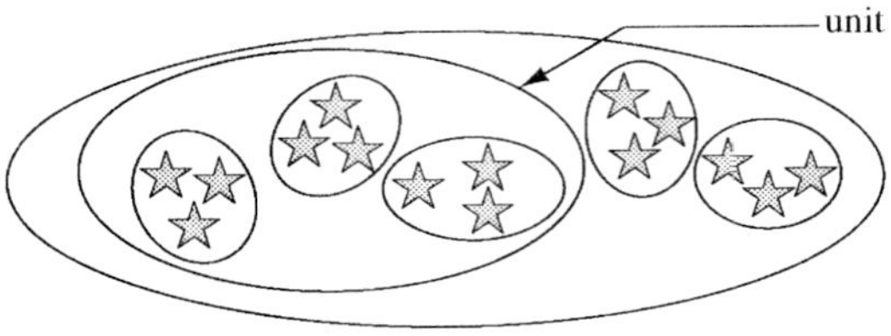

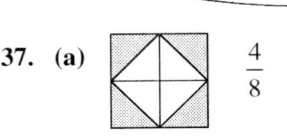

37. (a) $\frac{4}{8}$

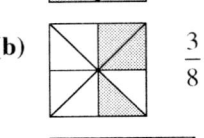

(b) $\frac{3}{8}$

(c) 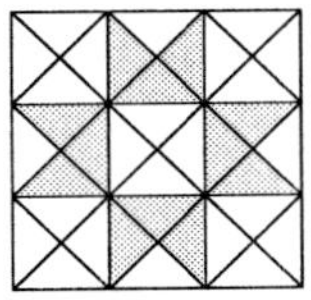 $\frac{12}{36}$

(d) 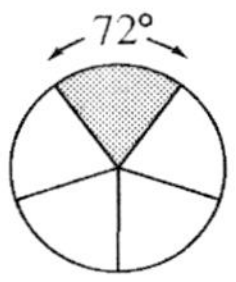 $\frac{1}{5}$

Copyright © 2015 Pearson Education, Inc.

(e) 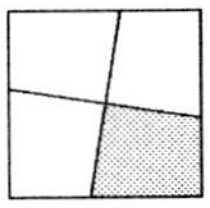 $\dfrac{1}{4}$

(f) 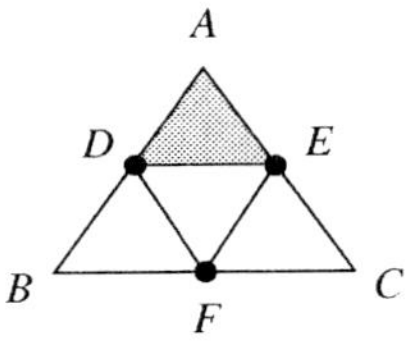 $\dfrac{1}{4}$

(g) 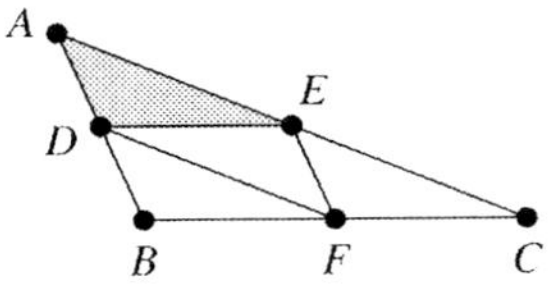 $\dfrac{1}{4}$

39. Add tick marks to show that the rope was originally 60 feet long.

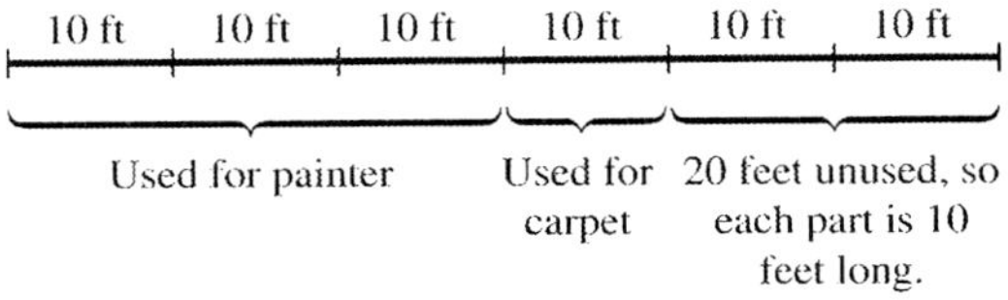

| 10 ft | 10 ft | 10 ft | 10 ft | 10 ft | 10 ft |

Used for painter   Used for   20 feet unused, so
                   carpet   each part is 10
                              feet long.

41. Your investigation should suggest this is a general property of Pascal's triangle. [Optional proof: Suppose the nth row of Pascal's triangle is separated to have k entries on the left and $n + 1 - k$ to the right of the divider, to look like $\binom{n}{0}, \binom{n}{1}, \cdots, \binom{n}{k-1} \Big| \binom{n}{k}, \cdots, \binom{n}{n}$. In Chapter 9 it is shown that $\binom{n}{k} = \dfrac{n(n-1)\cdots(n-k+1)}{k(k-1)\cdots 3 \cdot 2 \cdot 1}$, from which it follows $\dfrac{\binom{n}{k-1}}{\binom{n}{k}} = \dfrac{k}{n+1-k}$.]

43. (a) Since $\dfrac{3}{4} = \dfrac{18}{24}$, the tank contains 18 gallons.

(b) Since $\dfrac{1}{4} = \dfrac{6}{24}$, the tank contains 6 gallons.

(c) Since $\dfrac{5}{8} = \dfrac{15}{24}$, the tank contains 15 gallons.

45. (a) Since there are two possible outcomes (heads or tails) and only one of these is "successful," the probability is $\dfrac{1}{2}$.

(b) Since there are 12 face cards (jack, queen, and king in each of 4 suits) out of 52 cards, the probability is $\dfrac{12}{52}$ or $\dfrac{3}{13}$.

(c) Since there are 3 even numbers out of six possible outcomes, the probability is $\dfrac{3}{6}$ or $\dfrac{1}{2}$.

(d) Since there are 25 green marbles out of $20 + 30 + 25 = 75$ marbles, the probability is $\dfrac{25}{75}$ or $\dfrac{1}{3}$.

(e) There are 50 successful outcomes (20 red and 30 blue) out of the 75 marbles so the probability is $\dfrac{50}{75} = \dfrac{2}{3}$.

47. B.   **49.** A.   **51.** B.

Section 6.2
Addition and Subtraction of Fractions

Problem Set 6.2

1. (a) $\dfrac{1}{3} + \dfrac{1}{2} = \dfrac{5}{6}$ **(b)** $\dfrac{2}{3} + \dfrac{1}{2} = \dfrac{7}{6}$

3. (a)

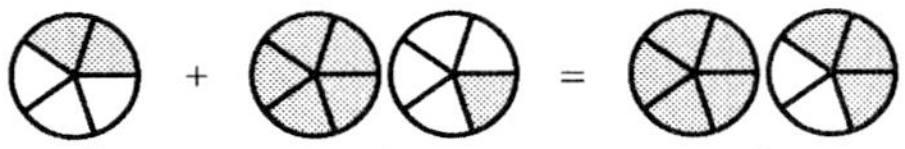

$\dfrac{2}{5} + \dfrac{6}{5} = \dfrac{8}{5}$

(b)

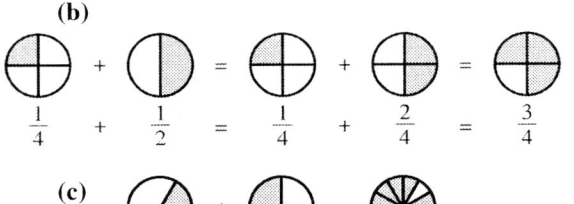

$\dfrac{1}{4} + \dfrac{1}{2} = \dfrac{1}{4} + \dfrac{2}{4} = \dfrac{3}{4}$

(c)

$\dfrac{2}{3} + \dfrac{1}{4} = \dfrac{11}{12}$

5. (a)

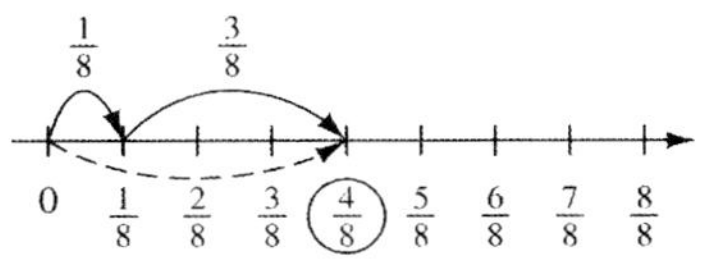

58   *Chapter 6 Fractions and Rational Numbers*

(b)

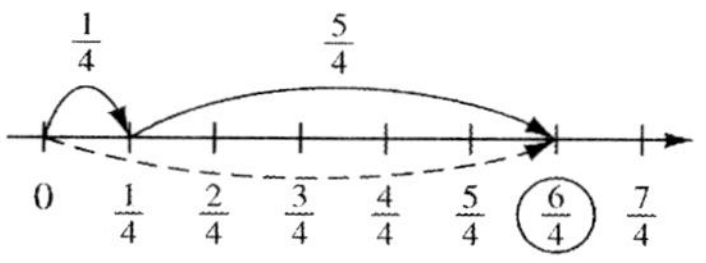

(c)

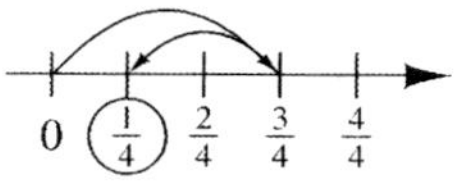

$$\frac{3}{4} + \frac{-2}{4} = \frac{1}{4}$$

7. (a) $\dfrac{2}{7} + \dfrac{3}{7} = \dfrac{2+3}{7} = \dfrac{5}{7}$

(b) $\dfrac{6}{5} + \dfrac{4}{5} = \dfrac{6+4}{5} = \dfrac{10}{5} = \dfrac{2}{1}$ or 2

(c) $\dfrac{3}{8} + \dfrac{11}{24} = \dfrac{9}{24} + \dfrac{11}{24} = \dfrac{9+11}{24} = \dfrac{20}{24} = \dfrac{5}{6}$

(d) $\dfrac{6}{13} + \dfrac{2}{5} = \dfrac{30}{65} + \dfrac{26}{65} = \dfrac{30+26}{65} = \dfrac{56}{65}$

9. (a) Since $9 = 2 \cdot 4 + 1$,
$$\frac{9}{4} = \frac{2 \cdot 4 + 1}{4} = 2 + \frac{1}{4} = 2\frac{1}{4}.$$

(b) Since $17 = 5 \cdot 3 + 2$,
$$\frac{17}{3} = \frac{5 \cdot 3 + 2}{3} = 5 + \frac{2}{3} = 5\frac{2}{3}.$$

(c) Since $111 = 4 \cdot 23 + 19$,
$$\frac{111}{23} = \frac{4 \cdot 23 + 19}{23} = 4 + \frac{19}{23} = 4\frac{19}{23}.$$

(d) Since $3571 = 35 \cdot 100 + 71$,
$$\frac{3571}{-100} = \frac{-3571}{100} = \frac{-(35 \cdot 100 + 71)}{100}$$
$$= -\left(35 + \frac{71}{100}\right) = -35\frac{71}{100}$$

11. (a) $\dfrac{5}{6} - \dfrac{1}{4} = \dfrac{7}{12}$

(b) Fraction strip:

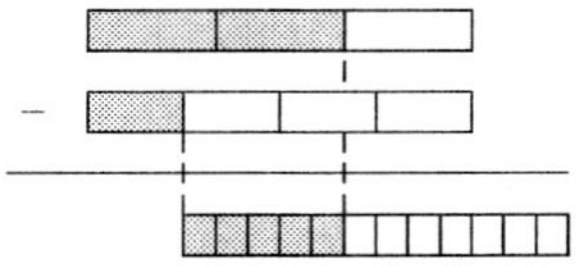

$$\frac{2}{3} - \frac{1}{4} = \frac{5}{12}$$

13. (a) $\dfrac{3}{4} - \dfrac{1}{3} = \dfrac{9}{12} - \dfrac{4}{12} = \dfrac{5}{12}$

(b) Number-line:

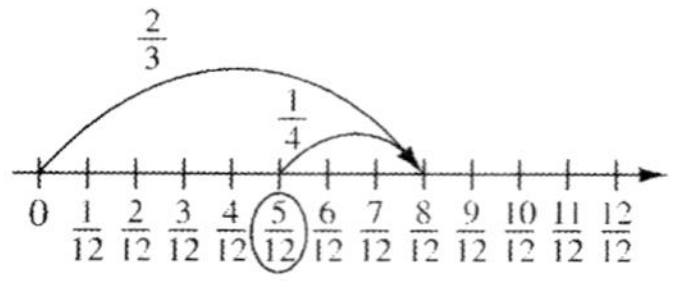

$$\frac{2}{3} - \frac{1}{4} = \frac{5}{12}$$

15. (a) $\dfrac{6}{8} - \dfrac{5}{12} = \dfrac{9}{12} - \dfrac{5}{12} = \dfrac{9-5}{12} = \dfrac{4}{12} = \dfrac{1}{3}$

(b) $\dfrac{1}{4} - \dfrac{14}{56} = \dfrac{1}{4} - \dfrac{1}{4} = 0$

(c) $\dfrac{137}{214} - \dfrac{-1}{3} = \dfrac{137 \cdot 3 - 214 \cdot (-1)}{214 \cdot 3} = \dfrac{625}{642}$

(d) $\dfrac{-23}{100} - \dfrac{198}{1000} = \dfrac{-230}{1000} - \dfrac{198}{1000} = \dfrac{(-230) - 198}{1000}$
$$= \dfrac{-428}{1000} = \dfrac{-107}{250}$$

17. (a) $\dfrac{2}{3} < \dfrac{3}{4}$ if, and only if, $\dfrac{3}{4} - \dfrac{2}{3} > 0$.
$$\frac{3}{4} - \frac{2}{3} = \frac{9}{12} - \frac{8}{12} = \frac{1}{12} > 0$$

(b) $\dfrac{4}{5} < \dfrac{14}{17}$ is, and only if, $\dfrac{14}{17} - \dfrac{4}{5} > 0$.
$$\frac{14}{17} - \frac{4}{5} = \frac{70}{85} - \frac{68}{85} = \frac{2}{85} > 0$$

(c) $\dfrac{19}{10} < \dfrac{99}{50}$ if, and only if, $\dfrac{99}{50} - \dfrac{19}{10} > 0$.
$$\frac{99}{50} - \frac{19}{10} = \frac{99}{50} - \frac{95}{50} = \frac{4}{50} = \frac{2}{25} > 0$$

19. Answers will vary.

21. Answers will vary. Dannea should have eaten half of the remaining 6 pieces not half of the original pizza.

Copyright © 2015 Pearson Education, Inc.

Solutions to Problem Set 6.3   **59**

23. Melanie's two-game average is indeed $\dfrac{3}{8}$, but she cannot determine this by adding the fractions $\dfrac{1}{3}$ and $\dfrac{2}{5}$ since their sum is $\dfrac{11}{15}$, as John correctly pointed out. When fractions are added according to Melanie's rule, it is sometimes know as taking the *mediant*, and is written as $\dfrac{1}{3} \oplus \dfrac{2}{5} = \dfrac{3}{8}$.

25. Since $\dfrac{1}{2} + \dfrac{1}{3} + \dfrac{1}{9} = \dfrac{9+6+2}{18} = \dfrac{17}{18}$, the will did not call for the distribution of all of the horses.

27. (a) $\dfrac{1}{10} + 6 \cdot \dfrac{1}{70} + 6 \cdot \dfrac{1}{21} + 3 \cdot \dfrac{1}{14} + 3 \cdot \dfrac{11}{105}$

$= \dfrac{1}{10} + \dfrac{6}{70} + \dfrac{2}{7} + \dfrac{3}{14} + \dfrac{11}{35}$

$= \dfrac{7+6+20+15+22}{70} = \dfrac{70}{70} = 1$

(b) $\dfrac{1}{10} + 6 \cdot \dfrac{1}{70} + 3 \cdot \dfrac{1}{14} = \dfrac{7+6+15}{70} = \dfrac{28}{70} = \dfrac{2}{5}$

(c) $\dfrac{1}{10} + 3 \cdot \dfrac{1}{70} = \dfrac{7+3}{70} = \dfrac{10}{70} = \dfrac{1}{7}$

29. (a) It terminates.

$\boxed{\tfrac{3}{4}}\ \boxed{\tfrac{2}{3}}\ \boxed{\tfrac{1}{2}}\ \boxed{\tfrac{1}{4}}$
$\boxed{\tfrac{1}{12}}\ \boxed{\tfrac{1}{6}}\ \boxed{\tfrac{1}{4}}\ \boxed{\tfrac{1}{2}}$
$\boxed{\tfrac{1}{12}}\ \boxed{\tfrac{1}{12}}\ \boxed{\tfrac{1}{4}}\ \boxed{\tfrac{5}{12}}$
$\boxed{0}\ \boxed{\tfrac{1}{6}}\ \boxed{\tfrac{1}{6}}\ \boxed{\tfrac{1}{3}}$
$\boxed{\tfrac{1}{6}}\ \boxed{0}\ \boxed{\tfrac{1}{6}}\ \boxed{\tfrac{1}{3}}$
$\boxed{\tfrac{1}{6}}\ \boxed{\tfrac{1}{6}}\ \boxed{\tfrac{1}{6}}\ \boxed{\tfrac{1}{6}}$
$\boxed{0}\ \boxed{0}\ \boxed{0}\ \boxed{0}$

(b) It terminates.

$\boxed{\tfrac{2}{7}}\ \boxed{\tfrac{4}{5}}\ \boxed{\tfrac{3}{2}}\ \boxed{\tfrac{5}{6}}$
$\boxed{\tfrac{18}{35}}\ \boxed{\tfrac{7}{10}}\ \boxed{\tfrac{2}{3}}\ \boxed{\tfrac{23}{42}}$
$\boxed{\tfrac{13}{70}}\ \boxed{\tfrac{1}{30}}\ \boxed{\tfrac{5}{42}}\ \boxed{\tfrac{1}{30}}$
$\boxed{\tfrac{16}{105}}\ \boxed{\tfrac{3}{35}}\ \boxed{\tfrac{3}{35}}\ \boxed{\tfrac{16}{105}}$
$\boxed{\tfrac{1}{15}}\ \boxed{0}\ \boxed{\tfrac{1}{15}}\ \boxed{0}$
$\boxed{\tfrac{1}{15}}\ \boxed{\tfrac{1}{15}}\ \boxed{\tfrac{1}{15}}\ \boxed{\tfrac{1}{15}}$
$\boxed{0}\ \boxed{0}\ \boxed{0}\ \boxed{0}$

(c) Yes. Let d be the least common denominator of the top row entries in a fractional DIFFY array. If we multiply each top row entry by d, we obtain integers and we may perform DIFFY on these integers. Each entry in the fractional DIFFY array will be $\dfrac{1}{d}$ times the corresponding entry in the integer array. The fractional array must terminate because the integer array will terminate with 0, 0, 0, 0, and of course $\dfrac{0}{d} = 0$.

(Note that even if negative integers are present, each row of the integer array after the first will contain only whole numbers, so the result for whole numbers still applies.)

31. The wall is $3\dfrac{1}{2} + \dfrac{5}{8} + \dfrac{5}{8} = 3\dfrac{1}{2} + \dfrac{10}{8} =$

$3\dfrac{1}{2} + \dfrac{5}{4} = 3\dfrac{2}{4} + 1\dfrac{1}{4} = 4\dfrac{3}{4}$ inches wide.

33. The original piece is 2 feet or 24 inches long. The remaining piece is $24 - 10\dfrac{1}{2} - \dfrac{1}{16} =$

$24 - 10\dfrac{8}{16} - \dfrac{1}{16} = 23\dfrac{16}{16} - 10\dfrac{8}{16} - \dfrac{1}{16} =$

$13\dfrac{8}{16} - \dfrac{1}{16} = 13\dfrac{7}{16}$ inches long.

35. A. $\dfrac{5}{8} + \dfrac{1}{8} = \dfrac{6}{8}$

37. D. $1 - \left(\dfrac{3}{8} + \dfrac{1}{4} \right) = 1 - \left(\dfrac{3}{8} + \dfrac{2}{8} \right) = 1 - \dfrac{5}{8} = \dfrac{3}{8}$

Section 6.3
Multiplication and Division of Fractions

Problem Set 6.3

1. $4 \times \dfrac{2}{5} = \dfrac{8}{5}$

3. (a) $4 \cdot \dfrac{3}{8} = \dfrac{12}{8}$

Copyright © 2015 Pearson Education, Inc.

60   *Chapter 6 Fractions and Rational Numbers*

(b) $-4 \cdot \dfrac{3}{8} = \dfrac{-12}{8}$

5. $\dfrac{4}{6} \times 5 = 5 \times \dfrac{4}{6} = \dfrac{20}{6} = 3\dfrac{2}{6}$

7. **(a)** $3 \times \dfrac{5}{2} = \dfrac{15}{2}$ **(b)** $\dfrac{2}{3} \times \dfrac{3}{2} = \dfrac{6}{6}$

   (c) $\dfrac{3}{4} \times 2 = \dfrac{6}{4}$

9. $2\dfrac{1}{4} \times 3\dfrac{1}{2} = \dfrac{9}{4} \times \dfrac{7}{2} = \dfrac{63}{8} = 7\dfrac{7}{8}$

The area is $7\dfrac{7}{8}$ square miles. This can be seen in the diagram below, which shows the plot divided into six full square miles, three $\dfrac{1}{4}$-square mile rectangular regions, and nine $\dfrac{1}{8}$-square mile rectangular regions. This gives a total area of $6 + \dfrac{3}{4} + \dfrac{9}{8} = 7\dfrac{7}{8}$ square miles .

11. **(a)** The reciprocal of $\dfrac{1}{8}$ is $\dfrac{8}{1}$ or 8.

   (b) The reciprocal of $5 = \dfrac{5}{1}$ is $\dfrac{1}{5}$.

   (c) The reciprocal of $1 = \dfrac{1}{1}$ is $\dfrac{1}{1}$ or 1 .

13. **(a)** $\dfrac{5}{4} \div \dfrac{3}{4} = \dfrac{5}{3}$

   (b) $\dfrac{7}{8} \div \dfrac{7}{11} = \dfrac{11}{8}$

   (c) $\dfrac{2}{5} \div \dfrac{7}{5} = \dfrac{2}{7}$

   (d) $\dfrac{4}{7} \div \dfrac{4}{21} = \dfrac{21}{7} = 3$

15. **(a)** $\dfrac{2}{3} \cdot \left(\dfrac{3}{4} + \dfrac{9}{12} \right) = \dfrac{2}{3} \cdot \left(\dfrac{9}{12} + \dfrac{9}{12} \right) = \dfrac{2}{3} \cdot \left(\dfrac{18}{12} \right)$

$= \dfrac{2 \cdot 18}{3 \cdot 12} = \dfrac{36}{36} = 1$

   (b) $\left(\dfrac{3}{5} - \dfrac{3}{10} \right) \div \dfrac{6}{5} = \left(\dfrac{6}{10} - \dfrac{3}{10} \right) \div \dfrac{6}{5} = \left(\dfrac{3}{10} \right) \div \dfrac{6}{5}$

$= \dfrac{3}{10} \cdot \dfrac{5}{6} = \dfrac{15}{60} = \dfrac{1}{4}$

   (c) $\left(\dfrac{2}{5} \div \dfrac{4}{15} \right) \cdot \dfrac{2}{3} = \left(\dfrac{2}{5} \cdot \dfrac{15}{4} \right) \cdot \dfrac{2}{3}$

$= \dfrac{30}{20} \cdot \dfrac{2}{3} = \dfrac{60}{60} = 1$

17. **(a)** $\dfrac{2}{5}x - \dfrac{3}{4} = \dfrac{1}{2}$

Add $\dfrac{3}{4}$ to each side:

$\dfrac{2}{5}x - \dfrac{3}{4} + \dfrac{3}{4} = \dfrac{1}{2} + \dfrac{3}{4}$

$\dfrac{2}{5}x = \dfrac{2}{4} + \dfrac{3}{4} = \dfrac{5}{4}$

Multiply each side by $\dfrac{5}{2}$:

$\dfrac{5}{2} \cdot \dfrac{2}{5}x = \dfrac{5}{2} \cdot \dfrac{5}{4}$

$x = \dfrac{25}{8} = 3\dfrac{1}{8}$

   (b) $\dfrac{2}{3}x + \dfrac{1}{4} = \dfrac{3}{2}x$

Subtract $\dfrac{2}{3}x$ from both sides:

$\dfrac{2}{3}x + \dfrac{1}{4} - \dfrac{2}{3}x = \dfrac{3}{2}x - \dfrac{2}{3}x$

$\dfrac{1}{4} = \dfrac{9}{6}x - \dfrac{4}{6}x = \dfrac{5}{6}x$

Multiply both sides by $\dfrac{6}{5}$:

$\dfrac{6}{5} \cdot \dfrac{1}{4} = \dfrac{6}{5} \cdot \dfrac{5}{6}x$

$\dfrac{6}{20} = \dfrac{3}{10} = x$

Copyright © 2015 Pearson Education, Inc.

19. First, divide each acre in sixths, say with one horizontal and two vertical segments through each square. Using grouping, we see the 3 acres can be grouped into 3 groups of $\frac{5}{6}$ of an acre and $\frac{3}{5}$ of a group. That is, Sean will need $3\frac{3}{5}$ hours (or 3 hours and 36 minutes) to complete the mowing job. We have shown that $3 \div \frac{5}{6} = 3\frac{3}{5}$.

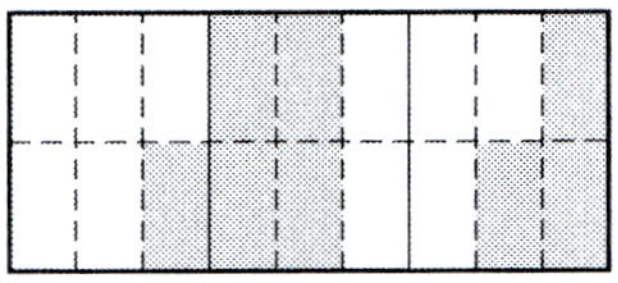

21. Gerry's equation is the same as $x \cdot \frac{5}{3} = 25$.

Multiply both sides of the equation by $\frac{3}{5}$, the reciprocal of $\frac{5}{3}$. Then we see that

$$x \cdot \frac{5}{3} \cdot \frac{3}{5} = 25 \cdot \frac{3}{5}, \text{ so that}$$

$$x = x \cdot 1 = x \cdot \frac{5}{3} \cdot \frac{3}{5} = 25 \cdot \frac{3}{5} = \frac{25 \cdot 3}{5} = 5 \cdot 3 = 15.$$

Therefore, Gerry needs 15 stepping stones.

23. Answers will vary. The problem really is asking for $\frac{2}{3} - \frac{1}{4}$.

25. Answers will vary. The problem asks about one-fourth of a class, not $\frac{1}{4}$ of $\frac{2}{3}$.

27. Answers will vary. This is a correct example, showing that $1\frac{3}{4} \div \frac{1}{2} = \frac{7}{4} \cdot 2 = \frac{14}{4} = 3\frac{2}{4} = 3\frac{1}{2}$ recipes can be made.

29. (a) Let r, g, and b denote the number of red, green, and blue marbles, respectively. Then

$$\frac{1}{5}(r + g + b) = r$$
$$b = \frac{1}{3}g$$
$$r = g - 10$$

Multiply both sides of the first equation by 5, then subtract r from each side to obtain $g + b = 4r$. Multiplying both sides of the second equation by 3, we have $g = 3b$. Substitute $3b$ for g in the first equation: $3b + b = 4r \Rightarrow 4b = 4r \Rightarrow b = r$. $g = 3b \Rightarrow g = 3r$. Substitute $3r$ for g in the third equation: $r = 3r - 10 \Rightarrow -2r = -10 \Rightarrow r = 5$. Then $b = 5$ and $3 \cdot 5 = 15 = g$. So there are 5 red marbles, 5 blue marbles, and 15 green marbles, for a total of 25 marbles.

(b) Let the five loops in the diagram each have the same number of marbles, with the R loop containing the red marbles. The remaining four loops are B and 4 G loops, which make it clear that the number of blue marbles in the B loop is $1/3$ of the number of green marbles in the three G loops. Since there are 10 fewer red marbles than green marbles, each loop contain 5 marbles. Then we see that there are 25 marbles in all, 5 red marbles, 5 blue marbles, and 15 green marbles.

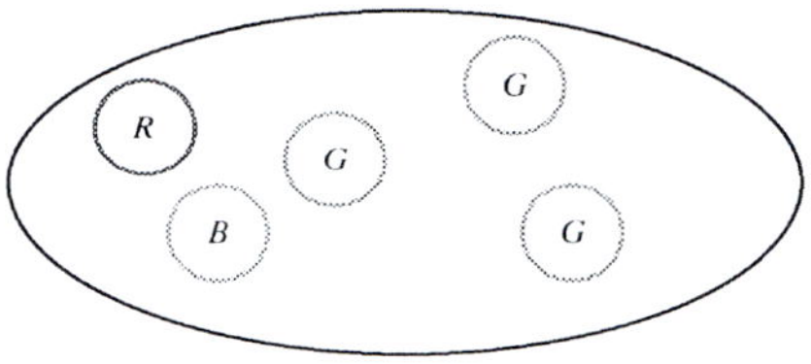

31. Make a cut perpendicular to the cuts already made, which is $\frac{1}{4}$ of the width of the cake from a side. The three original children are each left with $\frac{3}{4}$ of an original piece, which is $\frac{3}{4} \times \frac{1}{3} = \frac{1}{4}$ of the original cake. Each trimmed off piece of the cake is $\frac{1}{4} \times \frac{1}{3} = \frac{1}{12}$ of the cake. The fourth friend is given the three trimmed pieces, which is $\frac{1}{12} + \frac{1}{12} + \frac{1}{12} = \frac{3}{12} = \frac{1}{4}$ of the cake, as needed in an equal share.

62   *Chapter 6 Fractions and Rational Numbers*

33. (a) $\dfrac{2}{5} \div \dfrac{3}{4} = \dfrac{8}{15}$

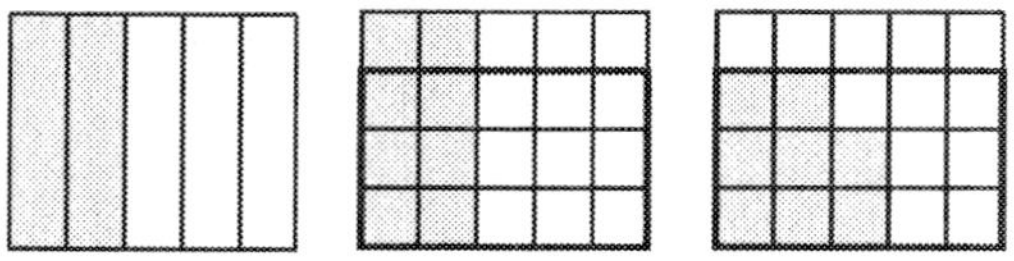

(b) $\dfrac{3}{5} \div \dfrac{5}{6} = \dfrac{18}{25}$

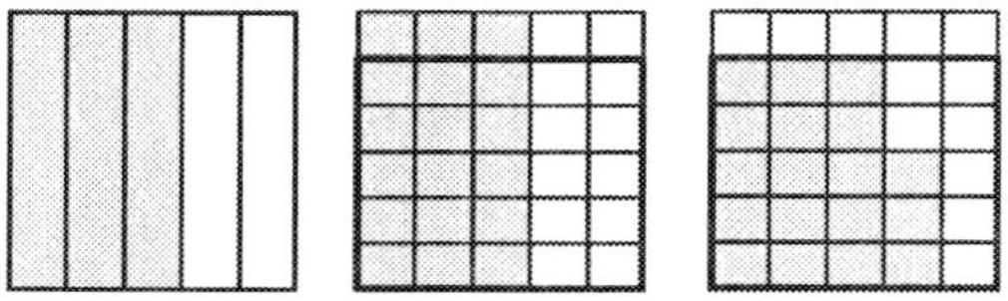

(c) $\dfrac{7}{8} \div \dfrac{1}{3} = \dfrac{21}{8} = 2\dfrac{5}{8}$

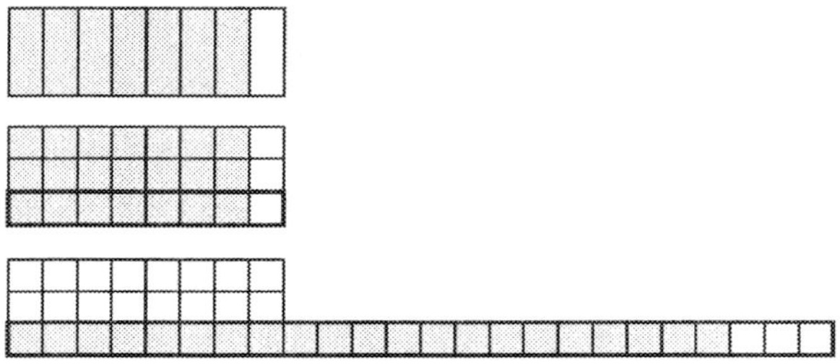

35. (a) The process terminates when one row consists of all ones.

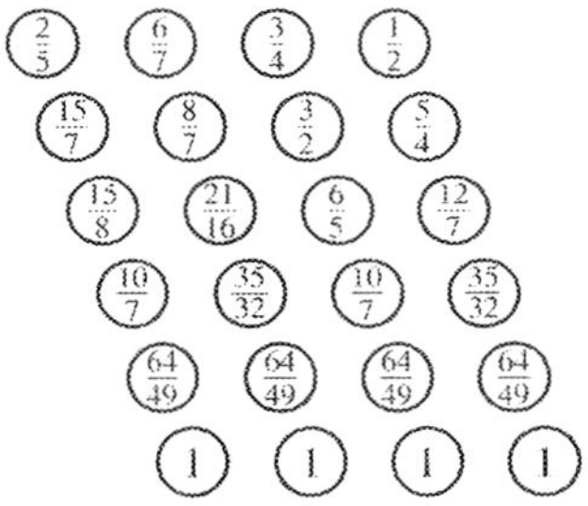

$\left(\tfrac{2}{5}\right)$ $\left(\tfrac{6}{7}\right)$ $\left(\tfrac{3}{4}\right)$ $\left(\tfrac{1}{2}\right)$

$\left(\tfrac{15}{7}\right)$ $\left(\tfrac{8}{7}\right)$ $\left(\tfrac{3}{2}\right)$ $\left(\tfrac{5}{4}\right)$

$\left(\tfrac{15}{8}\right)$ $\left(\tfrac{21}{16}\right)$ $\left(\tfrac{6}{5}\right)$ $\left(\tfrac{12}{7}\right)$

$\left(\tfrac{10}{7}\right)$ $\left(\tfrac{35}{32}\right)$ $\left(\tfrac{10}{7}\right)$ $\left(\tfrac{35}{32}\right)$

$\left(\tfrac{64}{49}\right)$ $\left(\tfrac{64}{49}\right)$ $\left(\tfrac{64}{49}\right)$ $\left(\tfrac{64}{49}\right)$

(1) (1) (1) (1)

(b) $\left(\tfrac{2}{7}\right)$ $\left(\tfrac{4}{5}\right)$ $\left(\tfrac{3}{2}\right)$ $\left(\tfrac{5}{6}\right)$

$\left(\tfrac{14}{5}\right)$ $\left(\tfrac{15}{8}\right)$ $\left(\tfrac{9}{5}\right)$ $\left(\tfrac{35}{12}\right)$

$\left(\tfrac{11}{75}\right)$ $\left(\tfrac{25}{24}\right)$ $\left(\tfrac{175}{108}\right)$ $\left(\tfrac{25}{24}\right)$

$\left(\tfrac{896}{625}\right)$ $\left(\tfrac{14}{9}\right)$ $\left(\tfrac{14}{9}\right)$ $\left(\tfrac{896}{625}\right)$

$\left(\tfrac{625}{576}\right)$ (1) $\left(\tfrac{625}{576}\right)$ (1)

$\left(\tfrac{625}{576}\right)$ $\left(\tfrac{625}{576}\right)$ $\left(\tfrac{625}{576}\right)$ $\left(\tfrac{625}{576}\right)$

(1) (1) (1) (1)

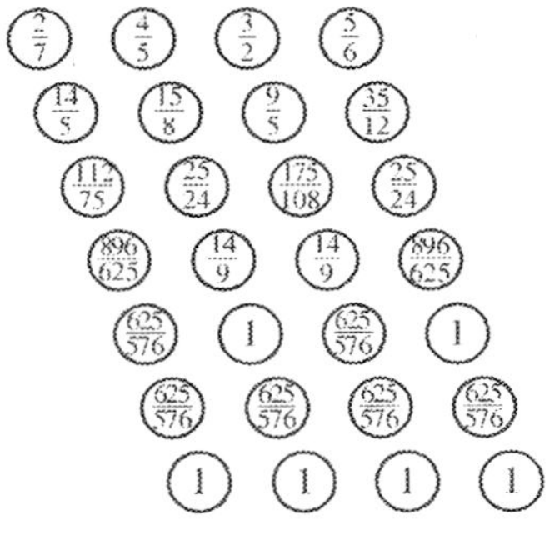

37. In each case, the number of votes must be rounded *up* to the next whole number.

3 members: $\dfrac{3}{4} \cdot 3 = \dfrac{9}{4} = 2\dfrac{1}{4}$, so 3 votes are needed.

4 members: $\dfrac{3}{4} \cdot 4 = \dfrac{12}{4} = 3$, so 3 votes are needed.

5 members: $\dfrac{3}{4} \cdot 5 = \dfrac{15}{4} = 3\dfrac{3}{4}$, so 4 votes are needed.

6 members: $\dfrac{3}{4} \cdot 6 = \dfrac{18}{4} = 4\dfrac{1}{2}$, so 5 votes are needed.

7 members: $\dfrac{3}{4} \cdot 7 = \dfrac{21}{4} = 5\dfrac{1}{4}$, so 6 votes are needed.

8 members: $\dfrac{3}{4} \cdot 8 = \dfrac{24}{4} = 6$, so 6 votes are needed.

39. Note that 14 ft = 168 inches.

$$168 \div 2\dfrac{1}{4} = \dfrac{168}{1} \div \dfrac{9}{4} = \dfrac{168}{1} \cdot \dfrac{4}{9}$$
$$= \dfrac{672}{9} = \dfrac{224}{3} = 74\dfrac{2}{3}$$

75 boards will be required, but the board on the end will need to be trimmed.

41. $35 \div \dfrac{3}{4} = \dfrac{35}{1} \cdot \dfrac{4}{3} = \dfrac{140}{3} = 46\dfrac{2}{3}$

Andre can make 46 aprons (with some material left over).

43. $2\dfrac{3}{4} \cdot \dfrac{6}{8} = \dfrac{11}{4} \cdot \dfrac{3}{4} = \dfrac{33}{16} = 2\dfrac{1}{16}$ cups

45. (a), (b)

Jennie spent 15 minutes picking strawberries from 8 plants. There are 48 plants altogether, so she picked $\dfrac{8}{48} = \dfrac{1}{6}$ of the strawberries. Therefore, it will take 6 times as long to pick all the strawberries. Note that 15 min = $\dfrac{1}{4}$ hr. It will take Jennie $6 \times \dfrac{1}{4} = \dfrac{3}{2} = 1\dfrac{1}{2}$ hours to pick all the strawberries.

Copyright © 2015 Pearson Education, Inc.

Solutions to Problem Set 6.4 **63**

47. B. 1 minute = 60 seconds. The average rate of

speed is $\dfrac{90 \text{ feet}}{60 \text{ sec}} = \dfrac{3}{2} = 1\dfrac{1}{2}$ feet per second.

Section 6.4
The Rational Number System

Problem Set 6.4

1. Commutative and associative properties for addition:

$$\left(3\dfrac{1}{5} + 2\dfrac{2}{5}\right) + 8\dfrac{1}{5} = (3 + 2 + 8) + \left(\dfrac{1}{5} + \dfrac{2}{5} + \dfrac{1}{5}\right)$$
$$= 13 + \dfrac{4}{5} = 13\dfrac{4}{5}$$

3. (a) $\dfrac{-4}{5}$

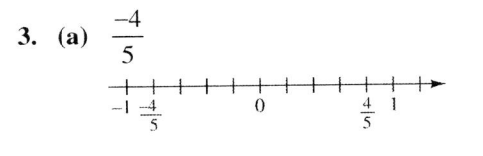

(b) $\dfrac{3}{2}$

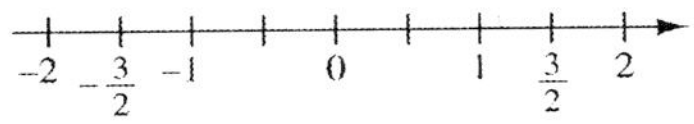

(c) $\dfrac{8}{3}$

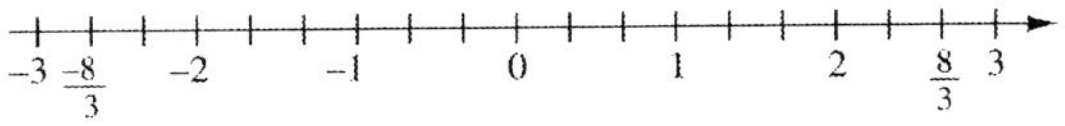

(d) $\dfrac{-4}{2}$ or -2

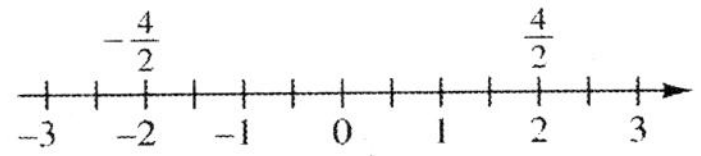

5. (a) $\dfrac{2}{5} - \dfrac{3}{4} = \dfrac{2 \cdot 4 - 5 \cdot 3}{5 \cdot 4} = \dfrac{-7}{20}$

(b) $\dfrac{-6}{7} - \dfrac{4}{7} = \dfrac{-6 - 4}{7} = \dfrac{-10}{7}$

(c) $\dfrac{3}{8} - \dfrac{1}{12} = \dfrac{9}{24} - \dfrac{2}{24} = \dfrac{7}{24}$

7. (a) $\dfrac{3}{5} \cdot \dfrac{7}{8} \cdot \left(\dfrac{5}{3}\right) = \dfrac{3 \cdot 7 \cdot 5}{5 \cdot 8 \cdot 3} = \dfrac{7}{8}$

(b) $\dfrac{-2}{7} \cdot \dfrac{3}{4} = \dfrac{-6}{28} = \dfrac{-3}{14}$

(c) $\dfrac{-4}{3} \cdot \dfrac{6}{-16} = \dfrac{-24}{-48} = \dfrac{1}{2}$

9. (a) $\dfrac{2}{3}$

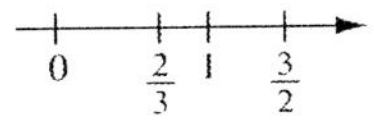

(b) $\dfrac{-9}{4}$

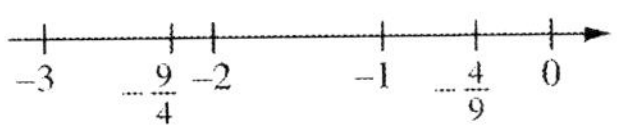

(c) $\dfrac{-11}{-4}$ or $\dfrac{11}{4}$

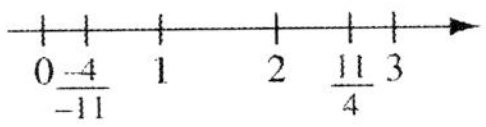

11. (a) $\dfrac{2}{3} \cdot \dfrac{4}{7} + \dfrac{2}{3} \cdot \dfrac{3}{7} = \dfrac{2}{3} \cdot \left(\dfrac{4}{7} + \dfrac{3}{7}\right) = \dfrac{2}{3} \cdot 1 = \dfrac{2}{3}$

(b) $\dfrac{4}{5} \cdot \dfrac{2}{3} - \dfrac{3}{10} \cdot \dfrac{2}{3} = \left(\dfrac{4}{5} - \dfrac{3}{10}\right) \cdot \dfrac{2}{3} = \dfrac{5}{10} \cdot \dfrac{2}{3}$
$$= \dfrac{1}{2} \cdot \dfrac{2}{3} = \dfrac{1 \cdot 2}{2 \cdot 3} = \dfrac{1}{3}$$

13. (a) Addition of rational numbers—definition

(b) Multiplication of rational numbers—definition

(c) Distributive property of multiplication over addition (for whole numbers)

(d) Addition of rational numbers—definition

(e) Multiplication of rational numbers—definition

15. (a)

$4x + 3 = 0$	Given
$4x = -3$	Definition of additive inverse
$\dfrac{1}{4}(4x) = \dfrac{1}{4}(-3)$	Multiply both sides by $\dfrac{1}{4}$.
$\left(\dfrac{1}{4} \cdot 4\right)x = \dfrac{1}{4}(-3)$	Associative property of multiplication
$1x = -\dfrac{3}{4}$	Multiplicative inverse property
$x = -\dfrac{3}{4}$	One is the multiplicative identity.

Copyright © 2015 Pearson Education, Inc.

64 *Chapter 6 Fractions and Rational Numbers*

(b)

$$x + \frac{3}{4} = \frac{7}{8} \qquad \text{Given}$$

$$\left(x + \frac{3}{4}\right) + \frac{-3}{4} = \frac{7}{8} + \frac{-3}{4} \qquad \text{Add } \frac{-3}{4} \text{ to both sides.}$$

$$x + \left(\frac{3}{4} + \frac{-3}{4}\right) = \frac{7}{8} + \frac{-3}{4} \qquad \text{Associative property of addition}$$

$$x + 0 = \frac{7}{8} + \frac{-3}{4} \qquad \text{Additive inverse}$$

$$x = \frac{7}{8} + \frac{-3}{4} \qquad \text{Additive identity, equivalence of fractions}$$

$$x = \frac{7}{8} + \frac{-6}{8} = \frac{1}{8} \qquad \text{Add.}$$

(c)

$$\frac{2}{3}x + \frac{4}{5} = 0 \qquad \text{Given}$$

$$\frac{2}{3}x = -\frac{4}{5} \qquad \text{Definition of additive inverse}$$

$$\frac{3}{2} \cdot \left(\frac{2}{3}x\right) = \frac{3}{2} \cdot \left(\frac{-4}{5}\right) \qquad \text{Multiply both sides by } \frac{3}{2}.$$

$$1x = \frac{3 \cdot (-4)}{2 \cdot 5} \qquad \text{Associative property of multiplication, multiplicative inverse property,}$$
$$\text{definition of multiplication of rational numbers}$$

$$x = \frac{-12}{10} \qquad \text{1 is the multiplication identity.}$$

$$x = -\frac{6}{5} \qquad \text{Equivalence of fractions}$$

(d)

$$3\left(x + \frac{1}{8}\right) = -\frac{2}{3} \qquad \text{Given}$$

$$3x + \frac{3}{8} = -\frac{2}{3} \qquad \text{Distributive property of multiplication over addition}$$

$$\left(3x + \frac{3}{8}\right) + \frac{-3}{8} = -\frac{2}{3} + \frac{-3}{8} \qquad \text{Add } \frac{-3}{8} \text{ to both sides.}$$

$$3x + \left(\frac{3}{8} + \frac{-3}{8}\right) = -\frac{2}{3} + \frac{-3}{8} \qquad \text{Associative property of addition}$$

$$3x + 0 = -\frac{2}{3} + \frac{-3}{8} \qquad \text{Additive inverse}$$

$$3x = -\frac{2}{3} - \frac{3}{8} \qquad \text{Additive identity}$$

$$3x = \frac{-25}{24} \qquad -\frac{2}{3} - \frac{3}{8} = \frac{-16}{24} - \frac{9}{24} = \frac{-25}{24}$$

$$\frac{1}{3} \cdot (3x) = \frac{1}{3} \cdot \left(\frac{-25}{24}\right) \qquad \text{Multiply both sides by } \frac{1}{3}.$$

$$1 \cdot x = \frac{1 \cdot (-25)}{3 \cdot 24} \qquad \text{Associative property of multiplication, multiplicative inverse property,}$$
$$\text{definition of multiplication of rational numbers}$$

$$x = \frac{-25}{72} \qquad \text{1 is a multiplicative identity.}$$

Copyright © 2015 Pearson Education, Inc.

17. Note that in each case the fractions need to be written with positive denominators.

   (a) $\dfrac{-4}{5} < \dfrac{-3}{4}$: $(-4) \cdot 4 = -16 < -15 = 5 \cdot (-3)$

   (b) $\dfrac{1}{10} > \dfrac{-1}{4}$: $1 \cdot 4 = 4 > -10 = 10 \cdot (-1)$

   (c) $\dfrac{-19}{60} > \dfrac{-1}{3}$:
   $(-19) \cdot 3 = -57 > -60 = 60 \cdot (-1)$

19. Answers will vary.

   (a) One answer is $\dfrac{1}{2}$, since $\dfrac{4}{9} < \dfrac{1}{2} < \dfrac{6}{11}$.

   (b) One answer is $\dfrac{2}{19}$, since

   $$\dfrac{1}{9} = \dfrac{2}{18} > \dfrac{2}{19} > \dfrac{2}{20} = \dfrac{1}{10}.$$

   (c) One answer is $\dfrac{28}{47}$ since $\dfrac{7}{12} = \dfrac{14}{24} = \dfrac{28}{48}$

   and $\dfrac{14}{23} = \dfrac{28}{46}$.

   (d) One answer is the mediant, $\dfrac{36}{145}$ (defined

   as $\dfrac{a+b}{c+d}$, where $c > 0$ and $d > 0$):
   $$\dfrac{141+183}{568+737} = \dfrac{324}{1305} = \dfrac{108}{435} = \dfrac{36}{145}$$
   $$\dfrac{141}{568} < \dfrac{36}{145} < \dfrac{183}{737}$$

21. (a) $\dfrac{104}{391}$ is approximately $\dfrac{100}{400} = \dfrac{1}{4}$.

   (b) $\dfrac{217}{340}$ is approximately $\dfrac{220}{330} = \dfrac{2}{3}$.

   (c) $\dfrac{-193}{211}$ is approximately $\dfrac{-200}{200} = -1$.

   (d) $\dfrac{453}{307}$ is approximately $\dfrac{450}{300} = \dfrac{3}{2} = 1\dfrac{1}{2}$.

23. (a) Combine like terms:
   $$\dfrac{1}{2} + \dfrac{1}{4} + \dfrac{3}{4} = \dfrac{1}{2} + 1 = \dfrac{3}{2}$$

   (b) Use the Distributive Property of Multiplication over addition:
   $$\dfrac{5}{2} \cdot \left(\dfrac{2}{5} - \dfrac{2}{10} \right) = \dfrac{5}{2} \cdot \dfrac{2}{5} - \dfrac{5}{2} \cdot \dfrac{2}{10}$$
   $$= 1 - \dfrac{5}{10} = 1 - \dfrac{1}{2} = \dfrac{1}{2}$$

   (c) Use equivalence of fractions:
   $$\dfrac{3}{4} \cdot \dfrac{12}{15} = \dfrac{3}{4} \cdot \dfrac{4}{5} = \dfrac{3}{5}$$

   (d) Use the definition of division:
   $$\dfrac{2}{9} \div \dfrac{1}{3} = \dfrac{2}{9} \cdot \dfrac{3}{1} = \dfrac{6}{9} = \dfrac{2}{3}$$

25. (a) What is his new total acreage?

   (b) How many miles does Janet live from Brian?

   (c) How many square yards of floor space are in the family room?

   (d) How many bottles of ginger ale will Clea fill?

27. The additional dotted lines partition the unit square into 16 congruent triangles, so each of

   these has area $\dfrac{1}{16}$.

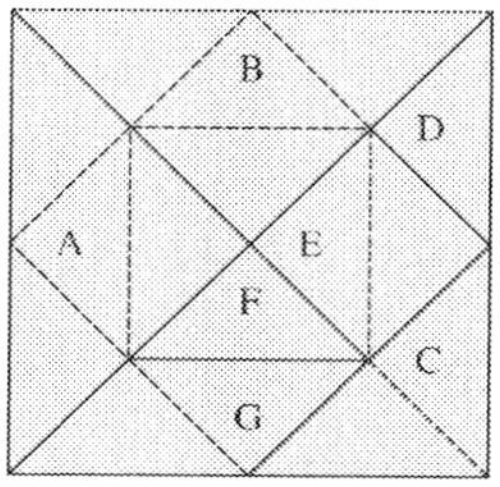

   Therefore, A has area $\dfrac{4}{16} = \dfrac{1}{4}$, B has area

   $\dfrac{4}{16} = \dfrac{1}{4}$, C has area $\dfrac{2}{16} = \dfrac{1}{8}$, D has area $\dfrac{1}{16}$,

   E has area $\dfrac{2}{16} = \dfrac{1}{8}$, F has area $\dfrac{1}{16}$, and G has

   area $\dfrac{2}{16} = \dfrac{1}{8}$.

66 *Chapter 6 Fractions and Rational Numbers*

29. It is likely that the student added numerators and added denominators to form the fraction $\frac{4}{8}$, which simplifies to $\frac{1}{2}$. It would be helpful to review the student's understanding of how a common denominator can be obtained, using a pie model with 15 equal sectors, for instance. Also, encourage the student to estimate and check for reasonableness of answers. For example, show the student with the pie model that $\frac{3}{5}$ is just a little smaller than $\frac{2}{3}$, so the answer to $\frac{1}{3} + \frac{3}{5}$ can be expected to be very close to $\frac{1}{3} + \frac{2}{3} = 1$. Also, since $\frac{3}{5}$ is larger than $\frac{1}{2}$, adding $\frac{1}{3}$ to $\frac{3}{5}$ must give an answer larger than $\frac{1}{2}$.

31. Since
$$\frac{2}{3}x + \frac{1}{2}x + \frac{1}{7}x + x = \left(\frac{2}{3} + \frac{1}{2} + \frac{1}{7} + 1\right)x$$
$$= \left(\frac{28}{42} + \frac{21}{42} + \frac{6}{42} + \frac{42}{42}\right)x$$
$$= \frac{97}{42}x$$
we have the equation $\frac{97}{42}x = 33$. Therefore,
$$x = \frac{42}{97} \cdot 33 = \frac{1386}{97}.$$

33.

250 cubits

180 cubits
or 1260 hands

$$\text{Steepness} = \frac{\text{run in hands}}{\text{rise in cubits}}$$
$$= \frac{1260}{250} = \frac{126}{25} = 5\frac{1}{25}$$

35. The diagram, as labeled below, shows that $\frac{2}{3}H = \frac{3}{4}P$, where H denotes the number of holes and P the number of pigeons. Therefore,

$H = \frac{3}{2} \cdot \frac{3}{4}P = \frac{9}{8}P$. Also, $H - P = 5$, so
$$5 = H - P = \frac{9}{8}P - P = \frac{1}{8}P \text{ and therefore}$$
$P = 8 \cdot 5 = 40$ and $H = P + 5 = 45$.

$H/3$	$H/3$	$H/3$	

$P/4$	$P/4$	$P/4$	$P/4$

37. (a) Each column, row, and diagonal adds up to $3x$.

(b) These computations give the Magic Square shown below.

$$x - z = \frac{1}{2} - \frac{1}{4} = \frac{2}{4} - \frac{1}{4} = \frac{1}{4}$$
$$x + y + z = \frac{1}{2} + \frac{1}{3} + \frac{1}{4} = \frac{6}{12} + \frac{4}{12} + \frac{3}{12} = \frac{13}{12}$$
$$x - y = \frac{1}{2} - \frac{1}{3} = \frac{3}{6} - \frac{2}{6} = \frac{1}{6}$$
$$x - y + z = \frac{1}{2} - \frac{1}{3} + \frac{1}{4} = \frac{6}{12} - \frac{4}{12} + \frac{3}{12} = \frac{5}{12}$$
$$x + y - z = \frac{1}{2} + \frac{1}{3} - \frac{1}{4} = \frac{6}{12} + \frac{4}{12} - \frac{3}{12} = \frac{7}{12}$$
$$x + y = \frac{1}{2} + \frac{1}{3} = \frac{3}{6} + \frac{2}{6} = \frac{5}{6}$$
$$x - y - z = \frac{1}{2} - \frac{1}{3} - \frac{1}{4} = \frac{6}{12} - \frac{4}{12} - \frac{3}{12} = -\frac{1}{12}$$
$$x + z = \frac{1}{2} + \frac{1}{4} = \frac{2}{4} + \frac{1}{4} = \frac{3}{4}$$

$\frac{1}{4}$	$\frac{5}{12}$	$\frac{5}{6}$
$\frac{13}{12}$	$\frac{1}{2}$	$-\frac{1}{12}$
$\frac{1}{6}$	$\frac{7}{12}$	$\frac{3}{4}$

(c) Since $x + y = 1$ and $x - y = -\frac{1}{3}$,
$$2x = (x + y) + (x - y) = 1 + \left(-\frac{1}{3}\right) = \frac{2}{3}, \text{ so}$$
$$x = \frac{1}{3}. \text{ Then } y = 1 - x = \frac{2}{3}.$$

(continued on next page)

Copyright © 2015 Pearson Education, Inc.

Solutions to Chapter 6 Review Exercises **67**

(continued)

Since $x - z = -\dfrac{5}{12}$,

$z = x + \dfrac{5}{12} = \dfrac{1}{3} + \dfrac{5}{12} = \dfrac{4}{12} + \dfrac{5}{12} = \dfrac{9}{12} = \dfrac{3}{4}.$

We may complete the Magic Square using

$x = \dfrac{1}{3},\ y = \dfrac{2}{3},$ and $z = \dfrac{3}{4}$.

$-\dfrac{5}{12}$	$\dfrac{5}{12}$	1
$\dfrac{7}{4}$	$\dfrac{1}{3}$	$-\dfrac{13}{12}$
$-\dfrac{1}{3}$	$\dfrac{1}{4}$	$\dfrac{13}{12}$

39. (a) If the solar year is $365\dfrac{1}{4}$ days long, there must be one extra day each 4 years. Having leap years on years divisible by 4 accomplishes this.

(b) An hour is $\dfrac{1}{24}$ days, a minute is $\dfrac{1}{24 \cdot 60}$ days, and a second is $\dfrac{1}{24 \cdot 60 \cdot 60}$ days. Therefore, a solar year is

$365 + \dfrac{5}{24} + \dfrac{48}{24 \cdot 60} + \dfrac{46}{24 \cdot 60 \cdot 60}$

$= 365\dfrac{20,926}{24 \cdot 60 \cdot 60}$ days long.

(c) $365\dfrac{20,952}{24 \cdot 60 \cdot 60} = 365\dfrac{97 \cdot 9 \cdot 24}{20 \cdot 20 \cdot 9 \cdot 24}$

$= 365\dfrac{97}{400} = 365 + \dfrac{97}{400}$

$= 365 + \dfrac{100}{400} - \dfrac{4}{400} + \dfrac{1}{400}$

$= 365 + \dfrac{1}{4} - \dfrac{1}{100} + \dfrac{1}{400}$

$= 365\dfrac{1}{4} - \dfrac{1}{100} + \dfrac{1}{400}$

(d) Leap years occur when the year is divisible by 4, except when the year is divisible by 100 when there is no leap year, except when the year is also divisible by 400 when there is a leap year. For example, 1900, is divisible by 4 and 100 but not 400, so it is not a leap year. Since 2000 is divisible by 4, 100, and 400 it was a leap year.

41. We desire steps of height 5" to 7". Since 3' 10" = 46" and $\dfrac{46}{7} = 6\dfrac{4}{7}$ and $\dfrac{46}{5} = 9\dfrac{1}{5}$, an acceptable staircase could be made with 7, 8, or 9 steps, including the step onto the deck itself. If we choose to use 8 steps, each step is $\dfrac{46}{8}'' = 5\dfrac{3}{4}''$ high. Since the tread of each stair is $\dfrac{1}{2}''$ thick, we will need 7 risers of width $5\dfrac{3}{4}'' - \dfrac{1}{2}'' = 5\dfrac{1}{4}''$. The deck is $\dfrac{3}{4}''$ thick, so the eighth (top) riser needs to have width $5\dfrac{3}{4} - \dfrac{3}{4} = 5''.$

43. Cut the 8-inch side in half and the 10-inch side in thirds to get six $3\dfrac{1}{3}$-by-4 inch rectangles.

45. A.   **47.** D.

Chapter 6 Review Exercises

1. (a) Two of the four regions are shaded; $\dfrac{2}{4}$

(b) All six of the six regions are shaded; $\dfrac{6}{6}$

(c) None of the four regions is shaded; $\dfrac{0}{4}$

(d) Five of the three regions are shaded; $\dfrac{5}{3}$

3. Answers will vary. Possibilities include $\dfrac{-6}{8}, \dfrac{-9}{12}, \dfrac{-12}{16}$.

5. (a) $\dfrac{27}{81} = \dfrac{9}{27} = \dfrac{3}{9} = \dfrac{1}{3}$

(b) $\dfrac{100}{825} = \dfrac{20}{165} = \dfrac{4}{33}$

(c) $\dfrac{378}{72} = \dfrac{189}{36} = \dfrac{63}{12} = \dfrac{21}{4}$

(d) $\dfrac{3^5 \cdot 7^2 \cdot 11^3}{3^2 \cdot 7^3 \cdot 11^2} = \dfrac{3^3 \cdot 1 \cdot 11^1}{1 \cdot 7^1 \cdot 1} = \dfrac{297}{7}$

Copyright © 2015 Pearson Education, Inc.

68 *Chapter 6 Fractions and Rational Numbers*

7. (a) 36 because $36 \div 9 = 4$, $36 \div 12 = 3$

(b) 18 because $18 \div 18 = 1$, $18 \div 6 = 3$, $18 \div 3 = 6$

9.

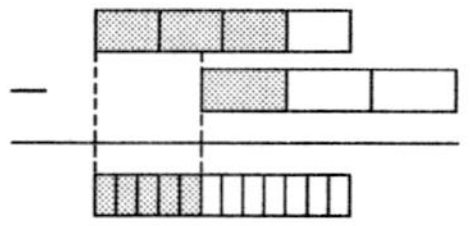

$$\frac{3}{4} - \frac{1}{3} = \frac{5}{12}$$

11. (a) $\dfrac{1}{3} + \dfrac{5}{8} - \dfrac{5}{6} = \dfrac{8}{24} + \dfrac{15}{24} - \dfrac{20}{24} = \dfrac{3}{24} = \dfrac{1}{8}$

(b) $\left(\dfrac{2}{3} - \dfrac{5}{4}\right) \div \dfrac{3}{4} = \left(\dfrac{8}{12} - \dfrac{15}{12}\right) \div \dfrac{3}{4} = \dfrac{-7}{12} \div \dfrac{3}{4}$

$$= \dfrac{-7}{12} \cdot \dfrac{4}{3} = \dfrac{-28}{36} = \dfrac{-7}{9}$$

(c) $\dfrac{4}{7} \cdot \left(\dfrac{35}{4} + \dfrac{-42}{12}\right) = \dfrac{4}{7} \cdot \left(\dfrac{35}{4} + \dfrac{-14}{4}\right)$

$$= \dfrac{4}{7} \cdot \dfrac{21}{4} = \dfrac{21}{7} = 3$$

(d) $\dfrac{123}{369} \div \dfrac{1}{3} = \dfrac{1}{3} \div \dfrac{1}{3} = 1$

13. Using a diagram, we see that we have 9 quarters of a pizza to share.

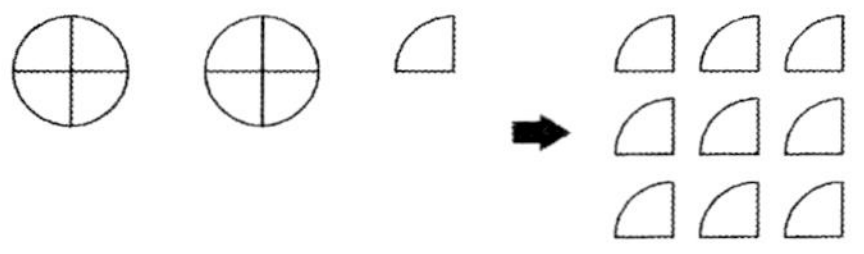

Since there are 9 quarters, each of the three should receive 3 of the quarters. That is,

$2\dfrac{1}{4} \div 3 = \dfrac{3}{4}$. This problem corresponds to the

sharing (partition model of division.)

15. $7\dfrac{1}{8} \div 1\dfrac{1}{4} = \dfrac{57}{8} \div \dfrac{5}{4} = \dfrac{57}{8} \cdot \dfrac{4}{5}$

$$= \dfrac{57 \cdot 4}{8 \cdot 5} = \dfrac{57}{2 \cdot 5} = \dfrac{57}{10}$$

The distance is $\dfrac{57}{10}$ miles or $5\dfrac{7}{10}$ miles.

17. (a)
$$3x + 5 = 11$$
$$3x + 5 + (-5) = 11 + (-5)$$
$$3x = 6$$
$$\dfrac{1}{3} \cdot (3x) = \dfrac{1}{3} \cdot 6$$
$$x = 2$$

(b)
$$x + \dfrac{2}{3} = \dfrac{1}{2}$$
$$x + \dfrac{2}{3} + \left(-\dfrac{2}{3}\right) = \dfrac{1}{2} + \left(-\dfrac{2}{3}\right)$$
$$x = \dfrac{1}{2} - \dfrac{2}{3} = \dfrac{-1}{6}$$

(c)
$$\dfrac{3}{5}x + \dfrac{1}{2} = \dfrac{2}{3}$$
$$\dfrac{3}{5}x + \dfrac{1}{2} + \left(-\dfrac{1}{2}\right) = \dfrac{2}{3} - \dfrac{1}{2}$$
$$\dfrac{3}{5}x = \dfrac{1}{6}$$
$$\dfrac{5}{3} \cdot \dfrac{3}{5}x = \dfrac{5}{3} \cdot \dfrac{1}{6} \Rightarrow x = \dfrac{5}{18}$$

(d)
$$-\dfrac{4}{3}x + 1 = \dfrac{1}{4}$$
$$-\dfrac{4}{3}x + 1 + (-1) = \dfrac{1}{4} + (-1)$$
$$-\dfrac{4}{3}x = -\dfrac{3}{4}$$
$$\left(-\dfrac{3}{4}\right) \cdot \left(-\dfrac{4}{3}\right)x = \left(-\dfrac{3}{4}\right) \cdot \left(-\dfrac{3}{4}\right) \Rightarrow x = \dfrac{9}{16}$$

19. Write $\dfrac{5}{6} = \dfrac{55}{66}$ and $\dfrac{10}{11} = \dfrac{60}{66}$. Then $\dfrac{56}{66} = \dfrac{28}{33}$

and $\dfrac{57}{66}$ are between $\dfrac{5}{6}$ and $\dfrac{10}{11}$. Other

answers may be given.

Copyright © 2015 Pearson Education, Inc.

Chapter 7 Decimals, Real Numbers, and Proportional Reasoning

Section 7.1
Decimals and Real Numbers

Problem Set 7.1

1. (a) $273.412 = 200 + 70 + 3 + \dfrac{4}{10} + \dfrac{1}{100} + \dfrac{2}{1000}$

$= 2 \cdot 10^2 + 7 \cdot 10^1 + 3 \cdot 10^0$

$\qquad + 4 \cdot 10^{-1} + 1 \cdot 10^{-2} + 2 \cdot 10^{-3}$

(b) $0.000723 = \dfrac{7}{10,000} + \dfrac{2}{100,000} + \dfrac{3}{1,000,000}$

$= 7 \cdot 10^{-4} + 2 \cdot 10^{-5} + 3 \cdot 10^{-6}$

3. (a) 0.21 **(b)** 0.235

(c) 0.278 **(d)** 0.302

5. 2.45

7. (a) $0.324 = \dfrac{324}{1000} = \dfrac{81}{250}; \quad 250 = 2 \cdot 5^3$

(b) $0.028 = \dfrac{28}{1000} = \dfrac{7}{250}; \quad 250 = 2 \cdot 5^3$

9. (a) $\dfrac{7}{20} = \dfrac{7 \cdot 5}{20 \cdot 5} = \dfrac{35}{100} = 0.35$

(b) $\dfrac{7}{16} = \dfrac{7 \cdot 5^4}{2^4 \cdot 5^4} = \dfrac{4375}{10,000} = 0.4375$

11. (a)
$$
\begin{array}{r}
0.875 \\
8\overline{)7.000} \\
\underline{64} \\
60 \\
\underline{56} \\
40 \\
\underline{40} \\
0
\end{array}
\qquad \dfrac{7}{8} = 0.875
$$

(b)
$$
\begin{array}{r}
1.325 \\
40\overline{)53.000} \\
\underline{40} \\
130 \\
\underline{120} \\
100 \\
\underline{80} \\
200 \\
\underline{200} \\
0
\end{array}
\qquad \dfrac{53}{40} = 1.325
$$

13. (a)
$$
\begin{array}{r}
0.833 \\
6\overline{)5.000} \\
\underline{4\,8} \\
20 \\
\underline{18} \\
20 \\
\underline{18} \\
2
\end{array}
\qquad \dfrac{5}{6} = 0.8\overline{3}
$$

(b)
$$
\begin{array}{r}
0.583 \\
12\overline{)7.000} \\
\underline{60} \\
1\,00 \\
\underline{96} \\
40 \\
\underline{36} \\
4
\end{array}
\qquad \dfrac{7}{12} = 0.58\overline{3}
$$

(c)
$$
\begin{array}{r}
0.571428 \\
7\overline{)4.000000} \\
\underline{3\,5} \\
50 \\
\underline{49} \\
10 \\
\underline{7} \\
30 \\
\underline{28} \\
20 \\
\underline{14} \\
60 \\
\underline{56} \\
4
\end{array}
\qquad \dfrac{4}{7} = 0.\overline{571428}
$$

70 *Chapter 7 Decimals, Real Numbers, and Proportional Reasoning*

15. **(a)** Let $x = 0.321321...$.
Then $1000x = 321.321321...$
$1000x - x = 321.321321... - 0.321321...$
$999x = 321$
$$x = \frac{321}{999} = \frac{107}{333}$$

(b) Let $x = 0.12414141...$
Then $100x = 12.414141...$
Then $10{,}000x = 1241.414141...$
$10{,}000x - 100x$
$\quad = 1241.414141... - 12.414141...$
$9900x = 1229$
$$x = \frac{1229}{9900}$$

(c) Let $x = 3.262626...$.
Then $100x = 326.2626726...$
$100x - x = 326.262626... - 3.262626...$
$99x = 323$
$$x = \frac{323}{99}$$

17. **(a)** $x = 0.3\overline{54}$
$1000x = 354.\overline{54}$
$\underline{- (10x = \quad 3.\overline{54})}$
$990x = 351$
$$x = \frac{351}{990} = \frac{39}{110}$$

(b) $x = 5.21\overline{6}$
$1000x = 5216.\overline{6}$
$\underline{- (100x = \quad 521.\overline{6})}$
$900x = 4695$
$$x = \frac{4695}{900} = \frac{313}{60}$$

19. Answers will vary. One answer is $\dfrac{1}{999} = 0.\overline{001}$.

21. Proof is by contradiction. Suppose $\sqrt{3}$ is rational. Then $\sqrt{3}$ can be represented by a fraction in reduced form. Let $\dfrac{u}{v}$ be this fraction. Then $\sqrt{3} = \dfrac{u}{v}$. Squaring each side we

have $3 = \dfrac{u^2}{v^2}$, and thus $3v^2 = u^2$. Hence, 3 is a prime factor of u^2, and thus of u. So $u = 3k$ for some integer k, and $3v^2 = u^2 = (3k)^2 = 9k^2$. This implies $v^2 = 3k^2$ and 3 is a factor of v^2 and hence of v. But this means that u and v have a common factor of 3 and $\dfrac{u}{v}$ is not in reduced form. We have reached a contradiction. Thus, $\sqrt{3}$ must be irrational.

Thus, the assumption that $3 - \sqrt{2}$ is rational must be false. So $3 - \sqrt{2}$ is irrational.

(b) Assume that $2\sqrt{2}$ is rational. Then $2\sqrt{2} = r$ where r is a rational number. This implies that $\sqrt{2} = \dfrac{r}{2}$. But $\dfrac{r}{2}$ is rational since the rational numbers are closed under division. This contradicts the fact that $\sqrt{2}$ is irrational. Thus, $2\sqrt{2}$ must be irrational.

23. Answers will vary. It should be possible to find a ruler with fractional markings.

25. Janeshia failed to realize the place value of the digits 5 and 6. She read 56 as if it were a decimal instead of a whole number.

27. **(a)** Let $x = 0.747474...$.
Then $100x = 74.747474...$
$100x - x = 74.747474... - 0.747474...$
$99x = 74$
$$x = \frac{74}{99}$$

(b) Let $x = 0.777...$.
Then $10x = 7.777...$
$10x - x = 7.777... - 0.777...$
$9x = 7$
$$x = \frac{7}{9}$$

Copyright © 2015 Pearson Education, Inc.

(c) Let $x = 0.235235235...$.

Then $1000x = 235.235235235...$

$1000x - x = 235.235235235... - 0.235235235...$

$999x = 235$

$x = \dfrac{235}{999}$

(d) (i) $0.\overline{a} = \dfrac{a}{9}$ **(ii)** $0.\overline{ab} = \dfrac{ab}{99}$

(iii) $0.\overline{abc} = \dfrac{abc}{999}$

29. Answers will vary.

(a) One example is $\sqrt{2} + \left(3 - \sqrt{2}\right) = 3$. $\sqrt{2}$ and $\left(3 - \sqrt{2}\right)$ are both irrational. (See the answer to problem 22.)

(b) One example is $\sqrt{3} + \sqrt{3} = 2\sqrt{3}$. $\sqrt{3}$ and $2\sqrt{3}$ are both irrational.

31. Answers will vary. For example, $\dfrac{\sqrt{2}}{2\sqrt{2}} = \dfrac{1}{2}$, and $\dfrac{1}{2}$ is rational.

33. (a) Since $5b^2$ is divisible by 5, its last digit is either 0 or 5. The last digit of a^2 is 0 or 5 only when the last digit of a is 0 or 5. In every case, both a and b are divisible by 5, which contradicts the assumption that the fraction $\dfrac{a}{b}$ is in lowest terms.

(b) $\sqrt{5}$ is an irrational number.

35. No part of the unit square remains unshaded, and at the nth step an additional shading of area $\dfrac{9}{10^n}$ is added.

37. (a) $\dfrac{1}{11} \approx 0.\overline{09}$ **(b)** $\dfrac{15}{22} \approx 0.6\overline{81}$

(c) $\dfrac{5}{21} \approx 0.\overline{238095}$ **(d)** $\dfrac{2}{13} \approx 0.\overline{153846}$

39. C. 94.3 is between 94 and 95 but closer to 94.

41. B. $\sqrt{26}$ is irrational.

Section 7.2
Computations with Decimals

Problem Set 7.2

1. (a)
$$\begin{array}{r} 32.174 \\ +\ 371.500 \\ \hline 403.674 \end{array}$$
(b)
$$\begin{array}{r} 371.500 \\ -\ 32.174 \\ \hline 339.326 \end{array}$$

3. (a)
$$\begin{array}{r} 37.1 \\ \times\ 4.7 \\ \hline 25\ 97 \\ 148\ 4 \\ \hline 174.37 \end{array}$$
(b)
$$\begin{array}{r} 3.71 \\ \times\ 0.47 \\ \hline 2597 \\ 1\ 484 \\ \hline 1.7437 \end{array}$$

5. (a) Estimate: $4 + 31 = 35$
Answer: $4.112 + 31.3 = 35.412$

(b) Estimate: $31 - 4 = 27$
Answer: $31.3 - 4.112 = 27.188$

7. (a) Using mental arithmetic:
$23.07 + 4.8 + 0.971 \approx 23 + 5 + 1 = 29$
Exact value: $23.07 + 4.8 + 0.971 = 28.841$

(b) Using mental arithmetic:
$41.5 - 6.48 + 13.013 \approx 41.5 - 6.5 + 13 = 48$
Exact value: $41.5 - 6.48 + 13.013 = 48.033$

9. $13.6 \times 3.199 = 43.5064$
$\$43.51$ will be showing on the cost dial.

11. (a) $34.796 \times 10^3 = 34,796$

(b) $34.796 \times 10^{-3} = 0.034796$

13. (a) $34,762 = 3.4762 \times 10^4$

(b) $4,256,000 = 4.256 \times 10^6$

(c) $0.009031 = 9.031 \times 10^{-3}$

(d) $0.000004320017 = 4.320017 \times 10^{-6}$

15. (a) [ON/AC] 1.27 [EXP] [(−)] 5 [×] 8.235 [EXP] [(−)] 6 [=] 1.05×10^{-10}

72 *Chapter 7 Decimals, Real Numbers, and Proportional Reasoning*

(b) $\boxed{\text{ON/AC}}$ 98613428 $\boxed{\times}$ 5746312 $\boxed{=}$
5.67×10^{14}

(c) $\boxed{\text{ON/AC}}$ 1.27 $\boxed{\text{EXP}}$ $\boxed{(-)}$ 5 $\boxed{\div}$ 98613428 $\boxed{=}$
1.29×10^{-13}

(d) $\boxed{\text{ON/AC}}$ 98613428 $\boxed{\div}$ 8.234 $\boxed{\text{EXP}}$ $\boxed{(-)}$ 6
$\boxed{=}$ 1.20×10^{13}

17. (a) $0.007, 0.017, 0.01\overline{7}, 0.027$

(b) $24.999, 25.312, 25.412, 25.4\overline{12}$

19. Answers will vary.

21. Remind Toni that, in rounding, we are only concerned with the digit immediately to the right of the place to which we are rounding. In this case the digit is 4, and we round down to 7.2. The idea basically is that this guarantees that 7.2447 is closer to 7.2 than to 7.3. Indeed, $|7.2447 - 7.2| = 0.0447$ and $|7.3 - 7.2447| = 0.0553$.

23. The student is making two fundamental errors. First, no estimated answers are given before commencing with the calculation. Second, the examples show all of the numbers have been aligned to the right, disregarding the position of the decimal. The position of the decimal apparently is determined by the first addend. The student needs a better understanding of place value in decimal numerals.

25. (b) Note that $21.06 + 3.21 = 24.27$,
$5.00 - 3.21 = 1.79$, and
$24.27 + 5.00 = 29.27$.

$$\begin{array}{ccc} \underline{21.06} & \underline{3.21} & \underline{1.79} \\ 24.27 & \underline{5.00} & \\ & \underline{29.27} & \end{array}$$

(c) Note that $2.415 - 0.041 = 2.374$,
$7.723 - 2.415 = 5.308$, and
$5.308 - 0.041 = 5.267$.

$$\begin{array}{ccc} \underline{2.374} & \underline{0.041} & \underline{5.267} \\ \underline{2.415} & 5.308 & \\ & \underline{7.723} & \end{array}$$

(d) Any two numbers whose sum is 3.142 may be placed in the second row. Once these numbers are chosen, the rest can be found. If we choose the numbers 2.414 and 0.728, then the other two numbers are $2.414 - 1.414 = 1$ and $0.728 - 1.414 = -0.686$.

$$\begin{array}{ccc} \underline{1} & \underline{1.414} & \underline{-0.686} \\ 2.414 & \underline{0.728} & \\ & \underline{3.142} & \end{array}$$

(e) Yes—(d) can be completed in more than one way, by finding any two numbers, both greater than 1.414, that add to 3.142.

27. (a) Each number is $\dfrac{2.321}{2.11} = \dfrac{11}{10}$ times its predecessor. The sequence is 2.11, 2.321, 2.5531, 2.80841, 3.089251.

(b) Since $r = -0.2808 \div 1.404 = -0.2$, the sequence is 35.1, −7.02, 1.404, −0.2808, 0.05616.

(c) If each term is r times its predecessor, then $6.01r^3 = 0.75125$.
Thus, $r^3 = \dfrac{0.75125}{6.01} = 0.125$ and
$r = \sqrt[3]{0.125} = 0.5$. The sequence is 6.01, 3.005, 1.5025, 0.75125, 0.375625.

29. (a) Note that $1.32 + 3.41 = 4.73$,
$3.41 + 7.10 = 10.51$, and
$1.32 + 7.10 = 8.42$.

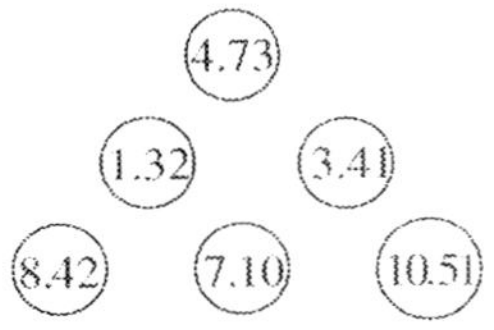

(b) Note that $0.710 + 3.111 = 3.821$,
$8.123 - 0.710 = 7.413$, and
$7.413 + 3.111 = 10.524$.

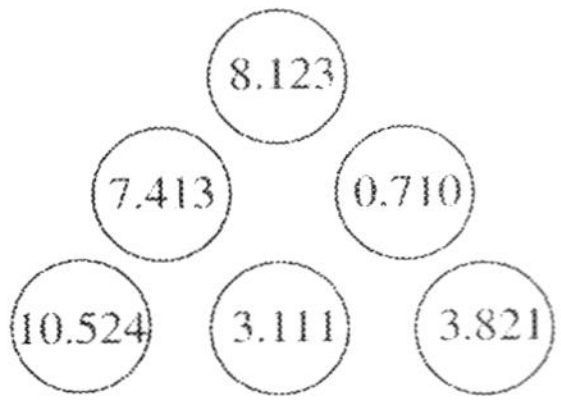

Copyright © 2015 Pearson Education, Inc.

(c) If a, b, and c are the numbers in the small circles, then
$2(a + b + c) = (a + b) + (b + c) + (a + c)$
$= 2.341 + 7.133 + 4.012 = 13.486$,
so the sum of a, b, and c is 6.743. Thus the number opposite 7.133 is
$6.743 - 7.133 = -0.39$ and the other missing numbers are
$6.743 - 2.341 = 4.402$ and
$6.743 - 4.012 = 2.731$.

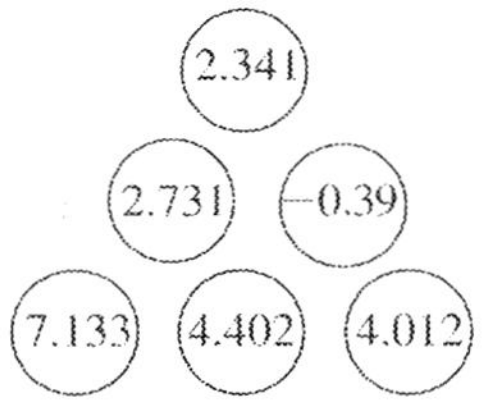

(2.341)
(2.731) (−0.39)
(7.133) (4.402) (4.012)

(d) The sum of the missing numbers is
$\frac{1}{2}(-7.141 + 0.517 + 0.003) = -3.3105$, so
the missing numbers are
$-3.3105 - (-7.141) = 3.8305$,
$-3.3105 - 0.517 = -3.8275$, and
$-3.3105 - 0.003 = -3.3135$.

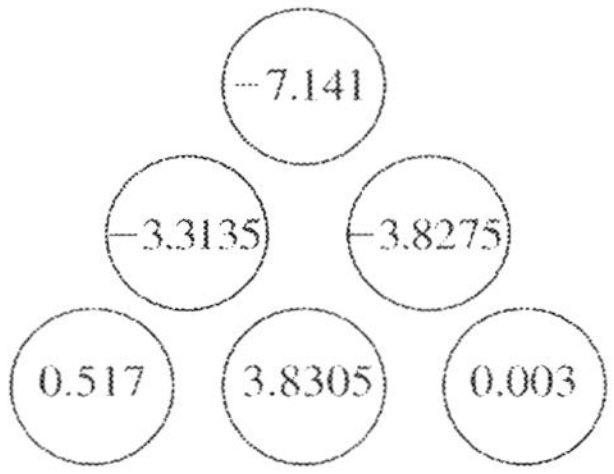

(−7.141)
(−3.3135) (−3.8275)
(0.517) (3.8305) (0.003)

31. $44.92 \div 4 = 11.23$, so each pair cost $11.23.

33. (a) It is helpful to draw a picture. Since
$24.75 + 2 \cdot 2.25 = 29.25$ and
$17.5 + 2 \cdot 2.25 = 22$, the drawing is as shown.

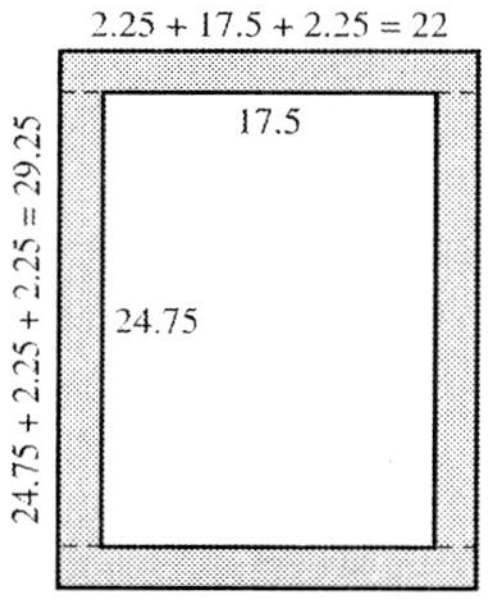

2.25 + 17.5 + 2.25 = 22

24.75 + 2.25 + 2.25 = 29.25

17.5

24.75

Thus, the area of the frame is
$2(22 \cdot 2.25) + 2(2.25 \cdot 24.75) = 210.375$
square inches. Note that other equations are possible.

(b) From the same drawing, it is apparent that the area of the picture is
$(17.5)(24.75) = 433.125$ square inches.

35. C. $4.8 < 4.86 < 4.9$

37. B.
$112 \times \$8.50 \times 0.01 = 952 \times 0.01 = \$95.20$

39. B. $43.23/3.93 = 11$

41. C.

Section 7.3
Proportional Reasoning

Problem Set 7.3

1. There are 10 girls, 14 boys, and
$10 + 14 = 24$ students.

(a) boys to girls $= \dfrac{14}{10} = \dfrac{7}{5}$

(b) girls to students $= \dfrac{10}{24} = \dfrac{5}{12}$

(c) boys to students $= \dfrac{14}{24} = \dfrac{7}{12}$

(d) girls to boys $= \dfrac{10}{14} = \dfrac{5}{7}$

(e) students to girls $= \dfrac{24}{10} = \dfrac{12}{5}$

(f) students to boys $= \dfrac{24}{14} = \dfrac{12}{7}$

3. (a) Yes. $2 \cdot 12 = 3 \cdot 8$

(b) Yes. $21 \cdot 36 = 28 \cdot 27$

(c) No. $7 \cdot 31 \neq 28 \cdot 8$

(d) Yes. $51 \cdot 95 = 85 \cdot 57$

5. (a)
$\dfrac{6}{14} = \dfrac{r}{21}$
$6 \cdot 21 = 14 \cdot r$
$r = \dfrac{6 \cdot 21}{14}$
$= 9$

(b)
$\dfrac{8}{12} = \dfrac{10}{r}$
$8 \cdot r = 12 \cdot 10$
$r = \dfrac{12 \cdot 10}{8}$
$= 15$

74 Chapter 7 Decimals, Real Numbers, and Proportional Reasoning

(c) $\dfrac{47}{3.2} = \dfrac{s}{7.8}$

$47 \cdot 7.8 = 3.2 \cdot s$

$s = \dfrac{47 \cdot 7.8}{3.2} = 114.5625$

7. (a) $3\dfrac{1}{2} \cdot 9.50 = 33.25$, so she earned \$33.25.

(b) $47.50 \div 9.50 = 5$, so she worked 5 hours.

9. Since there are 16 ounces in a pound, brand A is $\$0.43 \times 16 = \$6.88$ per pound, so brand B is the more expensive.

11. $s = kt$

$62.5 = k \cdot 7$

$k = \dfrac{62.5}{7}$

When $t = 10$,

$s = kt = 10 \cdot \dfrac{62.5}{7} = \dfrac{625}{7} \approx 89.2857.$

13. Let x be the speed in miles per hour.

Then $\dfrac{x}{90} = \dfrac{0.6}{1}$, so $x \cdot 1 = 90 \cdot 0.6 = 54$.

You can travel 54 miles per hour.

15. The 70¢ difference in the initial cost must be made up by traveling far enough to save 70¢. Since each quarter-mile saves 5¢, and $5 \times 14 = 70$, the break-even distance is 14 quarter miles, or $3\dfrac{1}{2}$ miles.

17. Measure with a metric ruler to determine $a, b, c, d, e, f,$ and g and form the ratios

$\dfrac{a}{1}, \dfrac{b}{2}, \dfrac{c}{3}, \dfrac{d}{4}, \dfrac{e}{5}, \dfrac{f}{6},$ and $\dfrac{g}{7}.$

Numerically the ratios are

$\dfrac{\frac{1}{2}}{1}, \dfrac{1}{2}, \dfrac{\frac{3}{2}}{3}, \dfrac{2}{4}, \dfrac{\frac{5}{2}}{5}, \dfrac{3}{6},$ and $\dfrac{\frac{7}{2}}{7}$

all of which are equal to $\dfrac{1}{2}$. Thus, any two of these ratios form a proportion. Moreover, note that $a = \dfrac{1}{2} \cdot 1$, $b = \dfrac{1}{2} \cdot 2$, $c = \dfrac{1}{2} \cdot 3$, $d = \dfrac{1}{2} \cdot 4$,

$e = \dfrac{1}{2} \cdot 5$, $f = \dfrac{1}{2} \cdot 6$ and $g = \dfrac{1}{2} \cdot 7$.

Thus, the height of each line segment is given by $h = \dfrac{1}{2} \cdot d$ where d is the distance of the line segment from 0; that is, h is proportional to d and the constant of proportionality is $\dfrac{1}{2}$.

19. Jo correctly multiplied 7×20 to find the weekly usage of 140 gallons. However, she multiplied the weekly rate of 140 by 30 to get the incorrect 4200-gallon monthly rate. The correct answer is $20 \times 30 = 600$ gallons per month. A similar error was made to get a yearly rate of $365 \times 140 = 51,100$ gallons, when the correct answer is $365 \times 20 = 7300$ gallons per year.

21. The photo is distorted because the two rectangles do not have proportional sides: $\dfrac{4}{6} \neq \dfrac{8}{10}$. Allison needs to crop her photo first, by cutting off a half-inch strip from the top and bottom to leave a 4-by 5- inch photo. This can be enlarged by doubling both the vertical horizontal dimensions to create a distortion-free 8- by 10-inch enlargement.

23. (a) B **(b)** C

(c) A **(d)** D

25. (a) $y = kx^2$

$27 = k(6^2) = 36k$

$k = \dfrac{27}{36} = \dfrac{3}{4}$

When $x = 12$, $y = kx^2 = \dfrac{3}{4}(12^2) = 108$.

(b) 27 to 108 or, more simply, 1 to 4.

(c) The value of y is multiplied by 4. Since y is proportional to x^2, if x is multiplied by any number, y is multiplied by the square of that number.

27. $\dfrac{100}{11.6} \div \dfrac{100}{11.8} = \dfrac{100}{11.6} \cdot \dfrac{11.8}{100} = \dfrac{1180}{1160} = \dfrac{59}{58}$

The ratio is 59 to 58.

Copyright © 2015 Pearson Education, Inc.

29. Compare the cost per ounce (or cost per gallon) in each case.

(a) $\dfrac{0.90}{32} < \dfrac{1.20}{40}$ because

$0.90 \cdot 40 < 32 \cdot 1.20$, so the better buy is 32 ounces for 90¢.

(b) $\dfrac{2.21}{1} < \dfrac{1.11}{0.5}$ because

$2.21 \cdot 0.5 < 1 \cdot 1.11$, so the better buy is a gallon for $2.21.

(c) $\dfrac{3.85}{16} < \dfrac{2.94}{12}$ because

$3.85 \cdot 12 < 16 \cdot 2.94$, so the better buy is a 16 ounce box for $3.85.

31. Since the ratio of boys to girls in Ms. Zombo's class is 3 to 2, $\dfrac{3}{5}$ of her students are boys and $\dfrac{2}{5}$ of her students are girls. Since the ratio of boys to girls in Mr. Stolarski's class is 4 to 3, $\dfrac{4}{7}$ of his students are boys and $\dfrac{3}{7}$ of his students are girls. Thus, there are 18 boys and 12 girls in Ms. Zombo's class, and 16 boys and 12 girls in Mr. Stolarski's class. The combined ratio is $\dfrac{18+16}{12+12} = \dfrac{34}{24} = \dfrac{17}{12}$, or 17 to 12.

33. They all seem to be about equal and to be approximately equal to the golden ratio, $\dfrac{1+\sqrt{5}}{2}$. If the measurements could be made absolutely accurately, the ratios would all exactly equal the golden ratio.

35. Let x be the number of gallons to travel 305 miles.

$$\frac{x}{305} = \frac{8.7}{192}$$
$$x \cdot 192 = 305 \cdot 8.7$$
$$x = \frac{305 \cdot 8.7}{192} \approx 13.82$$

It uses about 13.82 gallons.

37. Let x be the cost of 7 shirts.

$$\frac{x}{7} = \frac{59.97}{3}$$
$$x \cdot 3 = 7 \cdot 59.97$$
$$x = \frac{7 \cdot 59.97}{3} = 139.93$$

The cost will be $139.93.

39. Let x be the distance, in feet, between the rocks.

$$\frac{x}{\frac{22}{12}} = \frac{103}{3}$$
$$x \cdot 3 = \frac{22}{12} \cdot 103$$
$$x = \left(\frac{22}{12} \cdot 103\right) \div 3 = 62.9\overline{4}$$

The distance is $62.9\overline{4}$ ft.

---

41. (a)

		Number of Teeth in Cog						
		34	28	23	19	16	13	11
Number of Teeth in Chainring	24	0.71	0.86	1.04	1.26	1.50	1.85	2.18
	35	1.03	1.25	1.52	1.84	2.19	2.69	3.18
	51	1.50	1.82	2.22	2.68	3.19	3.92	4.64

(b) There are many nearly duplicated ratios. It would be more accurate to say the bike has 10 speeds.

(c) Answers will vary.

---

43. (a), (b) Use a proportion since the number of breakfasts per box is proportional to the size of the box. If there are x breakfasts in a 60-ounce box, then $\dfrac{x}{60} = \dfrac{8}{20}$ and

$$x = 60 \cdot \frac{8}{20} = 24.$$

Kerri should expect to have enough oatmeal for 24 days.

76 Chapter 7 Decimals, Real Numbers, and Proportional Reasoning

45. B. $\dfrac{25}{100} = \dfrac{1}{4}$ **47. B.**

Section 7.4
Percent

Problem Set 7.4

1. (a) $100 \cdot \dfrac{3}{16} = 18.75$, so $\dfrac{3}{16} = 18.75\%$.

   (b) $100 \cdot \dfrac{7}{25} = 28$, so $\dfrac{7}{25} = 28\%$.

   (c) $100 \cdot \dfrac{37}{40} = 92.5$, so $\dfrac{37}{40} = 92.5\%$.

   (d) $100 \cdot \dfrac{5}{6} = 83.\overline{3}$, so $\dfrac{5}{6} = 83.\overline{3}\%$.

3. (a) $100 \cdot 0.19 = 19$, so $0.19 = 19\%$.

   (b) $100 \cdot 0.015 = 1.5$, so $0.015 = 1.5\%$

   (c) $100 \cdot 2.15 = 215$, so $2.15 = 215\%$

   (d) $100 \cdot 3 = 300$, so $3 = 300\%$

5. (a) $70\% \times 280 = 0.7 \times 280 = 196$

   (b) $120\% \times 84 = 1.2 \times 84 = 100.8$

   (c) $38\% \times 751 = 0.38 \times 751 = 285.38$

   (d) $7\dfrac{1}{2}\% \times \$20,000 = 0.075 \times \$20,000$
   $$= \$1500$$

   (e) $0.02\% \times 27,481 = 0.0002 \times 27,481$
   $$= 5.4962$$

   (f) $1.05\% \times 845 = 0.0105 \times 845 = 8.8725$

7. (a) $105\% \times x = 50,400$
   $$1.05x = 50,400$$
   $$x = \dfrac{50,400}{1.05} = 48,000$$
   Michelle's former salary was $48,000.

   (b) Jason lost $12,000 - 11,160 = \$840$.
   $$\dfrac{840}{12,000} = 0.07 = 7\%$$
   Jason lost 7% of his investment.

9. (a) Guess: 50%; actual: 50%.

   (b) Guess: 50%; actual: 55%. To be 50%, only 5 of the 10 diagonal squares should be blue.

   (c) Guess: 25%; actual: 30%

   (d) Guess: 75%; actual: 68%

11. (a) $\dfrac{1}{8} = 0.125 = 12.5\%$

   (b) $\dfrac{3}{8} = 0.375 = 37.5\%$

   (c) $\dfrac{7}{18} = 0.3888\ldots \approx 39\%$

   (d) $\dfrac{1}{2} = 0.5 = 50\%$

13. There are $8 \cdot 5 = 40$ total small rectangles.

   (a) 50% of $40 = 0.5 \cdot 40 = 20$ small rectangles should be shaded.

   (b) 25% of $40 = 0.25 \cdot 40 = 10$ small rectangles should be shaded.

   (c) 20% of $40 = 0.2 \cdot 40 = 8$ small rectangles should be shaded.

   (d) 37.5% of $40 = 0.375 \cdot 40 = 15$ small rectangles should be shaded.

15. (a) $\dfrac{7}{28} = \dfrac{1}{4} = 25\%$ **(b)** $\dfrac{11}{33} = \dfrac{1}{3} = 33.\overline{3}\%$

   (c) $\dfrac{72}{144} = \dfrac{1}{2} = 50\%$

   (d) $\dfrac{44}{66} = \dfrac{2}{3} = 66.\overline{6}\%$

17. Answers will vary.

19. Patrick was looking the additive relationship when he determined his answer, but looking at the proportional (multiplicative relationship) Plant A grew by $\dfrac{3}{9} \approx 33\%$ compared to plant B which grew by just $\dfrac{3}{13} \approx 23\%$. Students need practice in thinking proportionally and using percentage change meaningfully.

Copyright © 2015 Pearson Education, Inc.

21. **(a)** Louis and Arnold prefer to think of saving money ("35% off") instead of focusing on how much is to be paid ("pay 65%").

   (b) 35% of $10 is $3.50. Subtract $3.50 from $10 to find the sale price of $6.50, which is the same as 65% of $10.

23. Yoshi has made an error often made with percent. Since $\frac{6}{3} = 2 = 200\%$, it is true that his new pay is 200% of his previous pay. However, it is incorrect to say that the *increase* in his pay is 200%. The increase was $3, which when compared to his previous pay of $3 is $\frac{3}{3} = 1 = 100\%$. Notice that if Yoshi was still paid $3 a day, he would be making 100% of his previous pay and have a 0% raise.

25. Yes. 5% off parts and 5% off labor results in a discount of 5% on the total job.

27. **(a)** For the customer, there is no difference. To see why, notice that the purchase price is multiplied by 1.08 to add on the tax, and 20% is taken off by multiplying by 0.8. Since $1.08 \times 0.8 = 0.8 \times 1.08$, it makes no difference to the customer if the sales tax is added before or after the coupon discount.

   (b) The cashier should redo the calculation. The store should remit only 8% of the sale price to the state, not 8% of the regular price.

29. She made $0.75 \times 16 = 12$ field goals out of $5 + 16 = 21$ attempts, so her percentage was $\frac{12}{21} \approx 0.5714 = 57.14\%$.

31. **(a)** Let x be the wholesale cost. Then $2x = x + 100\%x$ is the regular price, and the sale price is $0.8(2x) = 1.6x$. Then the store's profit is $1.6x - x = 0.6x$ and the percent profit is $\frac{0.6x}{x} = 0.60 = 60\%$.

   (b) Since 50% of $2x$ is x, there is no profit, or 0%.

33. The retail price for the heater would be $185 + (0.45) \times 185 = 185 \times 1.45 = 268.25$ dollars. Including tax you would have to pay

$268.25 + (0.05) \times 268.25 = 268.25 \times 1.05$
$\approx \$281.66$ for the water heater.

35. $11\% \times 158,000 = 0.11 \times \$158,000 = \$17,380$

37. **(a)** Let $P = 2500$, $r = 5.25$, $t = 1$, and $n = 7$. Then

$$P\left(1 + \frac{r}{100t}\right)^{nt} = 2500\left(1 + \frac{5.25}{100(1)}\right)^{7 \cdot 1}$$
$$= 2500(1.0525)^7 \approx 3576.80$$

The investment is worth $3576.80.

   (b) Let $P = 8000$, $r = 7$, $t = 2$, and $n = 10$. Then

$$P\left(1 + \frac{\frac{r}{t}}{100}\right)^{nt} = 8000\left(1 + \frac{\frac{7}{2}}{100}\right)^{10 \cdot 2}$$
$$= 8000(1.035)^{20} \doteq 15,918.31$$

The investment is worth $15,918.31.

39. Since $16,000 = P \cdot (1.06)^5$,

$$P = \frac{16,000}{(1.06)^5} \approx 11,956.13.$$

You would need to invest $11,956.13.

41. **(a)** Calculate successive powers of 1.05:
1.05, 1.1025, 1.1576, 1.2155, 1.2763, 1.3401, 1.4071, 1.4775, 1.5513, 1.6289, 1.7103, 1.7959, 1.8856, 1.9799, 2.0789, ...
Since $1.05^{15} > 2$, it takes 15 years.

   (b) Calculate successive powers of 1.07:
1.07, 1.1449, 1.2250, 1.3108, 1.4026, 1.5007, 1.6058, 1.7182, 1.8385, 1.9672, 2.1049, ... It takes 11 years.

   (c) Calculate successive powers of 1.14:
1.14, 1.2996, 1.4815, 1.6890, 1.9254, 2.1950, ... It takes 6 years.

   (d) Calculate successive powers of 1.20:
1.20, 1.44, 1.728, 2.0736, ...
It takes 4 years.

   (e) It seems pretty reasonable. $\frac{72}{5} = 14.4$, $\frac{72}{7} \approx 10.29$, $\frac{72}{14} \approx 5.14$, and $\frac{72}{20} = 3.6$, and these values are close to the ones we found in parts (a) through (d).

43. C. $11 \div 16 = 0.6875$, which is about 70%.

78 Chapter 7 Decimals, Real Numbers, and Proportional Reasoning

45. B. $0.7x = 48.99 \Rightarrow x = \dfrac{48.99}{0.7} \approx 70$

Chapter 7 Review Exercises

1. (a) $273.425 = 2 \cdot 10^2 + 7 \cdot 10^1 + 3 \cdot 10^0 + 4 \cdot 10^{-1}$
$$+ 2 \cdot 10^{-2} + 5 \cdot 10^{-3}$$

(b) $0.000354 = 3 \cdot 10^{-4} + 5 \cdot 10^{-5} + 4 \cdot 10^{-6}$

3. (a) $\dfrac{84}{175} = \dfrac{12}{25} = \dfrac{48}{100} = 0.48$

(b) $\dfrac{24}{99} = 0.\overline{24}$

(c) $\dfrac{7}{11} = \dfrac{63}{99} = 0.\overline{63}$

5. (a) Let $x = 10.\overline{363}$. Then
$$1000x = 10,363.\overline{363}.$$
$$1000x - x = 10,363.\overline{363} - 10.\overline{363}$$
$$999x = 10,353$$
$$x = \dfrac{10,353}{999} = \dfrac{3451}{333}$$

(b) Let $x = 2.1\overline{42}$. Then $10x = 21.\overline{42}$ and
$$1000x = 2142.\overline{42}$$
$$1000x - 10x = 2142.\overline{42} - 21.\overline{42}$$
$$990x = 2121$$
$$x = \dfrac{2121}{990} = \dfrac{707}{330}$$

7. (a) $0.\overline{2} = \dfrac{2}{9}$ **(b)** $0.\overline{36} = \dfrac{36}{99} = \dfrac{4}{11}$

9. (a) $31.47 + 3.471 + 0.0027 = 34.9437$

(b) $31.47 - 3.471 = 27.999$

(c) $31.47 \times 3.471 = 109.23237$

(d) $138.87 \div 23.145 = 6.0$

11. Since $\dfrac{4}{12} = 0.\overline{3}$, $\dfrac{5}{13} \approx 0.38$, and $\dfrac{2}{66} \approx 0.03$,
the numbers in order are 0.03, 0.33, $0.\overline{3}$,
0.3334, 0.38, or $\dfrac{2}{66}, 0.33, \dfrac{4}{12}, 0.3334, \dfrac{5}{13}$.

13. $3 - \sqrt{2}$ and $\sqrt{2}$ are irrational, but
$\left(3 - \sqrt{2}\right) + \sqrt{2} = 3$, which is rational.

15. (a) $8.25 \times 112.5 = 928.125 \text{ ft}^2$

(b) $928.125 \div 110 = 8.4375$ qts which rounds
to 9 qts.

17. She had 11 successes and $20 - 11 = 9$ failures,
so the ratio is 11 to 9.

19. Let x be the cost of 5 pounds of candy.
$$\dfrac{x}{5} = \dfrac{3.15}{2} \Rightarrow x = \dfrac{5 \cdot 3.15}{2} = 7.875$$
It would cost $7.88.

21. $y = kx \Rightarrow 7 = k \cdot 3 \Rightarrow k = \dfrac{7}{3}$
If $x = 5$, $y = kx = \dfrac{7}{3} \cdot 5 = \dfrac{35}{3}$.

23. (a) $\dfrac{5}{8} = 0.625 = 62.5\%$

(b) $2.115 = 211.5\%$

(c) $0.015 = 1.5\%$

25. $7.2\% \times \$49 = 0.072 \times \$49 \approx \$3.53$

27. $\dfrac{11}{20} = 0.55 = 55\%$

29. $\$3000\left(1 + \dfrac{8}{100(4)}\right)^{2 \cdot 4} = \$3000(1.02)^8$
$$\approx \$3514.98$$

Copyright © 2015 Pearson Education, Inc.

Chapter 8 Algebraic Reasoning, Graphing, and Connections with Geometry

Section 8.1
Variables, Algebraic Expressions, and Functions

Problem Set 8.1

1. (a) Constant; the number of feet never changes.

   (b) Variable; the number of hours of daylight increases and decreases during a year.

3. (a) Generalized variables

   (b) Relationship

5. (a) $2x - 3$ **(b)** $\dfrac{x}{2} + 2$

   (c) $x + 6 + x = 2x + 6$

7. (a) $2q - 10$

   (b) Assuming that he is successful in his attempt, he will weigh
   $q - 7(.25) = q - 1.75$ pounds.

   (c) Assuming that he is successful in his attempt, he will weigh $q + \dfrac{31}{8}$ pounds.

9. A writes x on a piece of paper, B multiplies by 5, C subtracts from 3, D cubes, E adds 4 then hands the paper to F. Therefore, there are 6 children, A, B, C, D, E, and F.
   Another correct answer has A and B as in previous sentence, C multiplies by -1, D adds 3, D cubes, F adds 4 then hands the paper to G. So, this solution has 7 children.

11. (a) Add $8x$ to both sides of the first equation to get $y = 8x + 2$. Subtract 4 from both sides of the second equation to get $y = 3x - 4$.

   (b) $8x + 2 = 3x - 4$

   (c) Subtract $3x$ from both sides to get $5x + 2 = -4$. Subtract 2 from both sides to get $5x = -6$. Divide both sides by 5 to get $x = -\dfrac{6}{5}$.

   (d) $y = 8x + 2 = 8\left(-\dfrac{6}{5}\right) + 2 = -\dfrac{38}{5}$.

13. (a) Not a function since element c in set A is associated with more than one element in set B and element d is not associated with any elements in set B.

   (b) A function, with range $\{p, q, r\} \subset B$.

15. (a)

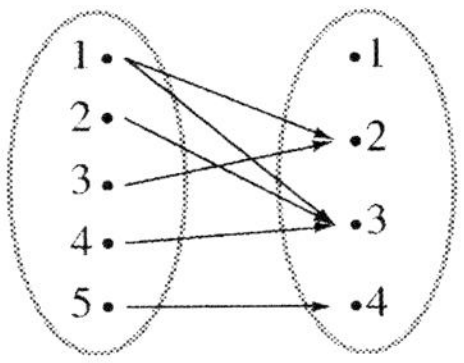

   Not a function with domain A, since $\{1\} \subset A$ is associated with more than one element of set B.

   (b)

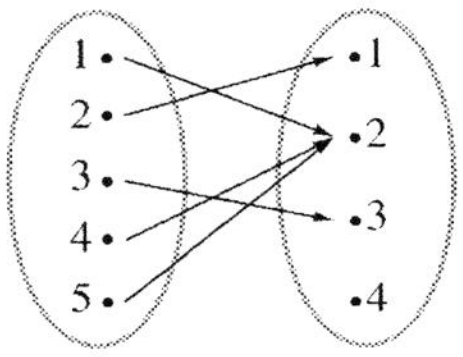

   A function with domain A, since every element of A has exactly one arrow that extends from that element to one of the elements in B.

   (c)

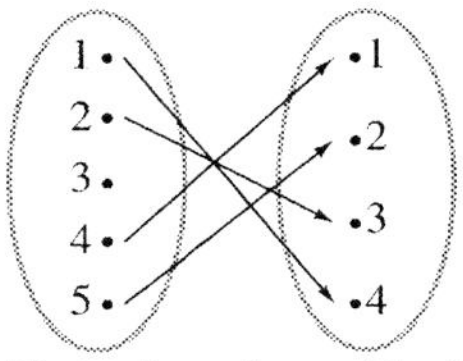

   Not a function with domain A, since $\{3\} \subset A$ is not associated with an element in B.

17. (a)

x	$f(x) = 2x - 2 = y$
-2	-6
-1	-4
0	-2
1	0
2	2
3	4

Copyright © 2015 Pearson Education, Inc.

80 *Chapter 8 Algebraic Reasoning and Representation*

(b)

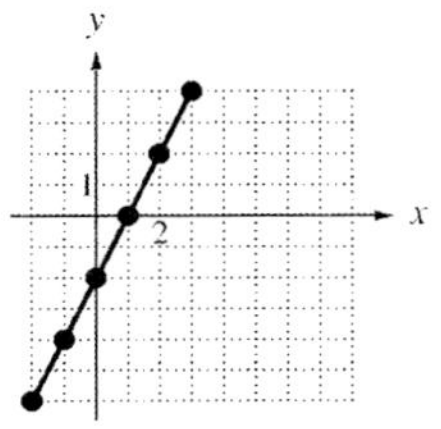

(c) Range $= \{ y \mid -6 \le y \le 4 \}$

19. (a) $h(2) = 2^2 - 1 = 3$

(b) $h(-2) = (-2)^2 - 1 = 3$

(c) Since $4^2 - 1 = (-4)^2 - 1 = 15$, the possible values are $t = 4$ or $t = -4$.

(d) $h(7.32) = 7.32^2 - 1 = 52.5824$

21. (a) She parked the car at the store about 1.5 hours past noon, or at 1:30 P.M.

(b) She shopped from 1.5 to 2.5 hours past noon, or $2.5 - 1.5 = 1$ hour

(c) The store was 10 miles away.

(d) The traffic was heavier coming home from the store, since the return trip took about half an hour and the trip to the store took only about 15 minutes.

23. (a)

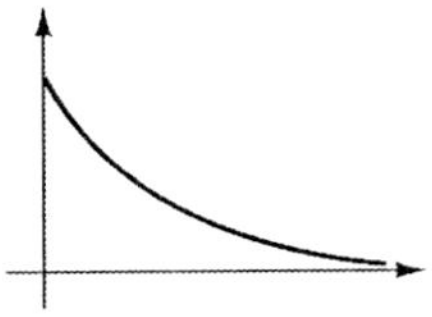

(b)

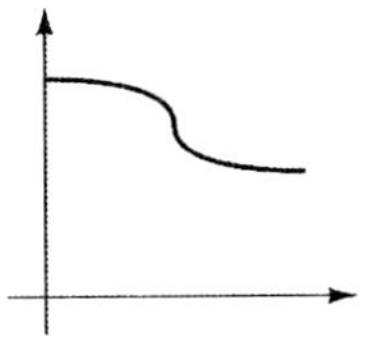

(c)

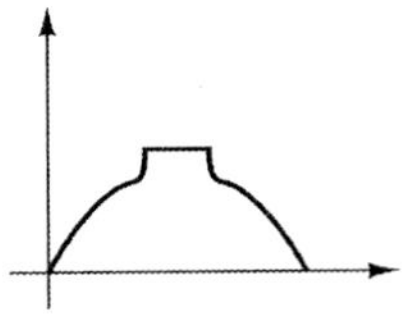

(d)

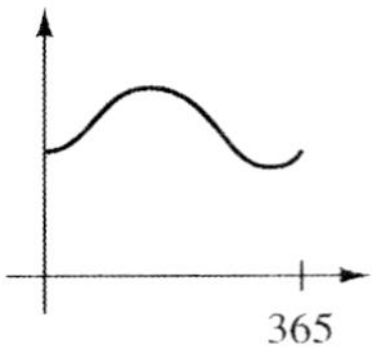

25. (a) Add 5: $y = x + 5$

(b) Double and add 2: $y = 2x + 2$

(c) Square and add 1: $y = x^2 + 1$

27.
$$
\begin{aligned}
F(0) &= f(g(0)) & F(1) &= f(g(1)) \\
&= f(0+3) & &= f(1+3) \\
&= f(3) & &= f(4) \\
&= 2 \times 3 & &= 2 \times 4 \\
&= 6 & &= 8 \\
F(2) &= f(g(2)) & F(3) &= f(g(3)) \\
&= f(2+3) & &= f(3+3) \\
&= f(5) & &= f(6) \\
&= 2 \times 5 & &= 2 \times 6 \\
&= 10 & &= 12
\end{aligned}
$$

29. Answers may vary. Show that the two expressions do not have the same value when, say, $a = 4$ and $b = 3$. For these values of the variables, $(4+3)^2 = 7^2 = 49$ but
$4^2 + 3^2 = 16 + 9 = 25$.
It can then be shown that
$$
\begin{aligned}
(4+3)^2 &= (4+3)(4+3) \\
&= 4 \cdot 4 + 4 \cdot 3 + 3 \cdot 4 + 3 \cdot 3 \\
&= 4 \cdot 4 + 2 \cdot (4 \cdot 3) + 3 \cdot 3.
\end{aligned}
$$
Eventually, it can be shown that
$$
\begin{aligned}
(a+b)^2 &= (a+b)(a+b) \\
&= (a+b)a + (a+b)b \\
&= a^2 + ba + ab + b^2 \\
&= a^2 + 2ab + b^2,
\end{aligned}
$$
showing the student how the general formula $(a+b)^2 = a^2 + 2ab + b^2$ is obtained.

31. Answers will vary.

33. Josiah focused in on the phrase "9 less", and so thought that he probably needed to subtract. He should be encouraged to plug numbers into the place of the variables to see if his answer was reasonable as a way to remedy this error. The correct answer is B.

Copyright © 2015 Pearson Education, Inc.

35. **(a)** Most students see the first row, and they jump to an answer of "adding the 2 each time."

(b) Have them model the problem with a function machine. Do it with counters and have them tell what happened inside the function machine. Then, they could try other examples and test their answers.

37. Let p denote the number of pencils and e the number of erasers. The total cost of p pencils, at 15 cents each, and e erasers at 6 cents each, is given by the expression $15p + 6e$. Since Huong spent 90 cents, p and e satisfy $15p + 6e = 90$. Dividing by 3 gives $5p + 2e = 30$, and this is equivalent to $p = \dfrac{(30 - 2e)}{5} = 6 - \dfrac{2}{5}e$. But p must be a positive whole number, so there are just two choices of e, either 5 or 10. The corresponding values of p are 4 and 2. In the first case Huong would have purchased just one more eraser than the number of pencils, so we conclude that Huong bought $e = 10$ erasers and $p = 2$ pencils.

39. **(a)** All 8 cubes have 3 painted faces.

(b) The 8 corner cubes have 3 painted faces. Each of the 12 edges of the large cube contains one cube with 2 painted faces. Each of the 6 faces of the large cube has one cube with 1 painted face. The one interior cube has no painted faces. Since $8 + 12 + 6 + 1 = 27 = 3^3$, all of the sugar cubes have been considered.

(c) The 8 corner cubes have 3 painted faces. Each of the 12 edges of the large cube contains $n - 2$ cubes with 2 painted faces, so $12(n - 2) = 12n - 24$ cubes have 2 painted faces. Each of the 6 faces of the large cube has $(n - 2)^2$ cubes with 1 painted face, so there are
$$6(n - 2)^2 = 6(n^2 - 4n + 4)$$
$$= 6n^2 - 24n + 24 \text{ cubes}$$
with one painted face. There are $(n - 2)^3 = n^3 - 6n^2 + 12n - 8$ interior cubes with no painted faces.

(d) The expressions, when added, must account for all n^3 sugar cubes. We see that
$$8 + (12n - 24) + (6n^2 - 24n + 24)$$
$$+ (n^3 - 6n^2 + 12n - 8)$$
$$= (8 - 24 + 24 - 8) + (12n - 24n + 12n)$$
$$+ (6n^2 - 6n^2) + n^3$$
$$= n^3.$$

41. The graphs are straight lines that cross when $t = 4.5$ hours and $y = 45$ miles.

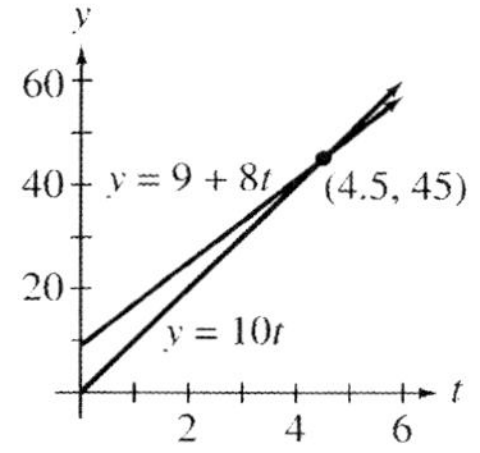

43. For $t \le 40$, Zal's pay is $\$18t$. For $t > 40$, Zal is paid $40 \times \$18 = \$720$ plus
$$\$18 + \frac{1}{2}(\$18) = \$27 \text{ for the hours beyond } 40.$$
This gives the expression $\$720 + (t - 40)(\$27)$ when $t > 40$.

45. A.

47. A. Since they each receive the total amount of money and share equally, they each received $x/4$ dollars.

49. A.

Section 8.2
Graphing Points, Lines, and Elementary Functions

Problem Set 8.2

1. **(a)–(d)**

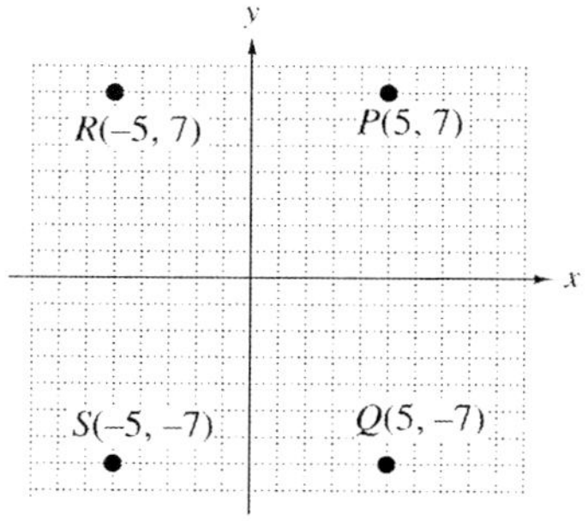

82 Chapter 8 Algebraic Reasoning and Representation

3.

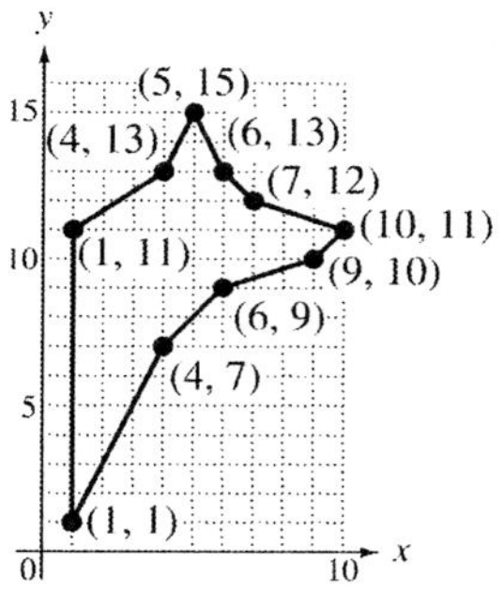

5. (a) $\sqrt{(4-(-2))^2+(13-5)^2} = \sqrt{6^2+8^2}$
$$= \sqrt{36+64}$$
$$= \sqrt{100} = 10$$

(b) $\sqrt{(8-3)^2+(8-(-4))^2} = \sqrt{5^2+12^2}$
$$= \sqrt{25+144}$$
$$= \sqrt{169} = 13$$

7. (a) $m = \dfrac{8-4}{3-1} = \dfrac{4}{2} = 2$

upward to the right

(b) $m = \dfrac{-6-5}{-2-(-2)} = \dfrac{-11}{0}$ (undefined)

vertical

(c) $m = \dfrac{-7-(-3)}{-4-(-2)} = \dfrac{-4}{-2} = 2$

upward to the right

9. $\dfrac{7-3}{4-b} = 2$
$$4 = 2(4-b)$$
$$2 = 4-b$$
$$b = 2$$

11. Substitute a for x and 3 for y:
$$2a + 3 \cdot 3 = 18$$
$$2a = 9$$
$$a = \dfrac{9}{2}$$

13. (a), (c)
For part (a), $3x + 5y = 12$.
If $x = -1$, then $-3 + 5y = 12$, or $y = 3$, so
$(-1, 3)$ is on the line.
If $y = 0$, then $3x = 12$, or $x = 4$, so $(4, 0)$ is
on the line.
For part (c), $5y - 3x = 15$.
If $x = 0$, then $5y = 15$, or $y = 3$, so $(0, 3)$ is
on the line.

If $y = 0$, then $-3x = 15$, or $x = -5$, so
$(-5, 0)$ is on the line.

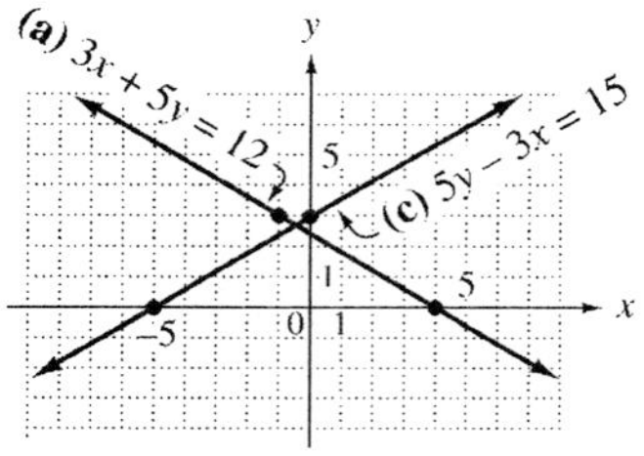

(a)

(b), (d)
For part (b), $6x = -10y + 12$.
If $x = -3$, then $-18 = -10y + 12$, or
$y = 3$, so $(-3, 3)$ is on the line.
If $y = 0$, then $6x = 12$, or $x = 2$, so $(2, 0)$ is
on the line.
For part (d), $6x + 10y = 24$.
If $x = -1$, then $-6 + 10y = 24$, or $y = 3$, so
$(-1, 3)$ is on the line.
If $y = 0$, then $6x = 24$, or $x = 4$, so $(4, 0)$ is
on the line.

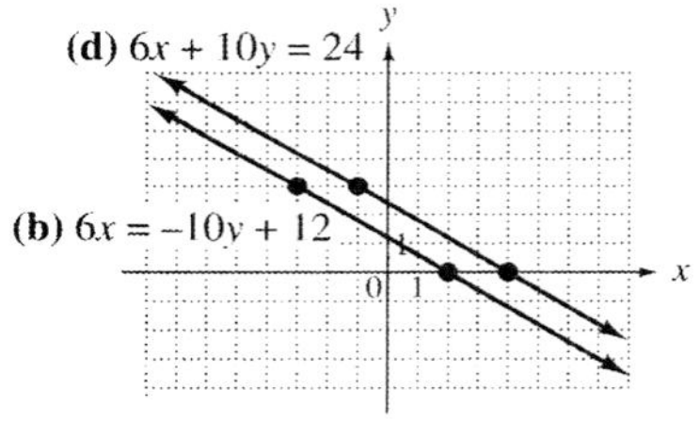

(e) Both lines include $(-1, 3)$ and $(4, 0)$.
Therefore, they are the same line.

15. (a)

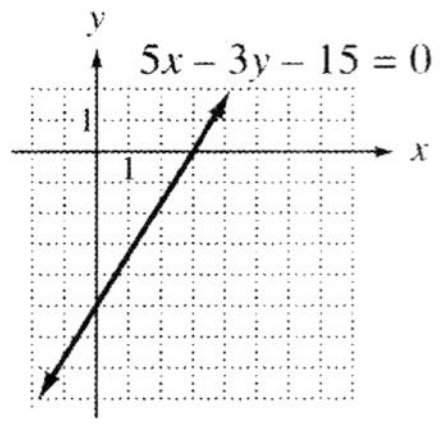

(b) Yes. The function is $y = \dfrac{5}{3}x - 5$.

(c) Yes. By solving the equation in part (c)
for x, we see that the function is
$$x = \dfrac{3}{5}y + 3.$$

Copyright © 2015 Pearson Education, Inc.

17. (a) $y = x^2$

x	y
–3	9
–2	4
–1	1
0	0
1	1
2	4
3	9

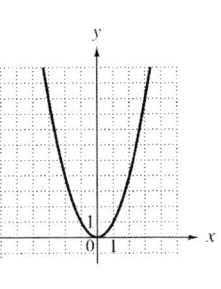

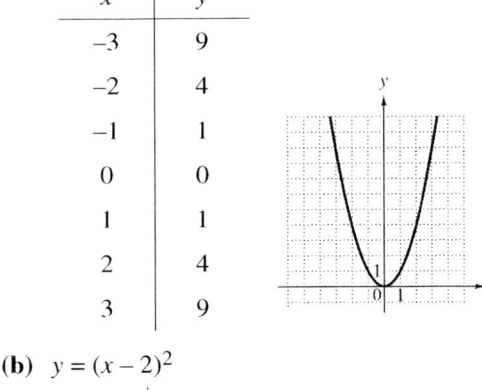

(b) $y = (x - 2)^2$

x	y
–3	25
–2	16
–1	9
0	4
1	1
2	0
3	1

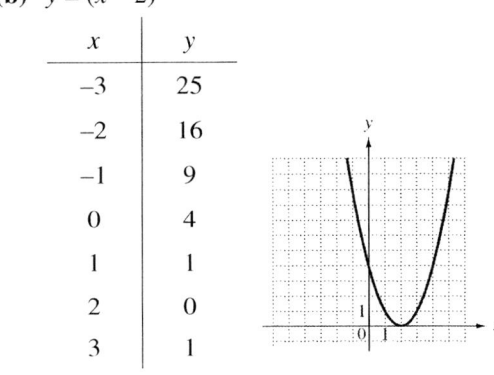

(c) $y = (x + 3)^2$

x	y
–3	0
–2	1
–1	4
0	9
1	16
2	25
3	36

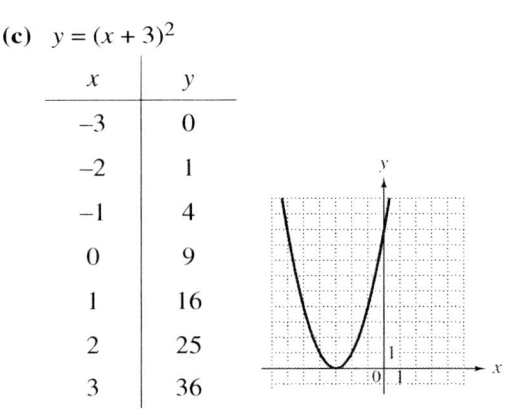

(d) They are horizontal shifts of the same curve.

19. (a)

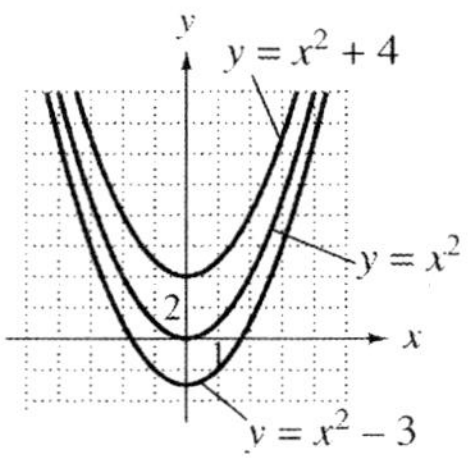

(b) The graphs of $y = x^2 + 4$ and $y = x^2 - 3$ are the same as the graph of $y = x^2$ except that the former is shifted up four units and the latter is shifted down three units.

21. (a) $y = 2x^2 - 4x + 10$

x	y
–3	40
–2	26
–1	16
0	10
1	8
2	10
3	16
4	26
5	40

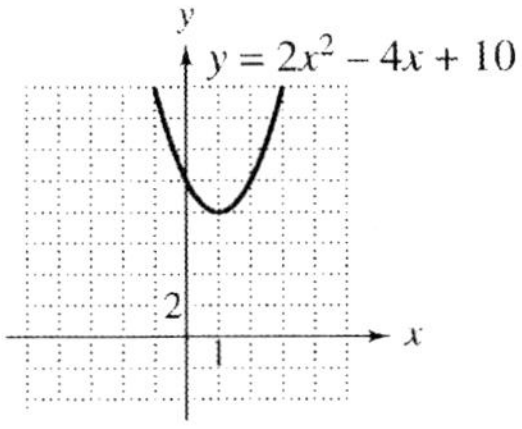

From the graph, the minimum value of $2x^2 - 4x + 10$ is 8. It occurs when $x = 1$.

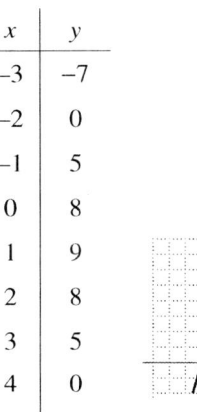

(b) $y = 2x - x^2 + 8$

x	y
–3	–7
–2	0
–1	5
0	8
1	9
2	8
3	5
4	0
5	–7

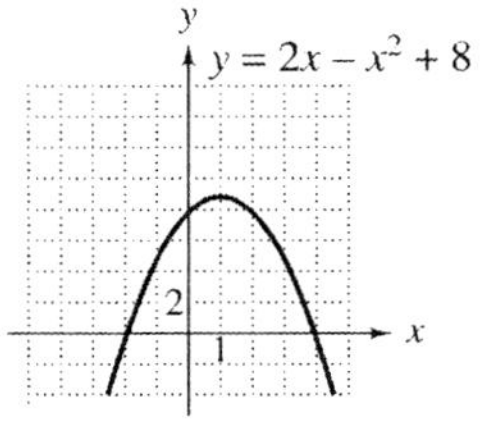

From the graph, the maximum value of $2x - x^2 + 8$ is 9. It occurs when $x = 1$.

23. Consider squares of size x, and graph the function $p(x) = 4x$ and $A(x) = x^2$ that respectively give the perimeter and area of the square as functions of x.

84 *Chapter 8 Algebraic Reasoning and Representation*

25. Note that B can be obtained by sliding A to the right 3 and up 5. Therefore, C can be obtained by sliding D to the right 3 and up 5. The coordinates of C are $(7 + 3, 0 + 5) = (10, 5)$. Hence $r = 10$ and $s = 5$.

27. There are 4 sections of length x and 3 sections of length w, so $4x + 3w = 24$, or $3w = 24 - 4x$. The pens form a $3w$ by x rectangle of area $(3w)x$.

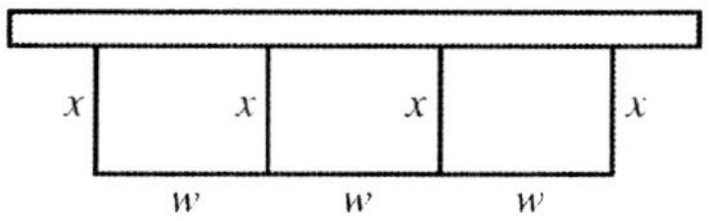

A substitution then shows that the area of the three pens is given by $f(x) = (24 - 4x)x$. Since x is a length, $x \geq 0$. Also, $4x$ must not be larger than the 24 feet of available fence, so $4x \leq 24$, or $x \leq 6$. The table of values of $f(x)$ for $x = 0$, 1, 2, 3, 4, 5, 6 suggests that the largest area occurs when $x = 3$. Then $3w = 24 - 12 = 12$, so $w = 4$. Thus each pen should measure 3 by 4 feet. The function can also be graphed.

x	$f(x)$
0	0
1	20
2	32
3	36
4	32
5	20
6	0

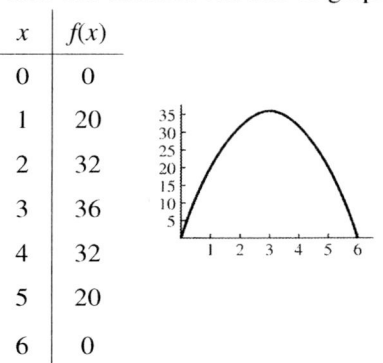

29. Since we are still doubling but starting with 25 cents, trying a few squares shows that the equation for the value on the nth square is $(.25)2^{n-1}$. The answer again is to take the money on the 64th day. (In fact, the money on a square is larger than \$1 million on day 23.)

31. (a) The run must be at least 12 times the 30 inch = 2.5 foot rise, so the minimum allowable run is $12 \times 2.5 = 30$ feet.

    (b) The rise is $\dfrac{1800}{5280}$ miles in a run of 7 miles, so the slope is $\dfrac{1800}{5280} \div 7 = 0.05$, or about 5%.

33. $L = mw + b$
    $10 = m \cdot 0 + b$, so $b = 10$.
    $14 = m \cdot 2 + b = 2m + 10$, so $m = 2$.
    The constants are $b = 10$ and $m = 2$.

35. 760 mi/hr = $0.2\bar{1}$ mi/sec $\doteq \dfrac{1}{5}$ mi/sec.

    Thus, $d \doteq \dfrac{1}{5}t$.

37. C. $C = (5, 3)$ and $D = (-2, -3)$, so the distance between C and D is given by

    $$\sqrt{\left(5 - (-2)\right)^2 + \left(3 - (-3)\right)^2}$$
    $$= \sqrt{(5 + 2)^2 + (3 + 3)^2} = \sqrt{49 + 36} = \sqrt{85}$$
    $$\approx 9$$

39. J

Section 8.3
Connections Between Algebra and Geometry

Problem Set 8.3

1. (a) Many answers are possible. An example is shown.
    When $x = 0$,
    $$4(0) + 2y = 6 \Rightarrow 2y = 6 \Rightarrow y = 3$$
    When $y = 0$,
    $$4x + 2(0) = 6 \Rightarrow 4x = 6 \Rightarrow x = \frac{6}{4} = 1.5$$
    Thus two points on the line are $(0, 3)$ and $(1.5, 0)$.

    (b) Answers may vary, since the points from part (a) will be used. However, any combination of points on the line will result in a slope of $m = -2$.
    $$m = \frac{0 - 3}{1.5 - 0} = \frac{-3}{1.5} = -2$$

    (c) $4x + 2y = 6$
    $$2y = -4x + 6$$
    $$y = -2x + 3$$
    The slope is -2 and the y-intercept is 3.

    (d) Regardless of the point chosen, the slope is still -2. The concept is that the slope is always the same constant.

Copyright © 2015 Pearson Education, Inc.

Solutions to Problem Set 8.3 **85**

3. First, find the slope of $3x - 5y + 45 = 0$ by solving for y:

$$0 = 3x - 5y + 45$$
$$5y = 3x + 45$$
$$y = \frac{3}{5}x + 9$$

The slope is $\frac{3}{5}$.

(a) $7x + ky = 21$

$$ky = -7x + 21$$
$$y = \frac{-7}{k}x + \frac{21}{k}$$

The slope must be $\frac{3}{5}$.

$$\frac{-7}{k} = \frac{3}{5} \Rightarrow 3k = -35 \Rightarrow k = -\frac{35}{3}$$

(b) $kx - 8y - 24 = 0$

$$-8y = -kx + 24$$
$$y = \frac{k}{8}x - 3$$

The slope must be $\frac{3}{5}$:

$$\frac{k}{8} = \frac{3}{5} \Rightarrow 5k = 24 \Rightarrow k = \frac{24}{5}$$

5. (a) $y = 2x - 3$

x	y
-3	-9
-2	-7
-1	-5
0	-3
1	-1
2	1
3	3

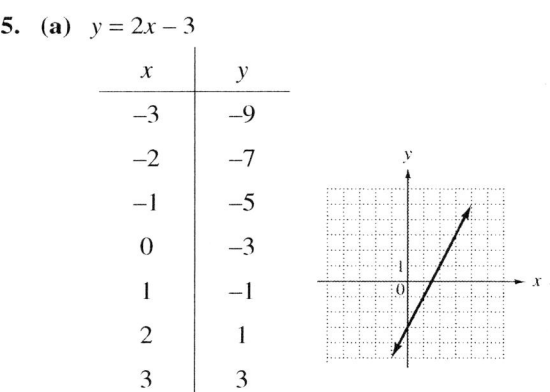

(b) $y = 0.5x + 2$

x	y
-3	0.5
-2	1
-1	1.5
0	2
1	2.5
2	3
3	3.5

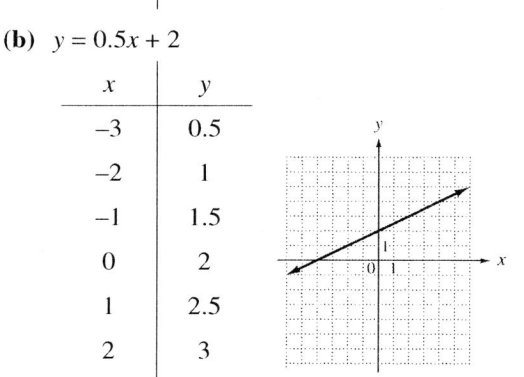

(c) $y = -3x$

x	y
-3	9
-2	6
-1	3
0	0
1	-3
2	-6
3	-9

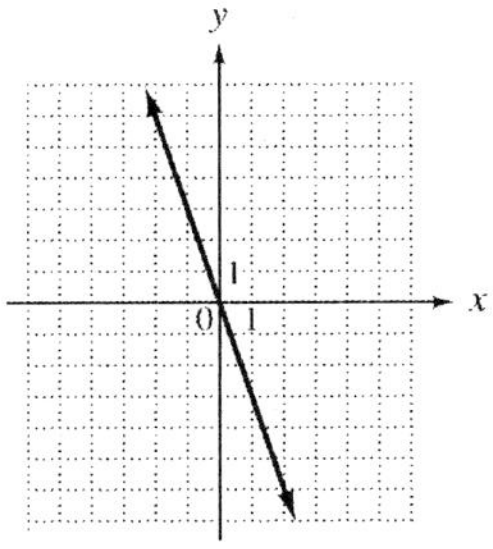

7. (a)

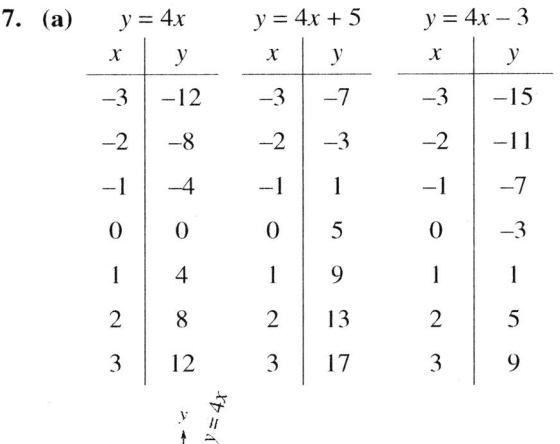

$y = 4x$		$y = 4x + 5$		$y = 4x - 3$	
x	y	x	y	x	y
-3	-12	-3	-7	-3	-15
-2	-8	-2	-3	-2	-11
-1	-4	-1	1	-1	-7
0	0	0	5	0	-3
1	4	1	9	1	1
2	8	2	13	2	5
3	12	3	17	3	9

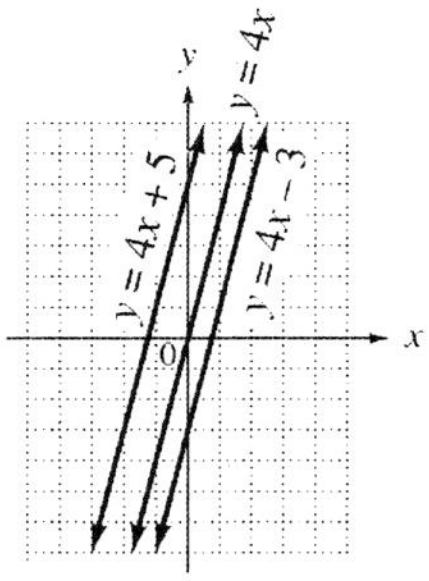

(b) The lines are parallel. This is expected because they are given in slope-intercept form, and all three have the same slope, 4.

9. (a)

$$(RS)^2 = \left(\sqrt{(7-1)^2 + (10-2)^2}\right)^2$$
$$= \left(\sqrt{6^2 + 8^2}\right)^2 = \left(\sqrt{36 + 64}\right)^2$$
$$= \left(\sqrt{100}\right)^2 = 100$$

(*continued on next page*)

Copyright © 2015 Pearson Education, Inc.

86 Chapter 8 Algebraic Reasoning and Representation

(*continued*)

$$(RT)^2 = \left(\sqrt{(5-1)^2 + (-1-2)^2}\right)^2$$

$$= \left(\sqrt{4^2 + (-3)^2}\right)^2 = \left(\sqrt{16+9}\right)^2$$

$$= \left(\sqrt{25}\right)^2 = 25$$

$$(ST)^2 = \left(\sqrt{(5-7)^2 + (-1-10)^2}\right)^2$$

$$= \sqrt{(-2)^2 + (-11)^2} = \left(\sqrt{4+121}\right)^2$$

$$= \left(\sqrt{125}\right)^2 = 125$$

Since

$$(RS)^2 + (RT)^2 = 100 + 25 = 125 = (ST)^2,$$

by the Pythagorean theorem, $\triangle RST$ is a right triangle.

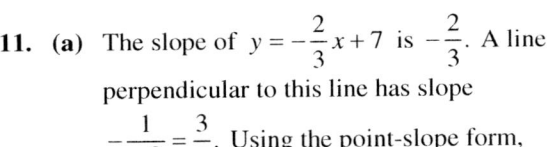

(b)

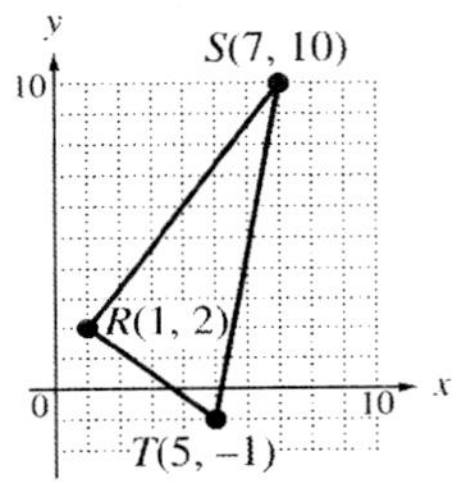

11. (a) The slope of $y = -\dfrac{2}{3}x + 7$ is $-\dfrac{2}{3}$. A line perpendicular to this line has slope

$-\dfrac{1}{-\frac{2}{3}} = \dfrac{3}{2}$. Using the point-slope form,

the equation of the perpendicular line through $(4, -1)$ is

$$y - (-1) = \frac{3}{2}(x-4)$$

$$y + 1 = \frac{3}{2}x - 6$$

$$y = \frac{3}{2}x - 7$$

(b) The slope of $2y - 6x - 5 = 0$ is $\dfrac{-6}{2} = 3$.

A line perpendicular to this line has slope

$-\dfrac{1}{3}$. Using the point-slope form, the

equation of the perpendicular line through $(0, 3)$ is

$$y - 3 = -\frac{1}{3}(x-0) \Rightarrow y = -\frac{1}{3}x + 3.$$

13. $RS = \sqrt{(1-5)^2 + (4-0)^2} = \sqrt{(-4)^2 + 4^2} = \sqrt{32}$

$TS = \sqrt{(7-5)^2 + (6-0)^2} = \sqrt{(2)^2 + 6^2} = \sqrt{40}$

Since two sides are not equal, then the triangle is not equilateral.

15. The slope of $\overline{AB}$ is $\dfrac{4 - (-2)}{2 - 1} = 6$, so the slope

of the perpendicular bisector is $-\dfrac{1}{6}$. The

midpoint of $\overline{AB}$ is $\left(\dfrac{1+2}{2}, \dfrac{4 + (-2)}{2}\right)$ or

$\left(\dfrac{3}{2}, 1\right)$. The equation of the perpendicular

bisector is $y - 1 = -\dfrac{1}{6}\left(x - \dfrac{3}{2}\right)$.

17. (a) By sketching the circle, it is clear that the equation of the tangent is $y = -3$.

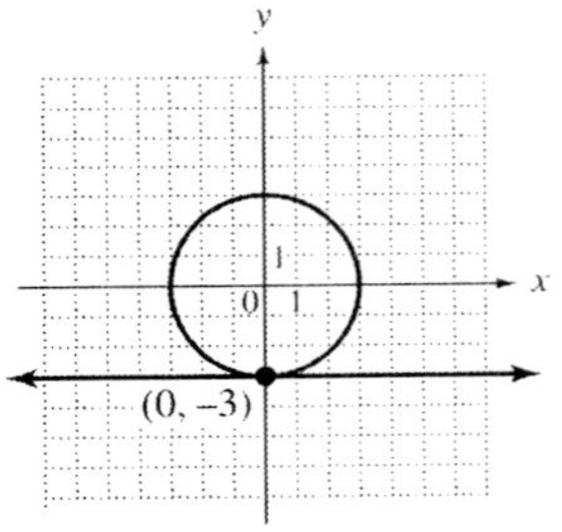

(b) By sketching the circle, it is clear that the equation of the tangent is $y = 3$.

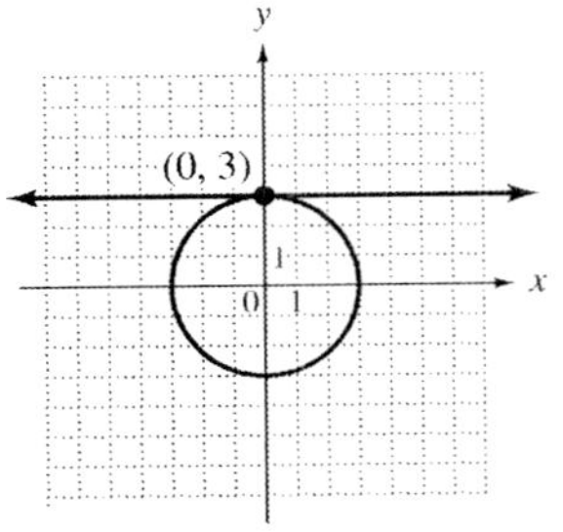

(c) The lines are parallel.

19. Responses will vary, but the conclusion is that l is parallel to n.

Copyright © 2015 Pearson Education, Inc.

21. All are true except (d). (e) is true when the two circles are the same.

a.

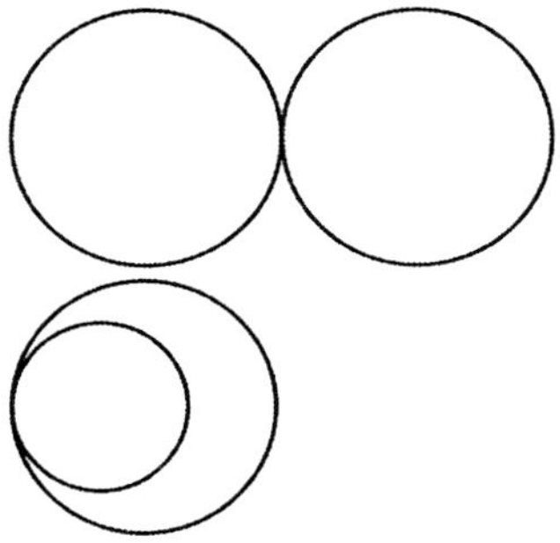

or

b.

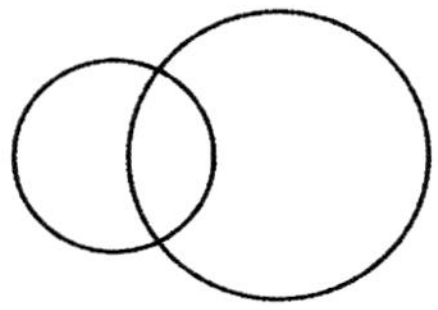

c.

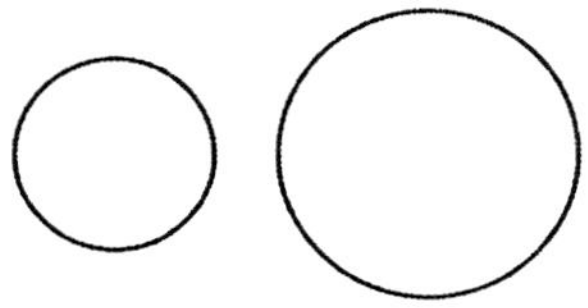

23. **(a)** A diameter is a chord which goes through the center of the circle.

(b) A tangent line to a circle is a line which passes through a point on the circle and is perpendicular to the diameter or radius at that point.

(c) Answers will vary.

25.

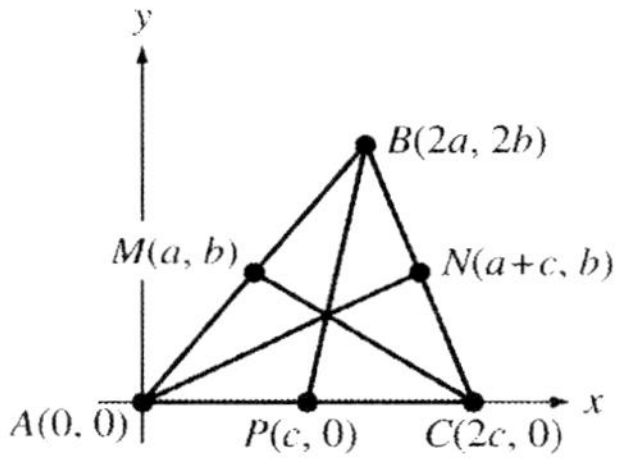

The midpoint M of AB is (a, b), the midpoint N of BC is $(a + c, b)$, and the midpoint P of AC is $(c, 0)$. So the equation of the median AN is

$$y = \frac{b}{a + c} x.$$

The equation of the median BP is

$$y = \frac{2b}{2a - c} (x - c),$$ and the equation of the

median CM is $y - b = \frac{b}{a - 2c} (x - a)$.

We can show that the point $G\left(\dfrac{2a + 2c}{3}, \dfrac{2b}{3}\right)$

lies on all three medians by verifying that the coordinates satisfy each equation.

$$y = \frac{b}{a + c} x \Rightarrow \frac{2b}{3} = \frac{b}{a + c}\left(\frac{2a + 2c}{3}\right)$$

$$y = \frac{2b}{2a - c}(x - c) \Rightarrow \frac{2b}{3} = \frac{2b}{2a - c}\left(\frac{2a + 2c}{3} - c\right)$$

$$y - b = \frac{b}{a - 2c}(x - a) \Rightarrow$$

$$\frac{2b}{3} - b = \frac{b}{a - 2c}\left(\frac{2a + 2c}{3} - a\right)$$

27. **(a)** By trial and error, these are 10 taxi segments. Notice that although there are infinitely many paths from A to B, the taxi segments correspond to the paths that accomplish the journey in 5 blocks—that is, no backtracking. Any such journey must include 3 eastward blocks and 2 northward blocks. Since these may be undertaken in any order, the number of paths may also be determined as the number of ways to arrange the symbols E, E, E, N, N. This is $C(5, 2) = 10$.

(b)

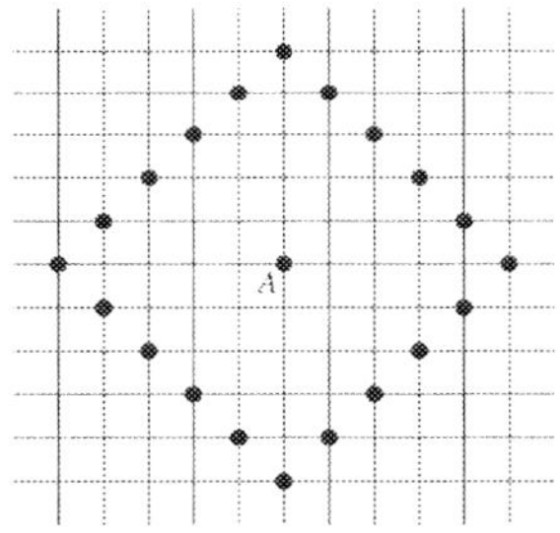

No, it does not look like a circle drawn with a compass. It looks like a square.

(c)

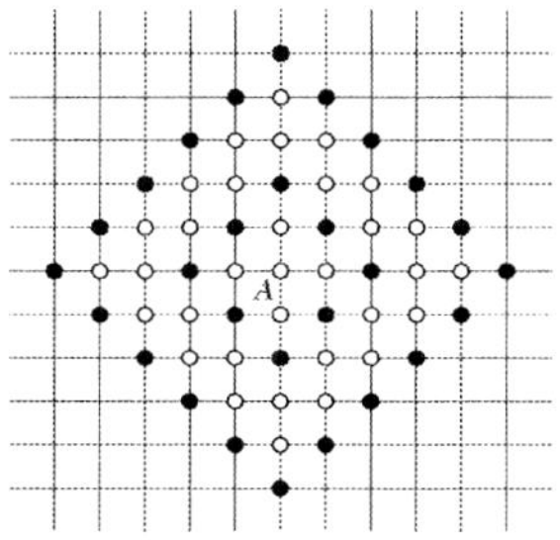

The points a particular distance away form a square oriented with one corner downward. The squares for different distances are nested inside one another.

29. A

Copyright © 2015 Pearson Education, Inc.

88 *Chapter 8 Algebraic Reasoning and Representation*

Chapter 8 Review Exercises

1. (a) $a + 5$ **(b)** $b < c$

(c) $c - b$ **(d)** $\dfrac{a+b+c}{3}$

3. (a) Multiply the input by 3 and add 2; that is, $y = 3x + 2$

(b) Multiply the input by the value one larger than the input; that is, $y = x(x + 1)$, or $y = x^2 + x$

(b) If $f(x) = 0$, then either $2x = 0$ or $x - 3 = 0$. Therefore, $x = 0$ or $x = 3$.

5. (a)

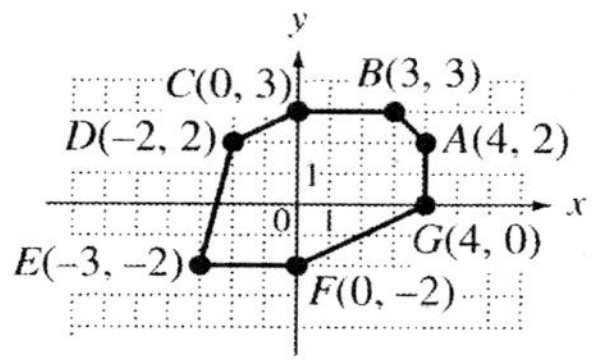

(b) B is in the first quadrant, C is on the positive y-axis, D is in the second quadrant, E is in the third quadrant, and F is on the negative y-axis.

(c) slope $\overline{BC} = \dfrac{0}{3} = 0$, slope $\overline{CD} = \dfrac{1}{2}$,

slope $\overline{DE} = \dfrac{4}{1} = 4$, slope $\overline{EF} = \dfrac{0}{3} = 0$,

slope $\overline{FG} = \dfrac{2}{4} = \dfrac{1}{2}$,

slope $\overline{GA}$ is undefined. Yes; $\overline{BC}$ is parallel to $\overline{EF}$, and $\overline{CD}$ is parallel to $\overline{FG}$.

7. (a) $y = 4 + 2(x - 3)$, by the point-slope form of the equation of a line. The equation can be rewritten in slope-intercept form as $y = 2x - 2$.

(b) $y = -1 + \dfrac{5 - (-1)}{(-2) - 6}(x - 6)$, by the two-point form of the equation of a line. The equation can be rewritten in slope-intercept form as $y = -\dfrac{3}{4}x + \dfrac{7}{2}$.

(c) $y = 3x - 4$, using the slope-intercept form of the equation of a line.

9. The altitude through R is perpendicular to $\overline{ST}$. The slope of $\overline{ST}$ is $\dfrac{6 - 0}{7 - 5} = 3$, so the slope of the altitude is $-\dfrac{1}{3}$. The altitude passes through $R(1, 4)$, so the equation of the altitude is $y - 4 = -\dfrac{1}{3}(x - 1)$.

11. (a)

x	$y = \left(\dfrac{1}{4}\right)^x$
0	1
1	0.25
2	0.0625
3	0.015625

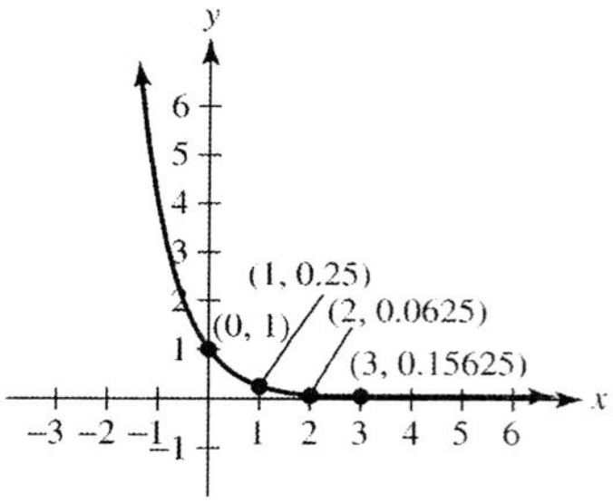

(b) Problem 25c of section 8.2 is a word problem whose math is identical with this one once you realize that $\$\dfrac{1}{4} = 25$ cents.

Copyright © 2015 Pearson Education, Inc.

Chapter 9   Geometric Figures

Section 9.1
Figures in the Plane

Problem Set 9.1

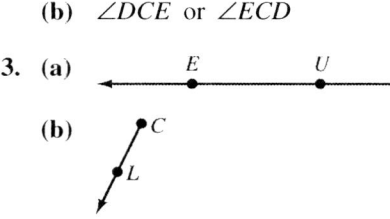

1. (a) $\overrightarrow{AB}$ or $\overrightarrow{BA}$

    (b) $\angle DCE$ or $\angle ECD$

3. (a)

    (b)

    (c)

5. (a) $\angle BAC$ will be a right angle if C is any of the circled points.

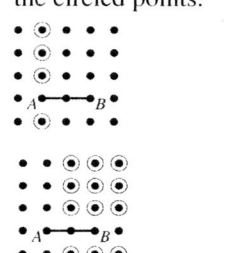

    (b)

    (c)

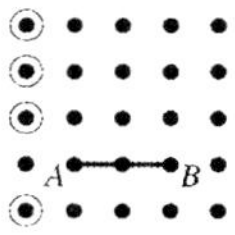

    (d)

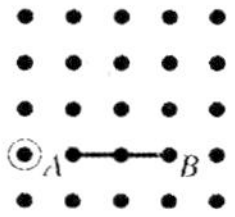

    (e)

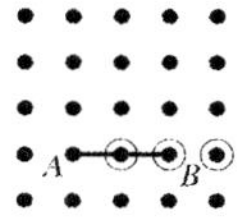

7. P, Q, and R are collinear.
    A possible drawing is shown.

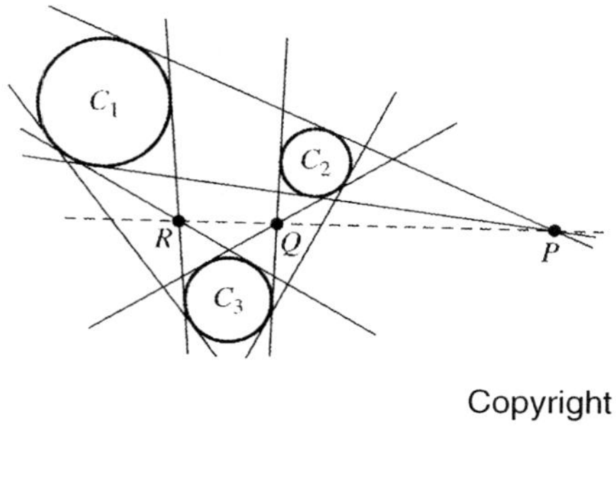

9. (a) $\overrightarrow{KL}$ and $\overrightarrow{MN}$ are parallel.

    (b) $\overrightarrow{LM}$ and $\overrightarrow{KN}$ are parallel.

    (c) $\overrightarrow{AB}$ is perpendicular to $\overrightarrow{KL}$ and $\overrightarrow{MN}$.

11. (a) For the figure shown in the textbook, $m(\angle A) \doteq 111°$, $m(\angle B) \doteq 88°$, $m(\angle C) \doteq 69°$, and $m(\angle D) \doteq 92°$. In general, opposite angles are supplementary: $m(\angle A) + m(\angle C) = 180°$ and $m(\angle B) + m(\angle D) = 180°$.

    (b) For the figure shown in the textbook, $m(\angle P) \approx 47°$, $m(\angle Q) \approx 26°$, $m(\angle R) \approx 26°$, and $m(\angle S) \approx 47°$. In general, $m(\angle P) = m(\angle S)$, and $m(\angle Q) = m(\angle R)$.

13. (a) One complete turn, or 360°

    (b) $\dfrac{1}{6}$ of a complete turn: $\dfrac{1}{6}(360°) = 60°$

    (c) $\dfrac{1}{30}$ of a complete turn: $\dfrac{1}{30}(360°) = 12°$

    (d) The hour hand moves 2 hours in 120 minutes. That is $\dfrac{2}{12}$ or $\dfrac{1}{6}$ of a complete rotation. $\dfrac{1}{6}(360°) = 60°$

    (e) The hour hand moves $\dfrac{5}{60}$ of a 1 hour in 5 minutes. The hour hand moves 30° in on hour, so it moves $\dfrac{5}{60}(30°) = 2.5°$ in 5 minutes.

90 Chapter 9 Geometric Figures

15. Since ∠1 and ∠4 are supplements of a known 60° angle, $m(\angle1) = m(\angle4) = 180° - 60° = 120°$.
By vertical angles, $m(\angle5) = 60°$. Since $l \parallel m$ and ∠8 corresponds to a known 60° angle along a transversal, $m(\angle8) = 60°$. Since $m(\angle8) + m(\angle11) = 180°$, $m(\angle11) = 180° - 60° - 70° = 50°$. By vertical angles, $m(\angle9) = 70°$, $m(\angle10) = m(\angle11) = 50°$, and $m(\angle12) = m(\angle8) = 60°$. Since $l \parallel m$ and ∠6 corresponds to ∠11 along a transversal, $m(\angle6) = m(\angle11) = 50°$. By vertical angles, $m(\angle3) = m(\angle6) = 50°$. Since ∠2 and ∠7 are supplements of ∠6, $m(\angle2) = m(\angle7) = 180° - 50° = 130°$. These results are summarized below.

Angle	∠1	∠2	∠3	∠4	∠5	∠6	∠7	∠8	∠9	∠10	∠11	∠12
Measure	120°	130°	50°	120°	60°	50°	130°	60°	70°	50°	50°	60°

17. (a) The measures of the interior angles of a triangle add up to 180°. Therefore, $m(\angle1) = 180° - 70° - 70° = 40°$.

(b) The measure of an exterior angle is the sum of the measures of its opposite interior angles, so
$60° = m(\angle2) + 20°$. Therefore, $m(\angle2) = 40°$.

19. (a) $x + x + 30° = 180°$ so $x = 75°$. The interior angles measure 75°, 75°, and 30°.

(b) $2y + y + 15° = 180°$ so $3y = 165°$, or $y = 55°$. The interior angles measure 55°, 110°, and 15°.

21. (a) No, because an obtuse angle has measure greater than 90°, and adding two such measures would exceed 180°, which is the sum of all three interior angle measures for any triangle.

(b) No, because a right angle has measure of 90°, so adding together the measures of two right angles and a third angle would exceed 180°.

(c) The sum of the three angle measures must be 180°, so each angle must be exactly 60°.

23. Since we are given that ∠APC is a right angle, $m\angle1 + m\angle2 = 90°$; similarly, we are given that ∠BPD is a right angle, so $m\angle3 + m\angle2 = 90°$. Therefore, solving for $m\angle1$ and $m\angle3$ in these equations, we have $m\angle1 = 90° - m\angle2$ and $m\angle3 = 90° - m\angle2$, respectively. By substitution, we have $m\angle1 = m\angle3$, so we conclude that $\angle1 \cong \angle3$.

25. Answers will vary.

27. Solutions will vary although asking them to draw two different segments of length one would work.

29. The confusion about parallelism is caused by the segments, as opposed to the lines $\overleftrightarrow{AB}$ and $\overleftrightarrow{CD}$, being examined to see if they intersect. Explain that Larisa should think of lines, rather than segments, to determine parallelism. You could then ask her if they intersect "off the paper."

31. (a) Sebastian assumed that all quadrilaterals contain at least one right angle. He also didn't realize that the triangle has a right angle.

(b) Answers will vary, but Sebastian needs to go back to definitions.

33. (a) 6 lines

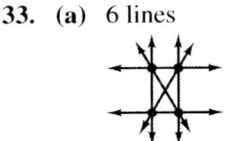

(b) 10 lines

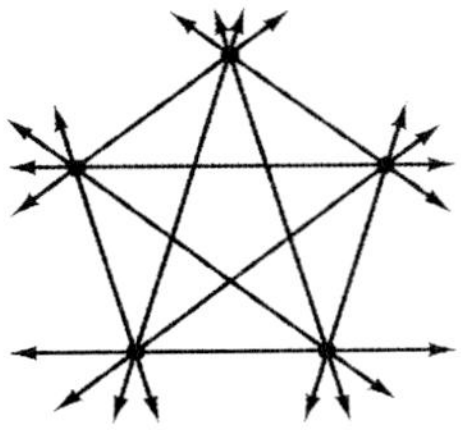

(c) Choosing any two points from the n available will determine a line. So an equivalent question is: How many ways can subsets of size 2 be chosen from a set of size n? $C(n, 2) = \dfrac{n(n-1)}{2}$ lines are determined by n points when no three points are collinear.

Copyright © 2015 Pearson Education, Inc.

Solutions to Problem Set 9.2   91

Section 9.2
Curves and Polygons in the Plane

Problem Set 9.2

1.
	(a)	(b)	(c)	(d)	(e)	(f)
Simple Curve	✓		✓		✓	✓
Closed Curve			✓	✓	✓	
Polygonal Curve	✓	✓	✓			
Polygon			✓			

3. (a) A nonsimple closed four-sided polygonal curve is a figure that has 4 line segments which when traced without lifting your pencil, at least one point will be touched more than once and the trace will end at the same point at which it started.

(b) A concave pentagon has 5 sides and has at least 2 vertices that cannot be connected with a line segment that is on or within the pentagon.

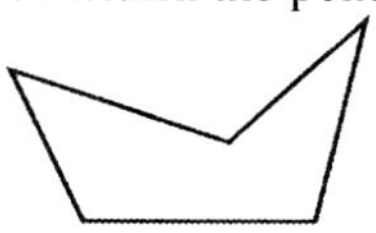

(c) An equiangular quadrilateral has 4 sides and all angles have the same measure. Any rectangle is acceptable.

(d) A convex octagon has 8 sides and all vertices can be connected with a line segment that is on or within the octagon.

35. The pencil turns through each interior angle of the triangle. Notice that each turn is in the same direction (counterclockwise). Since the pencil faces the opposite direction when it returns to the starting side, it has turned a total of 180°. This demonstrates that the sum of measures of the interior angles of a triangle is 180°.

37. $\triangle BCQ$ is a right triangle, so $m(\angle 2) + m(\angle 3) = 90°$. By the reflection property, $\angle 1 \cong \angle 2$ and $\angle 3 \cong \angle 4$. Therefore $m(\angle 5) = 180° - m(\angle 1) - m(\angle 2)$
$= 180° - 2m(\angle 2)$
and $m(\angle 6) = 180° - m(\angle 3) - m(\angle 4)$
$= 180° - 2m(\angle 3)$. Adding these equations gives $m(\angle 5) + m(\angle 6)$
$= 360° - 2m(\angle 2) - 2m(\angle 3)$
$= 360° - 2(m(\angle 2) + m(\angle 3))$
$= 360° - 2 \cdot (90°) = 180°$,
showing $\angle 5$ and $\angle 6$ are supplementary. By the alternate interior angles theorem, two lines are parallel if and only if the interior angles on the same side of a transversal are supplementary. Since $\angle 5$ and $\angle 6$ are supplementary, we then conclude that $\overline{AB}$ and $\overline{CD}$ are parallel.

39. A right triangle is formed with the three angles P, the angle of elevation, and 90°. So the angle of elevation is $90° - P$, using the fact that the three angles add up to 180°.

41. Answers will vary. The following diagram shows sample answers.

(a)

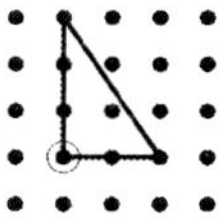

(b)

(c)

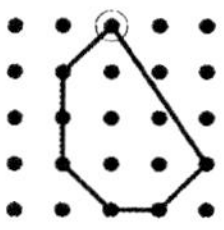

(d)

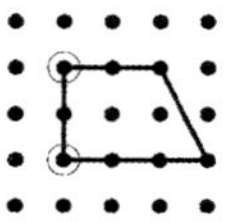

43. B. Since $p \parallel q$, $(m\angle 4 + m\angle 5) + (m\angle 6 + m\angle 7) = 180°$. We are given that $m\angle 4 = m\angle 5$ and $m\angle 6 = m\angle 7$, so substitution gives $2m\angle 5 + 2m\angle 6 = 180° \Rightarrow 2(m\angle 5 + m\angle 6) = 180° \Rightarrow m\angle 5 + m\angle 6 = 90°$. Since $\angle 8$ is the third angle in the triangle, and the sum of the angles in a triangle is 180°, $m\angle 8 = 90°$.

45. B. $90° + 45° + m\angle X = 180° \Rightarrow m\angle X = 45°$

Copyright © 2015 Pearson Education, Inc.

92 Chapter 9 Geometric Figures

5. (a)

(b)

The shaded region is always convex.

7. (a) 6 regions: the interior of the circle, the exterior of the square, and the four corner regions.

(b) 4 regions

9. The polygon has 6 sides, so the sum of the interior angle measures is
$(6 - 2)(180°) = 720°$. Therefore,
$$2x + 5x + 5x + 5x + 5x + 2x = 720°$$
$$24x = 720°$$
$$x = 30°.$$
Since $2x = 60°$ and $5x = 150°$, the angles measure 60°, 150°, 150°, 150°, 150°, and 60°.

11. The polygon has 8 sides, so the sum of the interior angle measures is
$(8 - 2)(180°) = 1080°$. Therefore,
$$y + 3y + 3y + 2y + y + 4y + y + 4y = 1080°$$
$$19y = 1080°$$
$$y \approx 56.84°.$$
Since $y \approx 56.84°$, $2y \approx 113.68°$, $3y \approx 170.53°$, and $4y \approx 227.37°$. The angles measure 56.84°, 170.53°, 170.53°, 113.68°, 56.84°, 227.37°, 56.84°, and 227.37°.

13. (a) The polygon has 6 sides, so the sum of the interior angle measures is
$(6 - 2)(180°) = 720°$.

(b) The polygon has 8 sides, so the sum of the interior angle measures is
$(8 - 2)(180°) = 1080°$.

15. (a) $\dfrac{(n - 2)(180°)}{n} = 175°$, so
$(n - 2)(180) = 175n$, or
$180n - 360 = 175n$. Then $5n = 360$, so
$n = 72$.

(b) The average of the interior angles will remain the same because the sum of the interior angles remains the same, $(n - 2)(180°)$, and there are still n interior angles so the average does not change.

17. (a) Yes. 24° corresponds to an exterior angle of the resulting polygon. Since $360 \div 24 = 15$, you will walk around a regular 15-gon.

(b) Each side of the regular 15-gon has length 10 paces.

19. (a) A rhombus is a quadrilateral with all four sides congruent, so the quadrilateral is *ACEG*.

(b) A rectangle is a quadrilateral with four right angles, so the quadrilateral is *BDFH*.

(c) An isosceles trapezoid is a quadrilateral with exactly one pair of sides parallel, and the other pair of sides congruent, so the quadrilateral is *ABDH*.

(d) A nonisosceles trapezoid is a quadrilateral with exactly one pair of sides parallel, and the other pair of sides not congruent, so the quadrilateral is *GADE*.

(e) A kite is a quadrilateral with exactly two distinct pairs of adjacent congruent sides, so the quadrilateral is *EFAD*.

21. In each case, the measure of the interior angle is given by $\dfrac{(n - 2)(180°)}{n}$, and the measures of the exterior angle and central angle are each given by $\dfrac{360°}{n}$ (or 180° minus the exterior measure).

	n	Interior Angle	Exterior Angle	Central Angle
(a)	7	$128\frac{4}{7}°$	$51\frac{3}{7}°$	$51\frac{3}{7}°$
(b)	8	$135°$	$45°$	$45°$

Copyright © 2015 Pearson Education, Inc.

Solutions to Problem Set 9.2

23.

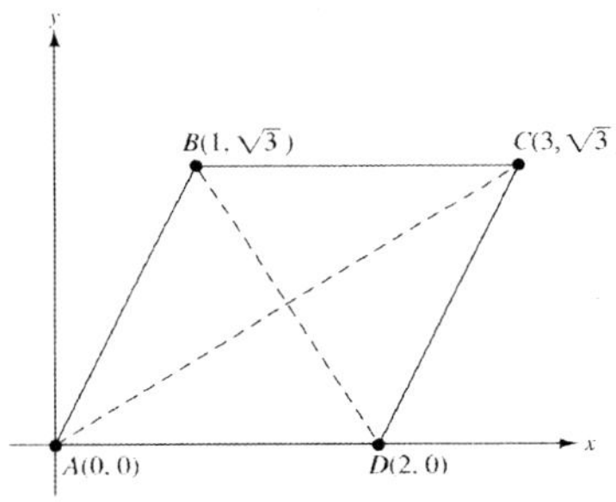

To show that the figure is a rhombus, we must show that the sides have equal lengths. Clearly, $AD = BC = 2$. Using the distance formula, we have

$$AB = \sqrt{(1-0)^2 + (\sqrt{3}-0)^2} = \sqrt{1+3} = 2 \text{ and}$$

$$DC = \sqrt{(3-2)^2 + (\sqrt{3}-0)^2} = \sqrt{1+3} = 2.$$

Therefore, $ABCD$ is a rhombus.

25. $Q = (a+c, b)$, so the square of the length of $\overline{OQ}$ is $(a+c)^2 + b^2$, and the square of the length of $\overline{PR}$ is $(a-c)^2 + b^2$. Thus, the sum of the squares of the length of diagonals is $2(a^2 + b^2 + c^2)$. The sum of the squares of the sides of the parallelogram is $(a^2 + b^2) + 2c^2 + a^2 + b^2$, which simplifies to $2(a^2 + b^2 + c^2)$, the same expression as the sum of the squares of the lengths of the diagonals.

27. (a) In the design with triangles, six angles of equal measure fill 360°. Therefore, each angle has measure $\frac{360°}{6} = 60°$. Similarly, $\frac{360°}{4} = 90°$ and $\frac{360°}{3} = 120°$ give the measures of the interior angles of a square and regular hexagon.

(b) The obtuse angles of the red trapezoid are congruent to the interior angles of the regular hexagon, so their measure is 120°. Two acute angles also cover 120°, so each acute angle has measure 60°.

(c) The obtuse angles of the blue rhombus are congruent to the interior angles of the regular hexagon, so their measure is 120°. Alternative, three angles fill 360°, so each has measure 120°. Two acute angles of

the blue rhombus cover 120° so each has measure 60°.

(d) The obtuse angles of the small rhombus, when combined with the 90° angle of the square and the 120° angle of the blue rhombus fill 360°. Therefore, the obtuse angle of the small rhombus has measure $360° - 90° - 120° = 150°$. The acute angle of the small rhombus, together with the 90° angle of a square, have the same total measure as the 120° obtuse angle of the large blue rhombus. Thus, the acute angle has measure $120° - 90° = 30°$.

29. Perhaps JaVonte was thinking that anything larger than a right angle was obtuse and then confused the size of the sides with the size of the angles. Talking with JaVonte about the definition of obtuse angle and being explicit about the size of the sides having nothing to do with whether a triangle is right, acute, or obtuse would be best.

31. Samuel is correct. It is also a square as well as being a rectangle. A review of the definition will help Samuel make sure that he masters this subtle point.

33. Martine is not correct. Here are two different counterexamples: The diagonal of a 7 by 24 rectangle has the same length as the diagonal of a 15 by 20 rectangle, and the diagonal of a 1 by $\sqrt{24}$ rectangle has the same length as the diagonal of a 3 by 4 rectangle.

35. (a) The boat can drift to any position inside the circle centered at A, where the radius of the circle is the length of the anchor rope.

(b) The boat is confined to the region in the overlap of circles at points A and C, whose radii are respective lengths of the anchor lines.

37. (a) By looking for patterns and forming a table, $n - 3$ diagonals are needed for an n-gon.

(b) $n - 2$ triangles

Copyright © 2015 Pearson Education, Inc.

94   *Chapter 9 Geometric Figures*

(c) After triangulating the polygon, all of the interior measure for the polygon is distributed among $n - 2$ triangles. So the sum of the interior angles of the n-gon is the same as the sum of the interior angles of $n - 2$ triangles. Each triangle has $180°$ as the sum of its three interior angles, so the sum of the interior angles of an n-gon equals $(n - 2) \cdot 180°$.

39. (a) By the total-turn theorem, the sum of the turn angles at the 5 points of the star is $720°$. Each exterior angle is the supplement of the corresponding interior angle, so
$$5 \cdot 180° = 720° + m(\angle 1) + m(\angle 2)$$
$$+ m(\angle 3) + m(\angle 4) + m(\angle 5).$$
This shows that
$$m(\angle 1) + \cdots + m(\angle 5) = 5 \cdot 180° - 720°$$
$$= 180°.$$

(b) $x + x + x + x + x = 180°$, so
$$x = \frac{180°}{5} = 36°.$$

41. Each new circle creates a new region each time it intersects a previously drawn circle. Since the new circle intersects each of the old circles in at most two points, this creates the following pattern:

	Number of
Circles	**Regions**
1	2
2	$2 + 2 \cdot 1 = 4$
3	$4 + 2 \cdot 2 = 8$
4	$8 + 2 \cdot 3 = 14$
5	$14 + 2 \cdot 4 = 22$
6	$22 + 2 \cdot 5 = 32$
7	$32 + 2 \cdot 6 = 44$
8	$44 + 2 \cdot 7 = 58$
9	$58 + 2 \cdot 8 = 74$
10	$74 + 2 \cdot 9 = 92$

43. (a) Extend side $\overline{AB}$ of the parallelogram, forming the exterior angles that are labeled $\angle a$ and $\angle b$. Then $m(\angle A) + m(\angle a) = 180°$, since $\angle A$ and $\angle a$ are adjacent supplementary angles. But $m(\angle a) = m(\angle B)$ because they are corresponding angles with respect to the

parallel sides $\overline{AD}$ and $\overline{BC}$. Therefore, $m(\angle A) + m(\angle B) = 180°$, which shows that $\angle A$ and $\angle B$ are supplementary.

(b) Part (a) shows that any two successive interior angles of a parallelogram are supplementary. Thus $\angle A$ and $\angle C$ are both supplementary to $\angle B$, so $\angle A \cong \angle C$. Similarly, $\angle B$ and $\angle D$ are both supplementary to $\angle A$, so $\angle B \cong \angle D$.

45. The square could fall through the hole. This is because a square's side length is less than the length of its diagonal. This cannot happen for a circle, because the hole's opening is slightly smaller than the diameter of the cover.

47. (a) *KLMN* is a parallelogram.

(b) *KLMN* is a rectangle.

(c) *ABCD* is a kite.

49. C. The slope of the side containing $(0, 2)$ and $(2, 0)$ is $\frac{0 - 2}{2 - 0} = -1$. The slope of the side containing $(-1, 1)$ and $(4, -4)$ is $\frac{-4 - 1}{4 - (-1)} = \frac{-5}{5} = 1$. It is obvious from the figure that the other two sides are not parallel, so the figure is a trapezoid.

51. C. Both figures have two obtuse angles.

Section 9.3
Figures in Space

Problem Set 9.3

1. (a) Polyhedron. It is a simple closed surface formed from planar polygonal regions.

(b) Not a polyhedron. It is not formed from planar polygonal regions.

(c) Polyhedron. Note that a polyhedron need not be convex.

3. (a) Pentagonal prism (and possibly a right pentagonal prism)

(b) Hexagonal pyramid

Copyright © 2015 Pearson Education, Inc.

Solutions to Problem Set 9.3

5. **(a)** 4. There are 4 faces, and each is in a different plane.

   (b) $\overline{AB}, \overline{AC}, \overline{AD}, \overline{BC}, \overline{BD}, \overline{CD}$

   (c) A, B, C, D

   (d) $\triangle ABD, \triangle BCD, \triangle ABC, \triangle ACD$

7. **(a)** Pentagonal right prism

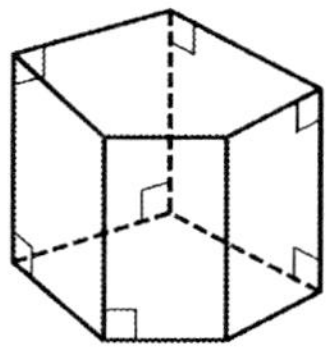

   (b) Oblique hexagonal prism

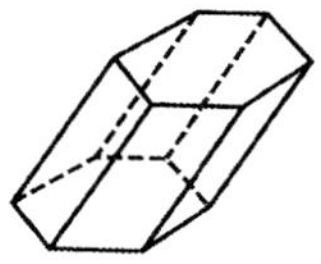

   (c) Octahedron

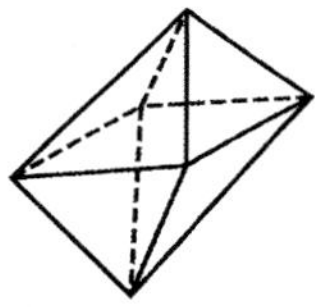

   (d) Right circular cone

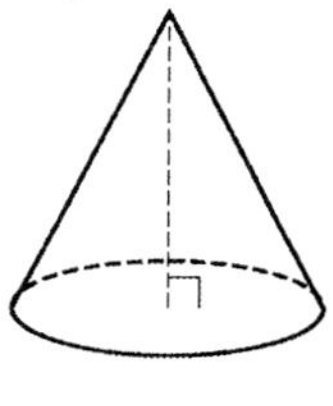

9. This problem may be solved by mental visualization or by actually constructing a model.

   (a), **(b)**, 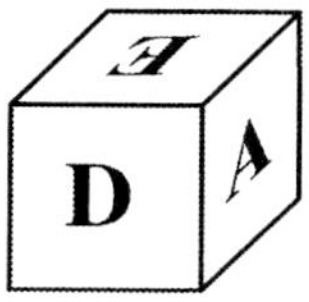 **(c)**

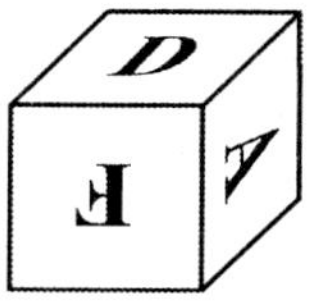

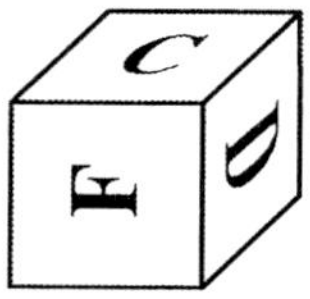

11. Several colorings will work, including this one.

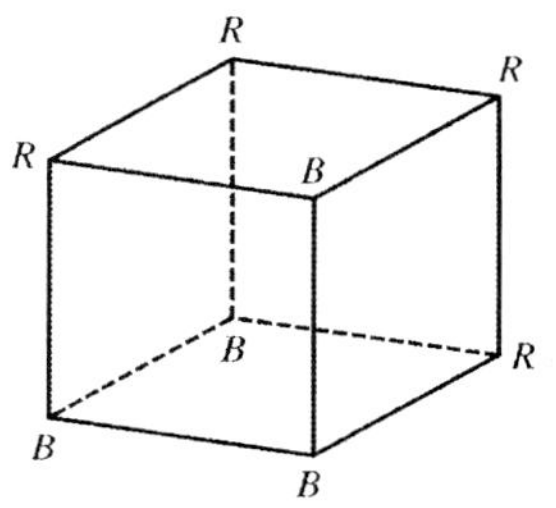

13. **(a)** $F = 7, V = 7, E = 12$, so $V + F = E + 2$ since $7 + 7 = 12 + 2$

   (b) $F = 10, V = 16, E = 24$, so $V + F = E + 2$ since $16 + 10 = 24 + 2$

   (c) $F = 20$. Each face is a triangle, so $2E = 3F$. That is, $E = 3 \cdot \frac{20}{2} = 30$. Each vertex is the endpoint of 5 edges, so $2E = 5V$, so $V = 2 \cdot \frac{30}{5} = 12$. Therefore, $V + F = E + 2$ since $12 + 20 = 30 + 2$.

15. **(a)** $F = 10, V = 7, E = 15$, so $V + F = E + 2$ since $7 + 10 = 15 + 2$

   (b) $F = 40, V = 22, E = 60$, so $V + F = E + 2$ since $22 + 40 = 60 + 2$

17. **(a)** $V = 10, F = 12, E = 20$
   $V + F = 22 = E + 2$, so Euler's formula holds.

   (b) $V = 12, F = 8, E = 18$
   $V + F = 20 = E + 2$, so Euler's formula holds.

   (c) Note that the square hole goes all the way through the figure.
   $V = 12, F = 12, E = 24$
   $V + F = 24 = E \neq E + 2$, so Euler's formula does not hold.
   The figure is not a polyhedron because it has a hole (it is not simple).

19. Answers will vary.

21. Answers will vary.

23. **(a)** Only (iv). The other hexominoes have squares that overlap when the hexomino is folded.

Copyright © 2015 Pearson Education, Inc.

96 Chapter 9 Geometric Figures

(b) Six of the hexominoes are found easily because they consist of a strip of four squares (which may be regarded as the "sides" of a cubical box), plus a flap for the top and a flap on the bottom. Note that the fifth one shown is (i) from part (a), and the second one shown is congruent to (iv) from part (a).

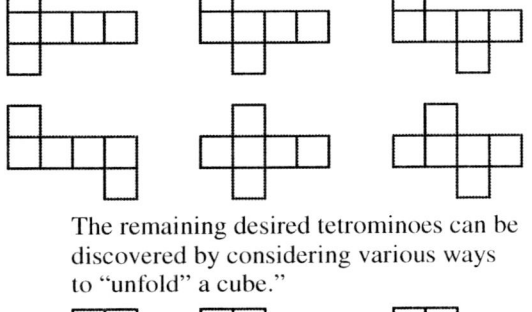

The remaining desired tetrominoes can be discovered by considering various ways to "unfold" a cube.

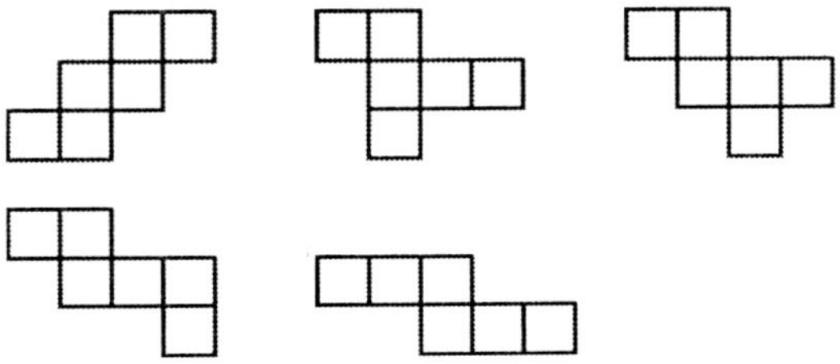

27. (a) Tianna is naming each prism by its face rather than its base. The shapes of the bases of a prism are used to name the prism.

(b) Show her that, except for a rectangular prism, there are just two shapes on the prism that are the bases. If she finds just the two shapes that are the same, she can identify the base. Also offer instruction about the fact that the bases of a prism must be congruent and are parallel.

29. Since a pentahedron has $F = 5$ faces, no one face could have 5 or more edges. Therefore each face is a quadrilateral or a triangle. Let Q be the number of quadrilateral faces and T be the number of triangular faces. Then $Q + T = 5$. Also, since each quadrilateral has 4 edges and each triangle has 3 edges, we have $4Q + 3T = 2E$. (The 2 factor is because each edge borders two faces of the polyhedron). Since $4Q$ and $2E$ are both even numbers, we see that $3T$, and hence T, must be even. That is T is 0, 2, or 4. Since it is clearly impossible to form a polyhedron with $Q = 5$ quadrilateral faces, this eliminates the $T = 0$ case. The remaining cases, $T = 2$ and $T = 4$, are shown below.

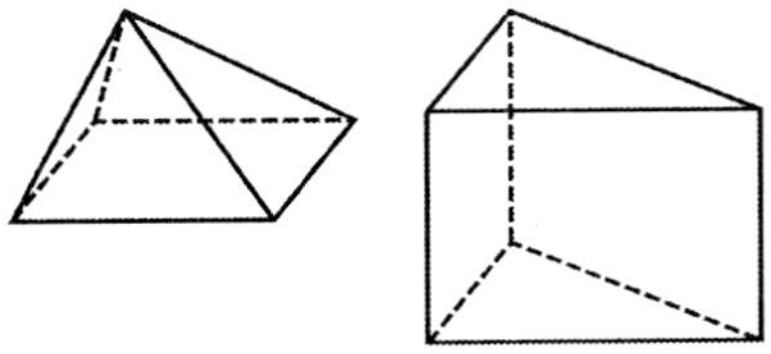

31. (a) 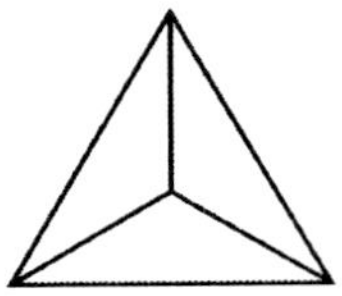 Tetrahedron   (b) Octahedron

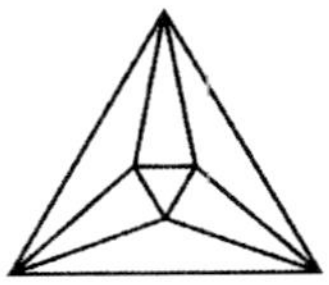

(c) Icosahedron

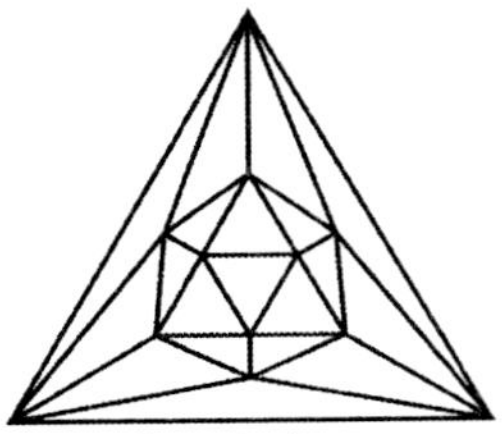

33. Fire = tetrahedron, earth = cube, air = octahedron, water = icosahedron, universe (world) = dodecahedron

25. (a) Polina is possibly picturing the triangle in her mind that serves as the "base" or bottom face for the shape. She is then thinking about the triangles that make up the "sides" or other faces and generalizing that for a geometric solid to have the same number of faces as vertices, then it must be the same shape all over the other geometric solids, such as a square pyramid, her mind might think of a square, and she isn't picturing that four faces or "sides" must come together to form the point or vertex at the top.

(b) Answers will vary, but a possible approach would be to use actual geometric solids for Polina to see and touch. Marking the vertices in some way may help her to see that a shape such as a hexagonal pyramid does have the same number of faces as vertices. Depending on Polina's background, her teacher could point out that, from the Euler Formula for Polyhedra, the solid must have one less edge than it does vertices or faces.

Chapter 9 Review Exercises

1. (a) $\overline{AC}$
   (b) $\overline{BD}$
   (c) $\overline{AD}$
   (d) $\angle ABC$ or $\angle CBA$ (or simply $\angle B$). Note that all other angles shown, when considered as a union of two rays, contain D.
   (e) $m(\angle BCD)$, $m(\angle DCB)$, or $m(\angle C)$, or 90°
   (f) $\overline{DC}$

3. (a) $180° - 37° = 143°$
   (b) $90° - 37° = 53°$

5. Since $\angle x$ and a 135° angle are adjacent supplementary angles, $x = 180° - 135° = 45°$. Since the sum of the measures of the angles in a triangle is 180°, $y = 180° - 102° - 45° = 33°$. Since $\angle y$ and $\angle z$ are adjacent supplementary angles, $z = 180° - y = 180° - 33° = 147°$. In summary, $x = 45°$, $y = 33°$, and $z = 147°$.

7. (a) No, because obtuse angles have measure greater than 90° and the sum of the three interior angles of a triangle is 180°. Two obtuse angles would produce a sum greater than 180°.
   (b) Yes, try angles of 100°, 100°, 100°, and 60°.
   (c) No, because acute angles have measure less than 90°, so if all the interior angles are acute the sum would be less than $4 \cdot 90° = 360°$. But the sum of the interior angles of a quadrilateral must be 360°.

9. Use the total-turn theorem. He turns through 360°.

11. Square right prism; triangular pyramid (or tetrahedron); oblique circular cylinder; sphere; hexagonal right prism.

13. (a)

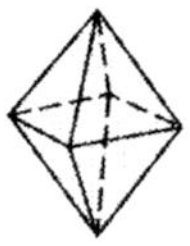

   (b) There are 6 vertices, 8 faces, and 12 edges, so $V = 6$, $F = 8$, and $E = 12$. Thus $V + F = 14$ and $E + 2 = 14$. Since these quantities are equal, Euler's formula holds.

35. Answers will vary, but include the following. In common with the cube, pyramid A, (i) is a polyhedron (ii) has a square face. In common with the cube, cylinder B (i) is a solid (ii) has a planar face (iii) is a common shape of food containers. In common with the cube, rectangle C has edges meeting at a right angle.

37. A

39. (a) Common properties include (i) each figure has many polyhedra, (ii) the figures look like they both have a perpendicular to a point on the opposite the base, and (iii) most importantly, the base of each consists of pentagons.
    (b) One is a (right) pyramid, the other is a (right) cylinder and so they are different. Said another way, there is a base for both but one has an opposing base and the other has an opposing vertex.

Chapter 10 Measurement: Length, Area, and Volume

Section 10.1
The Measurement Process

Problem Set 10.1

1. (a) Height, length, thickness, area, diagonal, weight
   (b) Length, diameter of cord
   (c) Height, length, depth
   (d) Length, height, width, perimeter (seating capacity), weight

3. (a) Since the figure can be made with 6 small isosceles triangles, it is 6 tga.
   (b) Since the figure can be made with 8 small isosceles triangles, it is 8 tga.
   (c) Since the figure can be made with 10 small isosceles triangles, it is 10 tga.

5. The volume of (i) is 6 cubes. The volume of (ii) is 5 cubes. The volume of (iii) is 5 cubes. The volume of (iv) is 7 cubes. In order of smallest to largest: (ii) (iii), (i) and (iv).

7. (a) Since the field is 360 feet long, its area is $360 \times 160 = 57,600$ square feet. There are 43,560 square feet in an acre, so the area of the field is $\frac{57,600}{43,560} = 1.3223...$
   That is about $1\frac{1}{3}$ acres.
   (b) The area of the soccer field is $110 \times 70 = 7700 \text{ m}^2$. Since 1 are $= 100 \text{ m}^2$, this is 77 ares. Since 1 ha $= 100$ a, the area is also 0.77 ha.

9. (a) $58,728 \text{ g} = 58.728 \times 10^3 \text{ g} = 58.728 \text{ kg}$
   (b) $632 \text{ mg} = 632 \times 10^{-3} \text{ g} = 0.632 \text{ g}$

11. (a) 3.5 kg   (b) 1200 kg

13. (a) 212 cm   (b) 50 mm

15. (a) About 28 cm by 22 cm
    (b) About 28 cm by 22 cm
    (c) About 2 cm
    (d) Answers will vary.

17. (a) $C = \frac{5}{9}(F - 32)$
    $C = \frac{5}{9}(68 - 32) = \frac{5}{9}(36) = 20$
    The temperature is 20° C.
    (b) $F = 1.8C + 32$
    $F = 1.8(13) + 32 = 55.4$
    The temperature is 55.4° F.
    (c) You would need to kow what the children are wearing and if the temperature is 0°C or 0°F.

19. 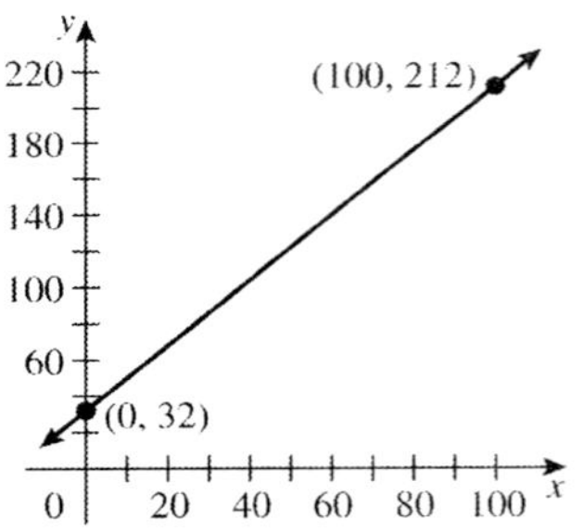 (a) The slope of the line is
    $\frac{212-32}{100} = \frac{180}{100} = \frac{9}{5} = 1.8$.
    The y-intercept is (0, 32), so the equation of the line is $y = 1.8x + 32$.
    (b) $F = 1.8C + 32$.

21. (a) 19 cm
    (b) Answers will vary. A size 8 men's shoe is about 28 cm long.
    (c) Answers will vary. A person 5 feet 7 inches tall is approximately 170 cm tall.
    (d) Answers will vary. The typical door in a house is about 203.5 cm tall and about 75.5 cm wide.
    (e) A yardstick is about 91.4 cm long.

23. (a) Majandra incorrectly associated the measurement of "feet" with the measurement of people only because that is the context in which most children hear that particular measurement discussed. She seems to have a good grasp on the largest and smallest lengths, but is having trouble distinguishing the middle amounts. The measurement of yards is a common problem for children in upper elementary.

(b) Answers may vary, but one approach is to let her physically mark the amounts of 6 feet and 6 yards using a ruler and a yardstick. This will guide her towards seeing the most reasonable table length. It is important, however, to make sure that she has some sort of personal measurement reference for the future since she might not always have access to a ruler or yardstick when she needs to estimate length. The teachers can use arm length, distance between objects, etc., to give Majandra a way to remember these measurement amounts.

25. (a) Corina correctly solved 18 cm = 180 mm and 4 m = 400 cm.

(b) Corina incorrectly solved the last two conversions. 40 mm = 4 cm and 3000 mm = 3 m.

(c) Answers may vary, however Corina's mistake is common because students are typically taught conversions going from a large metric measurement to a smaller metric measurements first. They quickly get into a bad habit of multiplying and just "adding zeros" to the end. When metric conversions going from small to large are given, many students just apply the same process of appending zeros and do not give much regard to their value or meaning. Corina needs to be shown with manipulatives what these amounts mean physically. The teacher can then guide her towards the correct answers.

(f) Counting the end zones, a football field is 120 yards long. Using the result of part (e), the field is

$$120 \text{ yards} \times \frac{91.4 \text{ cm}}{1 \text{ yard}} \times \frac{1 \text{ m}}{100 \text{ cm}}$$
$$\approx 110 \text{ m long}$$

27. (a) 1 jill = 2 jacks = 4 jiggers = 8 mouthfuls;
1 cup = 2 jills = 16 mouthfuls;
1 pint = 2 cups = 32 mouthfuls

(b) If one gallon is put in an empty pail, there is room for one bottle. If one bottle is put in, there is room for one bottle. Hence, each successive amount added to the pail leaves room for the same amount in the pail. Thus, there will be room for one more mouthful.

29.

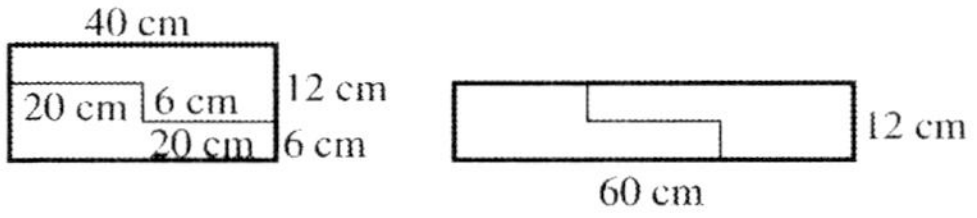

```
    40 cm
 ┌─────────┐
 │ 20 cm 6 cm │ 12 cm
 │ 20 cm 6 cm │
 └─────────┘
```

```
 ┌──────────────┐
 │              │ 12 cm
 │              │
 └──────────────┘
      60 cm
```

31. $\left(\frac{100 \text{ km}}{9 \text{ L}}\right)\left(\frac{3.7854 \text{ L}}{1 \text{ gal}}\right)\left(\frac{1 \text{ mi}}{1.6 \text{ km}}\right) \approx 26.3 \text{ mi/gal}$

35. (a) $\dfrac{22,300 \text{ kg}}{1.77 \text{ kg/L}} = 7682 \text{ L} \approx 4917 \text{ L}$

4917 liters were added.

(b) $\dfrac{22,300 \text{ kg}}{0.803 \text{ kg/L}} = 7682 \text{ L} \approx 20,089 \text{ L}$

20,089 liters should have been added.

37. The post is 60 in. = 5 ft long, so at $2.50 per foot costs $12.50. The tray frame requires four strips each 2 ft long, or 8 ft altogether. At $1 per foot, the strips cost $8.00. The tray bottom is $6.00. The total cost is
$12.50 + $8.00 + $6.00 = $26.50.

39. B.

Section 10.2
Area and Perimeter

Problem Set 10.2

1. The area is about 50 cm².

3. The dodecagon and square have the same area, equal to the sum of the areas of the same subregions in their dissections.

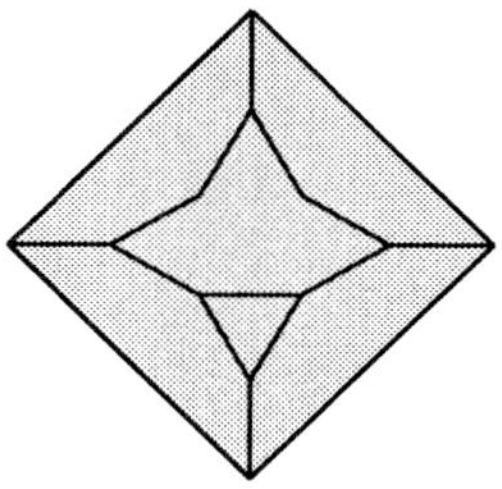

100 *Chapter 10 Measurement: Length, Area, and Volume*

5. (a) 12 units   **(b)** 8 units

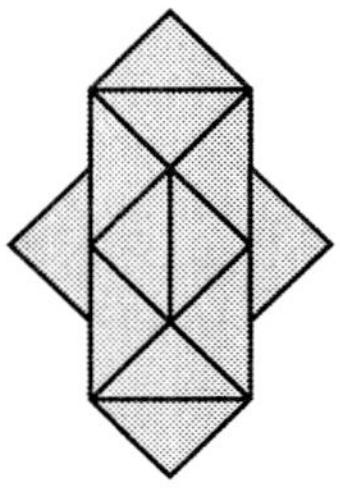

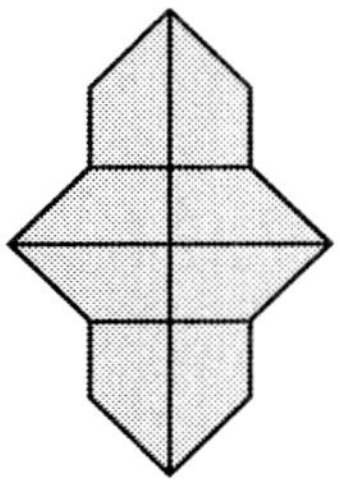

(c) 4 units

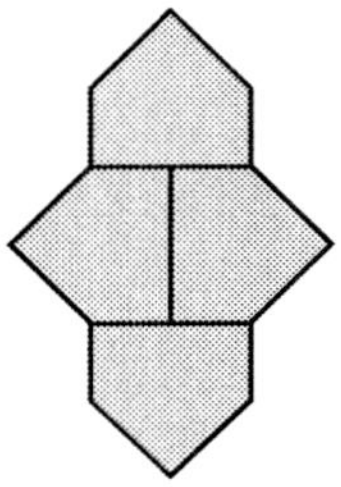

7. (a) $\frac{1}{2} \times 0.14 \text{ ft} \times 0.10 \text{ ft} = 0.007 \text{ ft}^2$

(b) $\frac{1}{2}(9 \text{ km} + 15 \text{ km}) \times 7 \text{ km} = 84 \text{ km}^2$

9. $\frac{1}{2} \times 12 \text{ m} \times 9 \text{ m} - \frac{1}{2} \times 4 \text{ m} \times 3 \text{ m} = 48 \text{ m}^2$

11. Let x = the width. Then $1 + x$ = the length.

(a) The perimeter is $2\big(x + (1 + x)\big) = 4x + 2$, so $4x + 2 = 20 \Rightarrow 4x = 18 \Rightarrow x = 4.5$. The width is 4.5 cm and the length is 5.5 cm.

(b) The area is $x(1 + x) = 6$, so

$$x^2 + x = 6 \Rightarrow x^2 + x - 6 = 0 \Rightarrow$$
$$(x + 3)(x - 2) = 0 \Rightarrow x = -3, \ x = 2$$

We disregard the negative solution since length must be positive.
The width is 2 cm. and the length is 3 cm.

13. (a) $0.085 \text{ mi}^2 \times \dfrac{27,878,400 \text{ ft}^2}{1 \text{ mi}^2}$
$$= 2,369,664 \text{ ft}^2$$

(b) $47,000 \text{ a} \times \dfrac{1 \text{ ha}}{100 \text{ a}} = 470 \text{ ha};$

$470 \text{ ha} \times \dfrac{10,000 \text{ m}^2}{1 \text{ ha}} = 4,700,000 \text{ m}^2$

(c) $5,800,000 \text{ m}^2 \times \dfrac{1 \text{ ha}}{10,000 \text{ m}^2} = 580 \text{ ha};$

$580 \text{ ha} \times \dfrac{1 \text{ km}^2}{100 \text{ ha}} = 5.8 \text{ km}^2$

15. (a) 1 cm by 24 cm; 2 cm by 12 cm; 3 cm by 8 cm; 4 cm by 6 cm. The dimensions can also be given in opposite order.

(b) The 4-cm by 6-cm rectangle has the smallest perimeter of 20 cm.

(c) The 1-cm by 24-cm rectangle has the largest perimeter of 50 cm.

17. (a) $A = \frac{1}{2}(81.2 \text{ m})(41.0 \text{ m}) = 1664.6 \text{ m}^2;$

$P = 49.0 \text{ m} + 68.1 \text{ m} + 81.2 \text{ m}$
$ = 198.3 \text{ m}$

(b) $A = \frac{1}{2}(87.6 \text{ in})(36.5 \text{ in}) = 1598.7 \text{ in}^2;$

$P = 36.5 \text{ in} + 87.6 \text{ in} + 94.9 \text{ in} = 219 \text{ in}$

(c) $A = \frac{1}{2}(5.4 \text{ cm})(2.8 \text{ cm}) = 7.56 \text{ cm}^2;$

$P = 5.4 \text{ cm} + 3.5 \text{ cm} + 8.0 \text{ cm}$
$ = 16.9 \text{ cm}$

19. (a) (i) $A = \frac{1}{2} \times 2 \times 2 = 2 \text{ cm}^2$

To find the perimeter, first use the Pythagorean theorem to find the length of the hypotenuse.

$\text{hypotenuse} = \sqrt{2^2 + 2^2} = \sqrt{8} = 2\sqrt{2}$
$\phantom{\text{hypotenuse}} \approx 2.83$
$P = 2 + 2 + 2\sqrt{2} = 4 + 2\sqrt{2} \approx 6.83 \text{ cm}$

(ii) $A = \frac{1}{2} \times 1 \times 4 = 2 \text{ cm}^2$

To find the perimeter, first use the Pythagorean theorem to find the length of the hypotenuse.

$\text{hypotenuse} = \sqrt{1^2 + 4^2} = \sqrt{17} \approx 4.12$
$P = 1 + 4 + \sqrt{17} = 5 + \sqrt{17} \approx 9.12 \text{ cm}$

Copyright © 2015 Pearson Education, Inc.

(iii) $A = \dfrac{1}{2} \times 0.5 \times 8 = 2 \text{ cm}^2$

To find the perimeter, first use the Pythagorean theorem to find the length of the hypotenuse.

$$\text{hypotenuse} = \sqrt{0.5^2 + 8^2} = \sqrt{64.25} \approx 8.02$$

$$P = 0.5 + 8 + \sqrt{64.25} = 8.5 + \sqrt{64.25}$$
$$\approx 16.52 \text{ cm}$$

(iv) $A = \dfrac{1}{2} \times 0.25 \times 16 = 2 \text{ cm}^2$

To find the perimeter, first use the Pythagorean theorem to find the length of the hypotenuse.

$$\text{hypotenuse} = \sqrt{0.25^2 + 16^2} = \sqrt{256.0625}$$
$$\approx 16.00$$

$$P = 0.25 + 16 + \sqrt{256.0625}$$
$$= 16.25 + \sqrt{256.0625} \approx 32.25 \text{ cm}$$

(b) The area of each triangle is the same, but the perimeter is getting longer each time. The triangles are becoming narrower, but longer, depending on which way you view them.

21. (a) The area is 12 square units since it has the same base and height as *ABCD*, so it has the same area.

(b) The area is 24 square units since it has the same base but twice the height as *ABCD*, so it has twice the area.

(c) 4 units, since 4 × 3 = 12.

23. 8 square units

25. The circumference (length) of a semi-circle of radius r is πr. Therefore, the inner three lanes have semi-circles of respective lengths 25π, 26π, and 27π. Moving outward, each lane is π meters (about 10.3 feet) longer, so the lanes are staggered π meters apart.

27. (a) If the radius of the circle is x, then the area of the circle is πx^2. The area of the square is $(2x)^2 = 4x^2$. So the ratio of the areas is $\dfrac{\pi x^2}{4x^2} = \dfrac{\pi}{4} \approx 78.5\%$.

(b) The perimeter is the length of the curve which bounds the region formed by removing the area inside the circle form the square. Thus, the boundary of the figure consists of two parts, that of the circle and that of the square. It is the boundary of the very light blue region. Since the radius of the circle is 6 cm, each side of the square is 12 cm. Thus, the answer is $4(12) + 2\pi(6) = 48 + 12\pi$ cm .

29. (a) $P = 2(l + w)$ and $A = lw$, so $l = \dfrac{4}{w}$ and

$$P = 2\left(\dfrac{4}{w} + w\right).$$

(b) Yes, the perimeter will become infinitely large as either w approaches zero or as w becomes very large itself. Thus, long, skinny rectangles can all have the same area. This is a different result from problem 28.

31. (a) The area of the small triangle is approximately $\dfrac{1}{2}(2)(1.7) = 1.7 \text{cm}^2$. The area of the large triangle is approximately $\dfrac{1}{2}(6)(1.7) = 5.1 \text{ cm}^2$.

The ratio is $\dfrac{5.1}{1.7} = 3$.

(b) The area of the square is approximately $2 \cdot 2 = 4 \text{ cm}^2$. The area under the trapezoid is approximately $\dfrac{1}{2}(4 + 8)(2) = 12 \text{ cm}^2$. Thus, the desired ratio is 3.

(c) The area of the regular pentagon can be calculated by dividing it into a trapezoid and a triangle and summing the two areas. The area is approximately

$$\dfrac{1}{2}(2 + 3.2)(1.9) + \dfrac{1}{2}(3.2)(1.2) \approx 6.9 \text{ cm}^2.$$

Similarly, the area of the large pentagon is approximately

$$\dfrac{1}{2}(10 + 7.2)(1.9) + \dfrac{1}{2}(7.2)(1.2) \approx 20.7 \text{ cm}^2$$

The ratio is approximately $\dfrac{20.7}{6.9} = 3$.

(d) Even without measuring, it is clear from the diagram in Example 2.6 that ratio of the area under the arch to the area of the rolling hexagon is exactly 3.

102 *Chapter 10 Measurement: Length, Area, and Volume*

(e) Given the results of parts (a), (b), (c), and (d) we conjecture that, if a polygonal arch is generated by rolling a regular n-gon along a line in this way, the ratio of the area under the polygonal arch to the area of the generating n-gon is exactly 3. In fact, it has been proven that the ratio is 3.

33. (a) Starting with the two triangles on top of each other, rotate the top one 180° about the midpoint of any side to form a parallelogram.

(b) Two triangles combine to form a parallelogram with base length the base of the triangle and height the height of the triangle. Therefore

2 · area (triangle) = area (parallelogram)
= bh.

35. (a) Put two copies of the trapezoid together as shown.

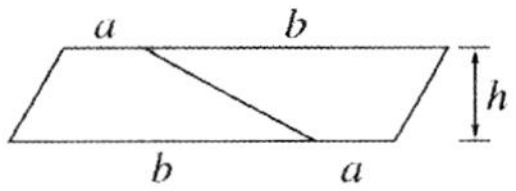

2 · area(trapezoid) = area(parallelogram)
= $h(a + b)$, so

area(trapezoid) = $\frac{1}{2}(a + b)\, h$

(b) Dissect the trapezoid as shown.

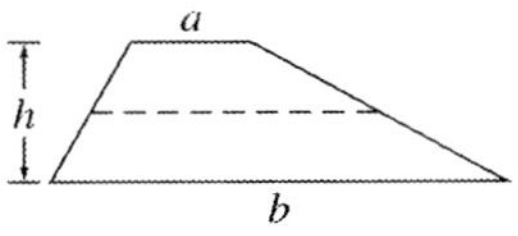

Rotate the bottom half 180° and place as shown.

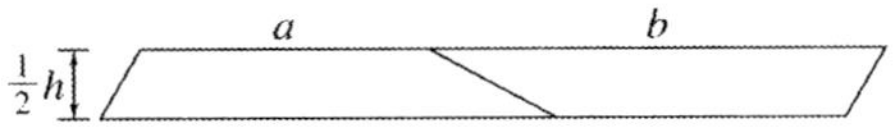

area(trapezoid) $= (a + b)\left(\frac{1}{2}h\right) = \frac{1}{2}(a + b)h$

37. (a) 4 m

(b), (c) Answers will vary.

(d) yes

39. (a) The perimeter of the bottom of the box is $2(11) + 2(7) = 36$ inches.

(b) Taneisha added the dimensions of the box to get $11 + 7 + 6 = 24$ inches.

(c) Answers may vary. Often when students are presented with a problem they know how to calculate, such as perimeter, but it is written in a different context, such as in this case, three dimensions instead of two are given, they don't know what to do with the third number (depth). So, they use it somewhere. Extra numbers that are not needed in problems often confuse students, and they can lose focus on what the task is. However, those extra numbers shouldn't be deleted from problems because that is unrealistic. Students should be instructed how to highlight what the question is asking them to do and then think about how to apply what they know. Taneisha could be shown a real box, asked to touch the perimeter of the bottom, and then apply the given numbers to find the answer.

41. March 14 is 3/14 and the first six digits of π are 3.14159. The Math Dept can enjoy pi(e).

43. Let P be the perimeter and A be the area.
$P = 6b$
Since the regular hexagon is composed of six equilateral triangles with sides of length b and altitudes of length $\frac{\sqrt{3}b}{2}$,

$$A = 6 \cdot \frac{1}{2}(b)\left(\frac{\sqrt{3}b}{2}\right) = \frac{3\sqrt{3}}{2}b^2.$$

45. (a) Dissect the vase as shown at the right.

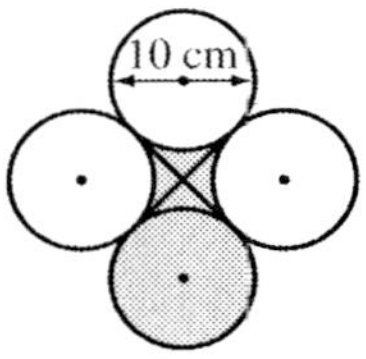

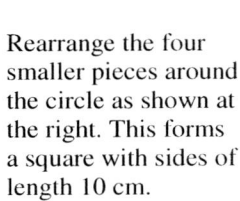

Rearrange the four smaller pieces around the circle as shown at the right. This forms a square with sides of length 10 cm.

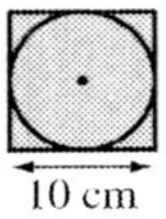

(b) area (vase) = area (square)
= 10 cm $\times$ 10 cm = 100 cm^2

Copyright © 2015 Pearson Education, Inc.

(c) Dissect the vase as shown at the right.

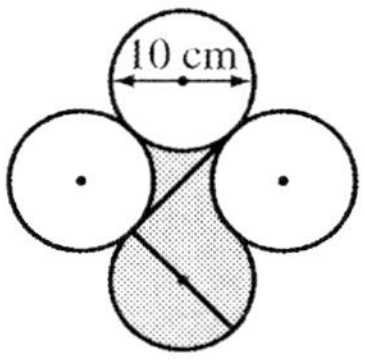

Rearrange the three pieces as shown.

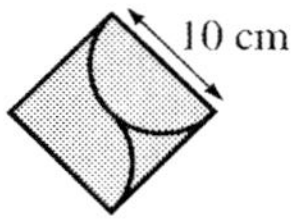

47. Draw a line through point B parallel to $\overline{AC}$. Let D be the intersection with the boundary line containing C. Since the triangles ABC and ADC have the same base AC and the same height, they have the same area. Thus, $\overline{AD}$ is a suitable new boundary line between the two fields.

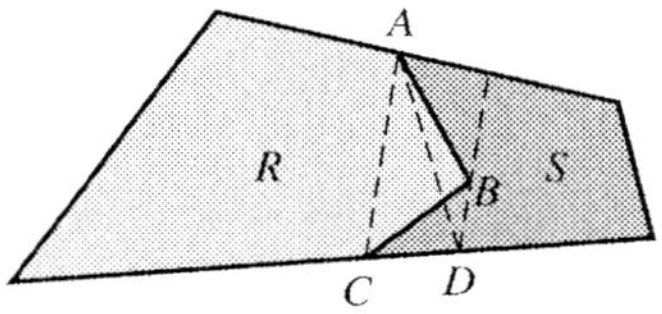

49. The areas of the rectangular portions of sidewalk total
(60 ft + 70 ft + 40 ft + 80 ft + 50 ft)(8 ft) = 2400 ft^2. The pieces formed with circular areas have total turning of 360°, so when placed together would form a circle with radius 8 ft, and area of $\pi(8 \text{ ft})^2 = 64\pi$ ft^2. Total area is
$(2400 + 64\pi)$ ft$^2 \approx 2601$ ft^2.

51. Since the circle has radius $\dfrac{9}{2}$, its area is
$\pi\left(\dfrac{9}{2}\right)^2$. This area is approximated with the
square of area 8^2, so $\pi\left(\dfrac{9}{2}\right)^2 \approx 8^2$.
Thus, $\pi \approx \dfrac{8^2}{\left(\dfrac{9}{2}\right)^2} \approx \left(\dfrac{16}{9}\right)^2$.

53. Draw $\overline{AP}, \overline{BP},$ and $\overline{CP}$.

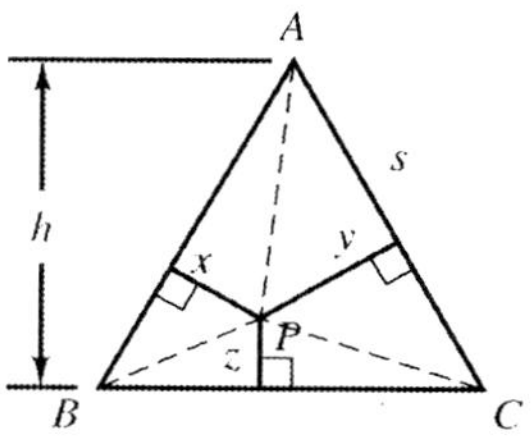

Then area$(\triangle ABC) = \dfrac{1}{2} sh$
$=$ area$(\triangle ABP) +$ area$(\triangle BPC) +$ area$(\triangle CPA)$
$= \dfrac{1}{2} sx + \dfrac{1}{2} sz + \dfrac{1}{2} sy = \dfrac{1}{2} s(x + y + z)$.
Therefore, $\dfrac{1}{2} sh = \dfrac{1}{2} s(x + y + z)$, so
$h = x + y + z$.
Alternate visual proof :

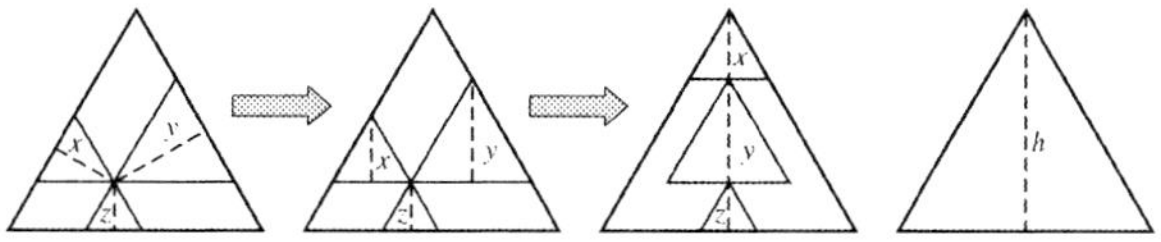

55. The lawn has an area 75 ft $\times$ 125 ft = 9375 ft^2.
Since 21 in. $= \dfrac{7}{4}$ ft, the lawn area is equivalent
to a rectangle 21 in. wide and
$9375 \div \dfrac{7}{4} = 5357.14....$ ft long. That is, Kelly
will walk about 5357 feet, a bit over a mile
(1 mile = 5280 ft).

57. Consider the carpet as a 6-ft by 4-ft rectangle with two semicircular ends of radius 2 ft. Then the carpet's area is
$(6 \text{ ft})(4 \text{ ft}) + \pi(2 \text{ ft})^2 \approx 36.57$ ft$^2 \approx 5266$ in^2,
so about 5266 in. of braid, or about 439 ft is required.

59. Since each tile measures $\dfrac{8}{12} = \dfrac{2}{3}$ foot on a side, the dimensions of the kitchen in tile units are $10 \div \dfrac{2}{3} = 15$ and $12 \div \dfrac{2}{3} = 18$. Therefore, the number of tiles needed is $15 \times 18 = 270$.

61. View one side of length 300 ft as a base. The altitude of the triangle is greatest if the angle is 90°. Thus $m(\angle A) = 90°$.

63. C.
$P = 2L + 2W \Rightarrow 32 = 2(10) + 2W \Rightarrow$
$32 = 20 + 2W \Rightarrow 12 = 2W \Rightarrow 6 = W$

65. D. By rearranging the lengths of the sides, we see that the answer $4a + 2b$.

104 *Chapter 10 Measurement: Length, Area, and Volume*

67. There lengths of two sides, x and y, must be computed from the other sides.

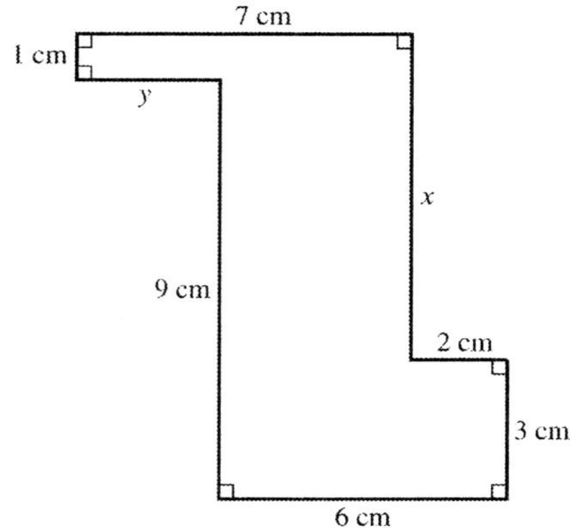

$x + 3 = 1 + 9 \Rightarrow x = 7$ cm.
$y + 6 = 7 + 2 \Rightarrow y = 3$ cm.
Thus, the perimeter is
$1 + 7 + 7 + 2 + 3 + 6 + 9 + 3 = 38$ cm.

Section 10.3
The Pythagorean Theorem

Problem Set 10.3

1. (a) By the Pythagorean theorem,
$$x^2 = 7^2 + 24^2 = 49 + 576 = 625.$$
Therefore, $x = \sqrt{625} = 25$.

(b) By the Pythagorean theorem,
$$x^2 + 8^2 = 17^2 \text{ or } x^2 = 17^2 - 8^2 = 225.$$
Therefore, $x = \sqrt{225} = 15$.

(c) By the Pythagorean theorem,
$$x^2 + 5^2 = 22^2 \text{ or}$$
$$x^2 = 22^2 - 5^2 = 484 - 25 = 459.$$
Therefore, $x = \sqrt{459} = 3\sqrt{51}$.

3. (a) By the Pythagorean theorem,
$$x^2 + (2x)^2 = (25)^2 \text{ or } 5x^2 = 625.$$
Therefore, $x = \sqrt{125} = 5\sqrt{5}$.

(b) Let y be the length of the bottom leg. By the Pythagorean theorem,
$$y^2 + 7^2 = 12^2, y^2 = 95, y = \sqrt{95}. \text{ By the}$$
Pythagorean theorem,
$$(x + 7)^2 + y^2 = 16^2 \text{ or } (x + 7)^2 = 161.$$
Therefore, $x = \sqrt{161} - 7$.

(c) Let y be the length of the leg of the smaller triangle. By the Pythagorean theorem, $y^2 + 6^2 = 10^2$, $y^2 = 64$, $y = 8$. By the Pythagorean theorem,
$$x^2 = (11 + 8)^2 + 6^2 = 361 + 36 = 397.$$
Therefore, $x = \sqrt{397}$.

5. (a) $x^2 = 10^2 + 15^2 = 325$, so $x = \sqrt{325}$;
$$y^2 = x^2 + 7^2 = 325 + 49 = 374, \text{ so}$$
$$y = \sqrt{374}.$$

(b) $x^2 = 1^2 + 1^2 = 2$, so $x = \sqrt{2}$;
$$y^2 = x^2 + 1^2 = 2 + 1 = 3, \text{ so } y = \sqrt{3}.$$

7. (a) height $= \sqrt{15^2 - 9^2} = 12$, so
area $= (20)(12) = 240$ square units

(b) base $= 2\sqrt{40^2 - 30^2} = 20\sqrt{7}$, so
$$\text{area} = \frac{1}{2}\left(20\sqrt{7}\right)(30)$$
$$= 300\sqrt{7} \text{ square units}$$

9. The length of the shortcut is
$$\sqrt{(50 \text{ ft})^2 + (100 \text{ ft})^2} \approx 112 \text{ ft. The distance}$$
along the sidewalk is 50 ft + 100 ft = 150 ft, so the children save about 150 ft – 112 ft = 38 ft.

11. (a) At 2 P.M. car A is 100 miles east, and car B is 40 miles north. By the Pythagorean theorem, the distance between the two cars is $\sqrt{(100 \text{ mi})^2 + (40 \text{ mi})^2} \approx 108$ mi. They are about 108 miles apart.

(b) At 3:30 P.M. car A is 175 miles east, and car B is 100 miles north. By the Pythagorean theorem, the distance between the two cars is
$$\sqrt{(175 \text{ mi})^2 + (100 \text{ mi})^2} \approx 202 \text{ mi.}$$
They are about 202 miles apart.

Copyright © 2015 Pearson Education, Inc.

Solutions to Problem Set 10.3   **105**

13. **(a)** Let s be the length of a side of the square. A diagonal of the square is a diameter of the circle.

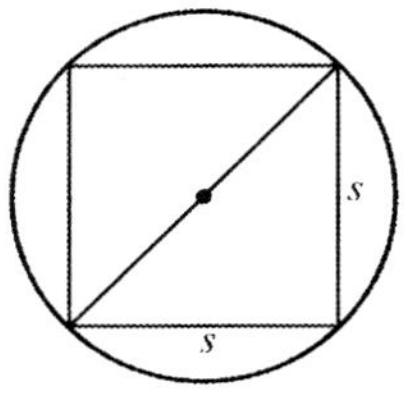

$s^2 + s^2 = 2^2, s^2 = 2.$ Thus $s = \sqrt{2}.$

(b) Let s be the length of a side of the cube. A diagonal through the cube is a diameter of the sphere. A diagonal along a face has length $\sqrt{s^2 + s^2} = \sqrt{2}s.$ Thus $\left(\sqrt{2}s\right)^2 + s^2 = 2^2,$ so $3s^2 = 4.$

Thus $s = \dfrac{2}{\sqrt{3}} = \dfrac{2\sqrt{3}}{3}.$

15. $r^2 + 24^2 = (18 + r)^2;$ since

$(18 + r)^2 = 18^2 + 36r + r^2,$

$r^2 + 24^2 = 18^2 + 36r + r^2,$ so

$r = \dfrac{24^2 - 18^2}{36} = 7.$

17. **(a)** Yes. $(14)^2 + \left(\sqrt{533}\right)^2 = 729 = (27)^2$

(b) No.

$\left(7\sqrt{2}\right)^2 + \left(4\sqrt{7}\right)^2 = 210 \neq 308 = \left(2\sqrt{77}\right)^2$

(c) Yes. $(9.5)^2 + (16.8)^2 = 372.49 = (19.3)^2$

19. **(a)**

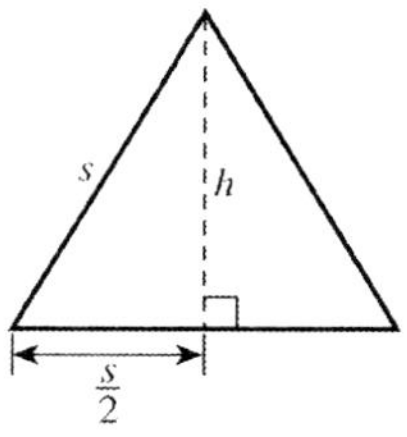

$\left(\dfrac{s}{2}\right)^2 + (h)^2 = s^2,$ by the Pythagorean

theorem. Therefore, $h = \dfrac{\sqrt{3}}{2}s.$

(b) The area is $\dfrac{1}{2}sh = \dfrac{1}{2}(s)\left(\dfrac{\sqrt{3}}{2}s\right)$

$= \dfrac{\sqrt{3}}{4}s^2$ square units.

(c) The regular hexagon is made up of six equilateral triangles, so the area is

$6\left(\dfrac{\sqrt{3}}{4}s^2\right) = \dfrac{3\sqrt{3}}{2}s^2$ square units.

(d)

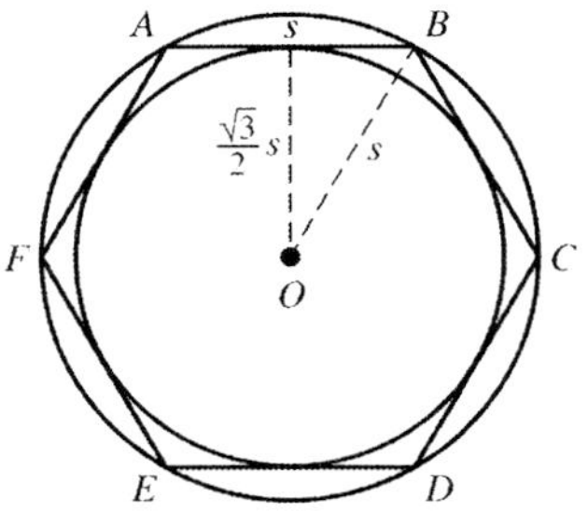

The inscribed circle has radius $h = \dfrac{\sqrt{3}}{2}s,$

so its area is $\pi\left(\dfrac{\sqrt{3}}{2}s\right)^2 = \dfrac{3}{4}\pi s^2.$

The circumscribed circle has radius $s,$ so its area is $\pi(s)^2 = \pi s^2.$ Therefore, the

area of the inscribed circle is $\dfrac{3}{4}$ the area

of the circumscribed circle.

21. **(a)** Consider the diagram of a right triangle with equilateral triangles drawn on the three sides as shown.

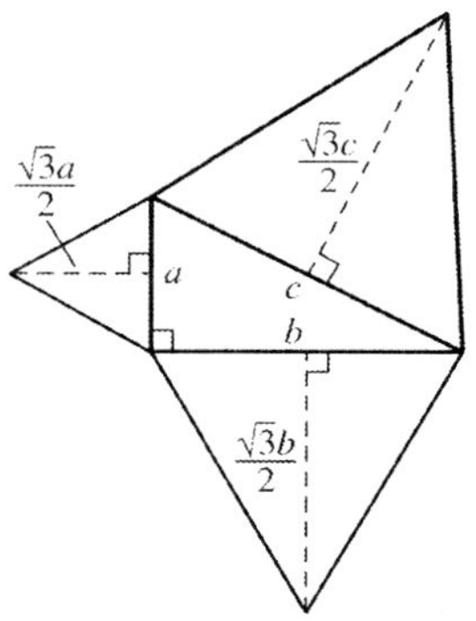

Using the Pythagorean theorem, we find that the altitudes of the three triangles are

$\dfrac{\sqrt{3}}{2}a, \dfrac{\sqrt{3}}{2}b,$ and $\dfrac{\sqrt{3}}{2}c.$

(*continued on next page*)

Copyright © 2015 Pearson Education, Inc.

106   Chapter 10 Measurement: Length, Area, and Volume

(continued)

Hence, the areas of the three triangles are

$$\left(\frac{\sqrt{3}}{2}a\right)\left(\frac{a}{2}\right) = \frac{\sqrt{3}}{4}a^2, \left(\frac{\sqrt{3}}{2}b\right)\left(\frac{b}{2}\right) = \frac{\sqrt{3}}{4}b^2,$$

and $\left(\frac{\sqrt{3}}{2}c\right)\left(\frac{c}{2}\right) = \frac{\sqrt{3}}{4}c^2$.

Since the inner triangle is a right triangle, $a^2 + b^2 = c^2$. Multiplying this equation through by $\frac{\sqrt{3}}{4} = \frac{\sqrt{3}}{2^2}$, gives

$\left(\frac{\sqrt{3}}{2}a\right)^2 + \left(\frac{\sqrt{3}}{2}b\right)^2 = \left(\frac{\sqrt{3}}{2}c\right)^2$ and this

shows that the sum of the areas of the two equilateral triangles on the sides of the right triangle equals the area of the equilateral triangle on the hypotenuse.

(b) Suggest that the student try to discover this by drawing an appropriate diagram and computing the required areas.

23. (a)

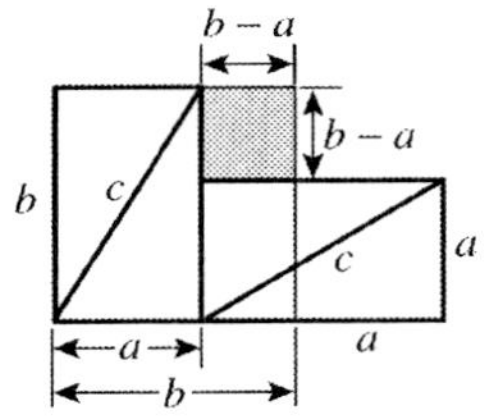

(b) Since the square and double square are tiled by the same five shapes, their areas are equal. The respective areas are c^2, and $a^2 + b^2$, so $c^2 = a^2 + b^2$.

25.

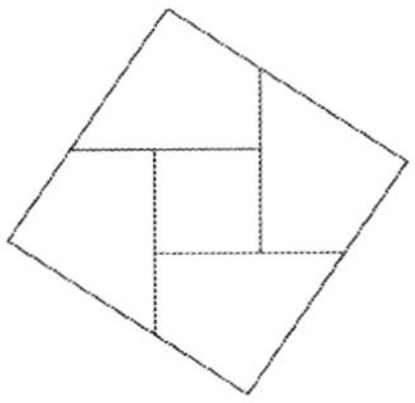

27. Sammy needs to understand that the area of the triangle is one-half the product of the length of the legs, not the hypotenuse. There may be confusion because the right triangle isn't "standing straight." In addition, he should always use units, in this case cm², in his answers.

29. Consider the following figure.

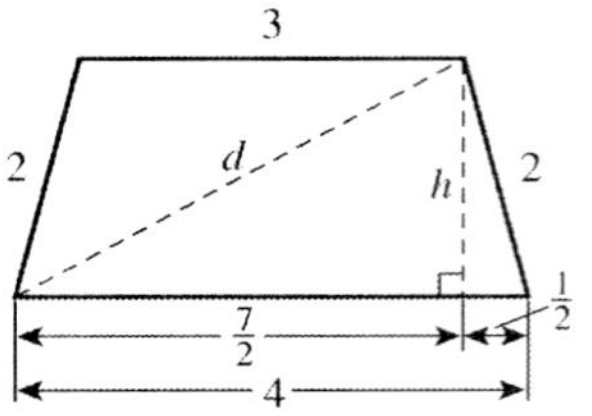

Using geometry, we can show that dropping an altitude to the bottom base divides the base as shown in the diagram above.

$h^2 + \left(\frac{1}{2}\right)^2 = 2^2$, so $h^2 = \frac{15}{4}$.

$d^2 = h^2 + \left(\frac{7}{2}\right)^2 = \frac{15}{4} + \frac{49}{4} = \frac{64}{4} = 16$

Therefore, the length of the diagonal is $d = 4$.

31. The radii of the circles were not given, so let them be denoted by R (larger circle) and r (smaller circle). The area of the annulus is then $\pi R^2 - \pi r^2 = \pi(R^2 - r^2) = \pi 100$ cm², since $R^2 - r^2 = 10^2$ by the Pythagorean theorem.

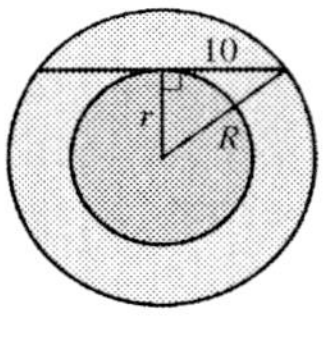

33. $\sqrt{(90\text{ ft})^2 + (90\text{ ft})^2} = 90\sqrt{2}$ ft or about 127 ft.

35. With lengths of 3, 4, and 5 one has a right triangle, since $3^2 + 4^2 = 25 = 5^2$.

37. Let s denote the length of a side of the octagon. Then $x^2 + x^2 = s^2$, so $s = \sqrt{2}x$. A side coming from the original square must also have length s, so we want $32 - 2x = \sqrt{2}x$. Therefore, $x = \frac{32}{2 + \sqrt{2}} \approx 9.373$, or about $9\frac{3}{8}$ inches.

39. G. The areas of the two squares are 28 cm² and 112 cm². Because of the Pythagorean theorem is the area of the shaded square, $121 - 28 = 93$ cm².

Copyright © 2015 Pearson Education, Inc.

Section 10.4
Volume

Problem Set 10.4

1. (a) surface area (b) volume
   (c) volume (d) surface area

3. (a) $V = Bh = \pi \times (10 \text{ m})^2 \times 4 \text{ m}$
       $= 400\pi \text{ m}^3 \approx 1257 \text{ m}^3$
   (b) $V = Bh = \pi (3 \text{ ft})^2 (7 \text{ ft}) = 63\pi \text{ ft}^3$
       $\approx 198 \text{ ft}^3$

5. (a) $V = \frac{1}{3}Bh = \frac{1}{3} \times \pi (5 \text{ cm})^2 \times (12 \text{ cm})$
       $= 100\pi \text{ cm}^3 \approx 314 \text{ cm}^3$
   (b) $V = \frac{1}{3}Bh = \frac{1}{3} \cdot \frac{2}{3}\pi(6 \text{ in})^2 (14 \text{ in})$
       $= 84\pi \text{ in}^3 \approx 264 \text{ in}^3$

7. Using Table 10.4, we have 2 cups = $\frac{1}{8}$ gal, so
   2 cups = $\frac{1}{8} \cdot 231 = 28\frac{7}{8}$ in^3. The volume of
   the hemisphere = $\frac{\frac{4}{3}\pi r^3}{2} = \frac{2}{3}\pi r^3$. So
   $\frac{2}{3}\pi r^3 = 28\frac{7}{8} \Rightarrow r^3 \approx 13.79 \Rightarrow r \approx 2.4$ in.

9. Have Ruth Ann construct a cone or pyramid
   out of cardboard. Also, build a rectangular box
   with a base of the same dimensions as the cone
   or pyramid. Fill the cone or pyramid with rice
   and then pour the rice into the box and see
   what fraction of the height the level of the rice
   reaches.

11. (a) The circumference of the cone is
    $\frac{3}{4} \cdot 2 \cdot \pi(4 \text{ in}) = 6\pi$ in, so the radius is
    $\frac{6\pi \text{ in}}{2\pi} = 3$ in.

    (b) $h^2 + (3 \text{ in})^2 = (4 \text{ in})^2$, so $h = \sqrt{7}$ in.

    (c) $V = \frac{1}{3}(\sqrt{7} \text{ in})(\pi (3 \text{ in})^2) \approx 25$ in^3

13. (a)

Radius (cm)	2	4	6	8
Volume (cm^3)	24π	96π	216π	384π

   (b) Samuel is correct if the height is fixed (in
       this case, 6 cm) but if we allow both the
       radius and the height to change, then the
       volume may increase or decrease as the
       radius increases.
   (c) Answers may vary. Have him try some
       examples, such as letting the radius
       increase from 2 cm to 4 cm, but
       decreasing the height from 6 cm to 1 cm.
       He should find that the volume decreases
       from 24π to 16π.

15. (a) Maurice correctly calculated the volume
        of the figure, 48. However, the correct
        answer is 48 ft^3, not 48 ft^2.
    (b) Maurice did not realize that the reason he
        was given only two number to multiply
        was because the area of the base had
        already been computed. He multiplied the
        numbers without thought as to the
        significance of the numbers. Maurice then
        thought that, since he didn't physically
        multiply three numbers, he shouldn't label
        it with a "little three."
    (c) Answers may vary. Maurice does not
        seem to fully understand what it means to
        label a number with "squared" or
        "cubed." It is important that students are
        given the correct vocabulary when
        calculating math so they learn the concept
        and can internalize what the procedure
        requires them to do.

17. The respective volumes are
    $V_{\text{cone}} = \frac{1}{3}(\pi r^2)(2r) = \frac{2}{3}\pi r^3$,
    $V_{\text{sphere}} = \frac{4}{3}\pi r^3$,
    $V_{\text{cylinder}} = (\pi r^2)(2r) = 2\pi r^3 = \frac{6}{3}\pi r^3$ so the
    volume ratios are 2 to 4 to 6, or 1 to 2 to 3.

41. B. The length of the hypotenuse in the lower
    triangle is $\sqrt{3^2 + 4^2} = \sqrt{25} = 5$. In the upper
    triangle, we have
    $3^2 + y^2 = 5^2 \Rightarrow y^2 = 25 - 9 = 16 \Rightarrow y = 4$.

108   *Chapter 10 Measurement: Length, Area, and Volume*

19. The volume of the ring is
$$\pi\left(\frac{9}{16}\text{ in}\right)^2\left(\frac{5}{4}\text{ in}\right) - \pi\left(\frac{1}{2}\text{ in}\right)^2\left(\frac{5}{4}\text{ in}\right)$$
$\approx 0.26\text{ in}^3$, so the ring weighs about
$$\left(\frac{6\text{ oz}}{1\text{ in}^3}\right)(0.26\text{ in}^3) = 1.56\text{ oz}.$$

21. The volume of a box is
$(4\text{ in})(5\text{ in})(8\text{ in}) = 160\text{ in}^3$, so the volume of two boxes is 320 in^3. The volume of a "tub" is $\pi(3\text{ in})^2(10\text{ in}) \approx 283\text{ in}^3$, so two boxes is a better buy.

23. **(a)** The scale factor is $\dfrac{5}{13}$, so the weight is
$$\left(\frac{5}{13}\right)^3(106.75) \approx 6.07\text{ pounds}.$$

   (b) The volume of the pyramid is about
$$\frac{1}{3}(177\text{ feet})(45\text{ acres})\left(\frac{43,560\text{ ft}^2}{1\text{ acre}}\right)$$
$$= 115,651,800\text{ ft}^3 \approx 116\text{ million ft}^3.$$

   (c) The length of a side of the cube is about
$(472,000,000\text{ ft}^3)^{1/3} \approx 779\text{ ft}.$

25. D. Since 2 feet = 24 inches, the new volume is
$(24-4) \times 10 \times 10 = 2000\text{ in}^3.$

Section 10.5
Surface Area

Problem Set 10.5

1. **(a)** $SA = 2 \cdot \dfrac{1}{2}(20+15)(12)\text{ cm}$
$$\quad\quad + (2)(13+15+12+20)\text{ cm}$$
$$= 540\text{ cm}^2$$

   (b) $SA = 2 \times \dfrac{1}{2}(6 \times 8)\text{ m} + (6+10+8)\text{ m} \times 5\text{ m}$
$$= 2 \times 24\text{ m}^2 + 24\text{ m} \times 5\text{ m}$$
$$= 168\text{ m}^2$$

3. **(a)** slant height $= \sqrt{(40\text{ m})^2 + (30\text{ m})^2} = 50\text{ m}$
$$SA = (60\text{ m})^2 + 4 \cdot \frac{1}{2}(60\text{ m})(50\text{ m})$$
$$= 9600\text{ m}^2$$

   (b) slant height $= \sqrt{(10\text{ ft})^2 - (6\text{ ft})^2} = 8\text{ ft}$
$$SA = (12\text{ ft})^2 + \frac{1}{2}(4)(12\text{ ft})(8\text{ ft}) = 336\text{ ft}^2$$

5. **(a)** $SA = 4\pi(2200\text{ km})^2 = 19,360,000\pi\text{ km}^2$
$$\approx 6.08 \times 10^7\text{ km}^2$$
$$V = \frac{4}{3}\pi(2200\text{ km})^3 \approx 4.46 \times 10^{10}\text{ km}^3$$

   (b) $SA = \dfrac{1}{2}[4\pi(6\text{ cm})^2] + \pi(6\text{ cm})^2$
$$= 108\pi\text{ cm}^2 \approx 339\text{ cm}^2$$
$$V = \frac{1}{2}\left[\frac{4}{3}\pi(6\text{ cm})^3\right] = 144\pi\text{ cm}^3$$
$$\approx 452\text{ cm}^3$$

7. **(a)** volume   **(b)**   volume

   (c) surface area   **(d)**   surface area

9. $V = \pi(3.25\text{ cm})^2 11\text{ cm} \approx 365\text{ cm}^3$
$$= 365\text{ mL} \approx \frac{365}{30}\text{ fl oz} = 12.2\text{ fl oz}$$

11. Let r be the radius of the sphere, so r is the radius of the cylinder and $2r$ is the height of the cylinder.
   area(sphere) $= 4\pi r^2$ square units;
   area(cylinder) $= 2 \cdot \pi r^2 + (2\pi r)(2r)$
$$= 6\pi r^2\text{ square units}$$
   Thus, $\dfrac{\text{area(sphere)}}{\text{area(cylinder)}} = \dfrac{4\pi r^2}{6\pi r^2} = \dfrac{2}{3}.$

13. **(a)** Since the diameter of the 16″ pizza is 2 times that of the 8″ pizza, the area of the 16″ pizza is $2^2 = 4$ times that of the 8″ pizza by the similarity principle. Therefore, it will feed 4 people.

   (b) Since 1.4^2 is about 2.0, the area of one 14″ pizza is nearly 2 times the area of one 10″ pizza by the similarity principle. Hence, one 14″ pizza is nearly the same amount of pizza as two 10″ pizzas, but costs $2.00 less. It is better to buy one 14″ pizza.

Copyright © 2015 Pearson Education, Inc.

15. Observe that the scale factor from I to II is $\frac{18}{6} = 3$, the scale factor from II to III is $\frac{15}{30} = \frac{1}{2}$, and hence the scale factor from I to III is $\frac{3}{2}$.

	I	II	III
Height	6	18	9 cm
Perimeter of Base	10	30	15 cm
Lateral Surface Area	40	360	90 cm^2
Volume	$\frac{80}{27}$	80	10 cm^3

17. **(a)** Doubling the sides increases the volume by a factor of 8, so the new cube holds 8 liters.

   (b) Since the scale factor k must satisfy $k^3 = 2$, $k = 2^{1/3}$. Thus, the length of each side is $2^{1/3} \cdot 10$ cm ≈ 12.6 cm.

19. If r is the radius, then
$$SA = 2\pi r^2 + (2\pi r)(3r) = 8\pi r^2 = 392 \Rightarrow$$
$$r^2 = \frac{49}{\pi} \Rightarrow r = \frac{7}{\sqrt{\pi}} \text{ cm and the height is}$$
$$\frac{21}{\sqrt{\pi}} \text{ cm.}$$

21. **(a)**

$$8\sqrt{2} \doteq 11.3 \text{ cm}$$
$$8\sqrt{3} \doteq 13.9 \text{ cm}$$

   (b) The area of each face: $(8)(8) = 64$ cm^2.
$$\frac{1}{2}(8)(8) = 32 \text{ cm}^2$$
$$\frac{1}{2}(8)(8) = 32 \text{ cm}^2$$
$$\frac{1}{2}(8)\left(8\sqrt{2}\right) = 32\sqrt{2} \text{ cm}^2$$
$$\frac{1}{2}(8)\left(8\sqrt{2}\right) = 32\sqrt{2} \text{ cm}^2$$

Total area $= 64 + 32 + 32 + 32\sqrt{2} + 32\sqrt{2}$
$$= \left(128 + 64\sqrt{2}\right) \text{ cm}^2$$
$$\approx 218.51 \text{ cm}^2$$

   (c) The volume of the cube is $(8 \text{ cm})^3 = 512 \text{ cm}^3$ and each pyramid has volume $\frac{512}{3} \text{cm}^3$.

23. Jerry was really writing $(4\pi + 24\pi)$, which is the sum of the area of the top of the can and the lateral area. (He forgot to include the area of the bottom of the can.) It is not unusual for children to compute the surface area by using only those parts of the surface that they can actually see.

25. Inez is correct if $r > 1$. Pick values of $r < 1$ to show Inez that her assertion is false.

27. The slant height of the cone is the hypotenuse of a right triangle with legs r and $2r$. That is, the slant height is $s = \sqrt{r^2 + (2r)^2} = \sqrt{5}r$. Thus, the respective surface areas are
$$S_{\text{cone}} = \pi r^2 + \frac{1}{2}\left(\sqrt{5}r\right)(2\pi r)$$
$$= \frac{\left(1 + \sqrt{5}\right)}{2} 2\pi r^2 = 2\tau(\pi r^2)$$
$$S_{\text{sphere}} = 4\pi r^2,$$
$$S_{\text{cylinder}} = 2(\pi r^2) + (2r)(2\pi r) = 6\pi r^2$$
So the ratio of surface areas are 2τ to 4 to 6, or equivalently, τ to 2 to 3.

29. **(a)** Each edge is the hypotenuse of a right triangle with legs of length s. Thus each edge has length $\sqrt{s^2 + s^2} = s\sqrt{2}$, so $ACEG$ is a regular tetrahedron.

   (b) The volume of
$$ACDE = \frac{1}{3}\left(\frac{1}{2}\right)(s)(s)(s) = \frac{1}{6}s^3.$$

   (c) Notice that $ACEG$ is formed by cutting four such congruent tetrahedrons, as in part (b), from the cube. Hence, the volume of $ACEG$ is $s^3 - 4 \cdot \frac{1}{6}s^3 = \frac{1}{3}s^3$.

110 *Chapter 10 Measurement: Length, Area, and Volume*

(d) The scale factor is $\dfrac{b}{\sqrt{2}}$, so the volume is

$$\left(\frac{b}{\sqrt{2}}\right)^3 \frac{1}{3} = \frac{b^3\sqrt{2}}{12}.$$

31. E. 100 square meters since $1\ cm^2 = 1\ meter^2$.

33. The surface area of the can is twice the area of the base plus the lateral area. The area of the top of the can is $\pi r^2 = \pi\left(4^2\right) = 16\pi\ cm^2$. The area of the side of the can is $2\pi rh = 2\pi(4)(3) = 24\pi\ cm^2$. Therefore, the surface area of the can is $2\left(16\pi\right) + 24\pi = 56\pi\ cm^2$.

35. A. Assuming that the roof is flat, the surface area of the front is $\left(c \times b\right)$, and the surface area of each side is $\left(a \times b\right)$, so the total surface area is $\left(c \times b\right) + 2\left(a \times b\right)$.

Chapter 10 Review Exercises

1. **(a)** Centimeters **(b)** Millimeters

  (c) Kilometers **(d)** Meters

  (e) Hectares **(f)** Square kilometers

  (g) Milliliters **(h)** Liters

3. $(60\ cm)(40\ cm)(35\ cm) = 84{,}000\ cm^3 = 84\ L$

5. Dissect the trapezoid by a horizontal line through M and rotate the bottom half by $180°$ as shown in the figure.

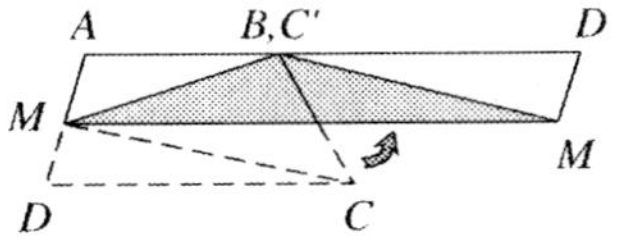

The triangle has half the area of the parallelogram $AD'M'M$ since they both have the same base and height, so it also has half the area of the trapezoid $ABCD$.

7. **(a)** 1 ft 4 in = 16 in, 4 ft = 48 in, so the area is $(16\ in)(48\ in) = 768\ in^2 = 5\dfrac{1}{3}ft^2$.

(b) Dissect the figure with a diagonal from the bottom left vertex to the top right vertex to form two triangles. The area is
$$\frac{1}{2}(8\ m)(9\ m) + \frac{1}{2}(3\ m)(6\ m) = 45\ m^2.$$

(c) $\dfrac{1}{2}(5\ cm + 7cm)(3\ cm) = 18\ cm^2$

9. **(a)** $A = (3\ ft)(4\ ft) + \dfrac{1}{2}\cdot \pi \cdot (1.5\ ft)^2$
$$\approx 15.5\ ft^2;$$
$$P = 4\ ft + 3\ ft + 4\ ft + \frac{1}{2}\cdot 2\pi(1.5\ ft)$$
$$= 11\ ft\ + \pi(1.5\ ft) \approx 15.7\ ft$$

(b) $A = \dfrac{3}{4}\cdot \pi(3\ m)^2 = \dfrac{27}{4}\pi\ m^2 \approx 21.2\ m^2;$
$$P = 3\ m + 3\ m + \frac{3}{4}\cdot 2\pi(3\ m)$$
$$= 6\ m + \frac{9}{2}\pi\ m \approx 20.1\ m$$

11. The height of the cone is
$$\sqrt{(35\ cm)^2 - (10\ cm)^2} = \sqrt{1125}\ cm$$
$$\approx 33.5\ cm$$

13. $P = 3 + \sqrt{2} + \sqrt{10} + \sqrt{5} + 1 + 5$
$$= 9 + \sqrt{2} + \sqrt{10} + \sqrt{5} \approx 15.8\ units$$

15. $V(\text{sphere}) = \dfrac{4}{3}\pi(10\ m)^3$ and $V(\text{four cubes}) = 4(10\ m)^3$. Since $\pi > 3$, then $\dfrac{4}{3}\pi > 4$, showing that the sphere has the larger volume. Also, a sphere of radius 10 m can be inscribed in a cube with sides of length 20 m, which is the same as eight cubes with sides of length 10 m put together.

Copyright © 2015 Pearson Education, Inc.

Chapter 11 Transformations, Symmetries, and Tilings

**Section 11.1
Rigid Motions and Similarity
Transformations**

Problem Set 11.1

1. (a) Not a rigid motion; distances between particular cards will change.

   (b) Yes, pieces will be in the same places relative to each other.

   (c) No, distances between particular pieces almost certainly will have changed.

   (d) Yes, the painting has not changed, only its location.

   (e) No, distances between bits of dough will have increased.

3. The figure is moved 3 units to the right.

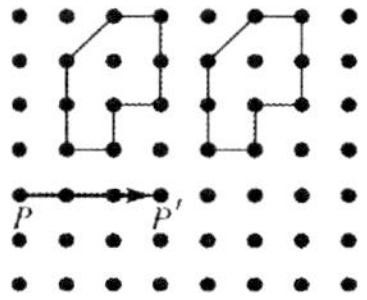

5. (a) The given triangle, $\triangle A'B'C'$, must be 3 units to the right and 2 units up from $\triangle ABC$.

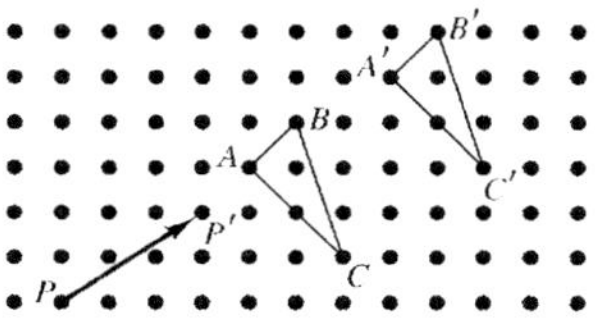

   (b) The rigid motion is a translation two units down and three units to the left. The translation arrow is from P' to P. This translation does the opposite of the translation that takes $\triangle ABC$ to $\triangle A'B'C'$.

7.

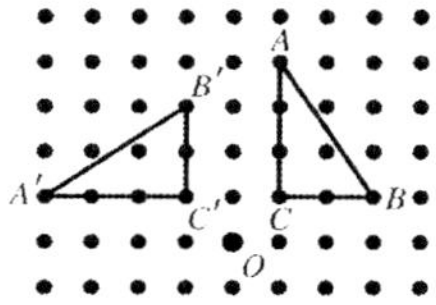

9. (a) The center of rotation is the same distance from A as from A', so it is on the perpendicular bisector of $\overline{AA'}$. (Similarly, it is on the perpendicular bisector of $\overline{BB'}$, but this turns out to be the same line.) By using trial and error among the points on the perpendicular bisector, we find that the only point that can produce the desired transformation of both points simultaneously is point O.

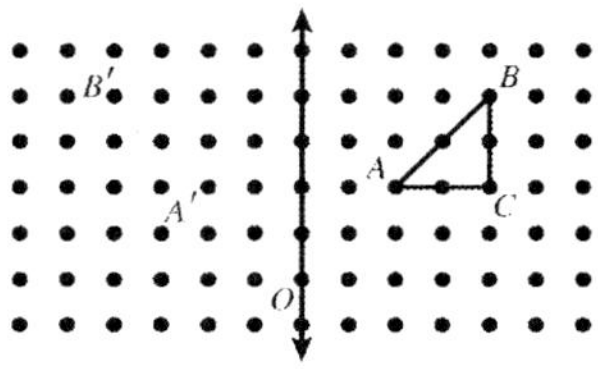

   (b) 90° (counterclockwise)

   (c)

11. (a) A rigid motion that has no fixed points is a translation or a glide-reflection.

    (b) A rigid motion that has exactly one fixed point is a rotation about the fixed point by an angle between 0° and 360°.

    (c) A rigid motion that has at least two fixed points and a non-fixed point is a reflection.

    (d) A rigid motion that has only fixed points is the identity transformation.

13.

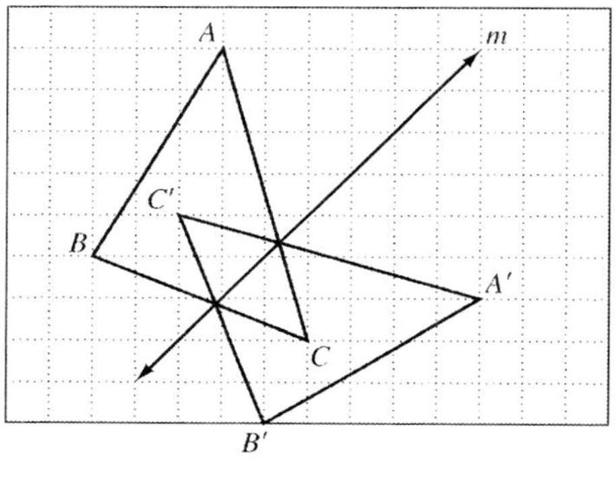

Copyright © 2015 Pearson Education, Inc.

112 *Chapter 11 Transformations, Symmetries, and Tilings*

15. **(a)** *A* is on *m*.

(b) *A* and *B* are each on *m*, so $m = \overleftrightarrow{AB}$.

(c) It takes *D* to *C*.

17. **(a), (b), (c)** The line of reflection is determined by the midpoints of $\overline{BB'}$ and $\overline{EE'}$. Since the reflection of *B'* across this line is 4 units to the right and 4 units below *B*, the slide arrow must point 4 units right and 4 units down.

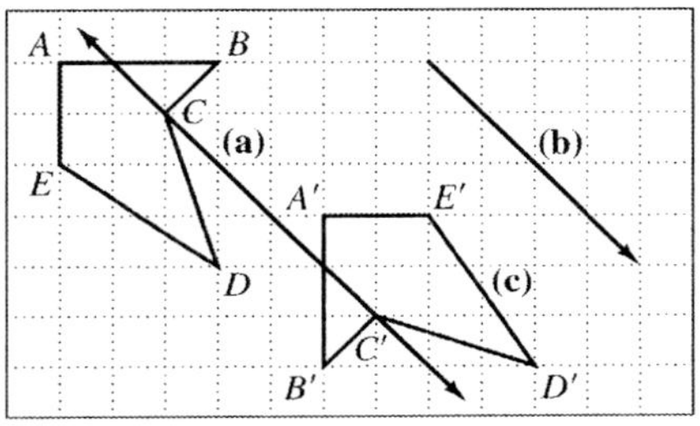

19. **(a)** Translate right 8 units.

(b) *j*   **(c)** *l*

(d) Rotate counterclockwise $90°$ about point *R*

(e) *j*   **(f)** *k*

21. **(a)** The center is at *P*, the intersection of $\overleftrightarrow{DG}$ and $\overleftrightarrow{EH}$ The scale factor is $\dfrac{GI}{DF} = \dfrac{6}{4} = \dfrac{3}{2}$.

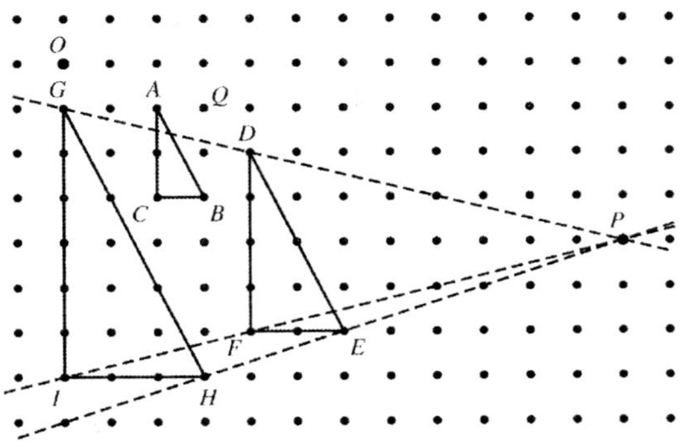

(b) The center is at the intersection of $\overleftrightarrow{AG}$ and $\overleftrightarrow{BH}$ which is the point *Q* one unit to the right of *A* in the above figure. The scale factor is $\dfrac{GI}{AC} = \dfrac{6}{2} = 3$.

(c) In $\triangle DEF$, each side is 2 times the length of the corresponding side in $\triangle ABC$, so the perimeter of $\triangle DEF$ is $2\left(3 + \sqrt{5}\right) = 6 + 2\sqrt{5}$. Similarly, the perimeter of $\triangle GHI$ is $3\left(3 + \sqrt{5}\right) = 9 + 3\sqrt{5}$.

(d) $4\left(3 + \sqrt{5}\right) = 12 + 4\sqrt{5}$

(e) Area $\triangle ABC = 1$
Area $\triangle DEF = 4$
Area $\triangle GHI = 9$
The areas are equal to the scale factor squared.

23. We need to rotate the figure $90°$ counterclockwise so that, for example, $\overline{A_1B_1}$ will be parallel to $\overline{A'B'}$. Then we can perform a size transformation whose center is found by intersecting two lines such as $\overleftrightarrow{B_1B'}$ and $\overleftrightarrow{D_1D'}$ Answers will vary; one possibility is the following. Rotate $90°$ counterclockwise about *B*. Then perform a size transformation centered at *P* with scale factor 2.

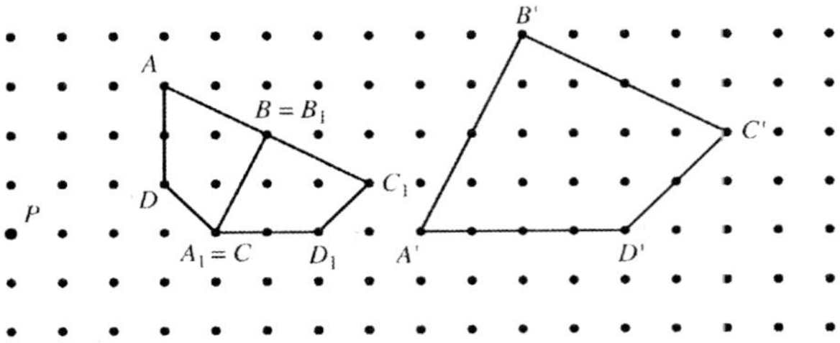

25. **(a)** Ephraim and DeVonte used a $180°$ rotation to transform their figure.

(b) Answers will vary. It might help to trace the figure and use a Mira to arrange the pattern blocks into the mirror image.

27. Under a similarity transformation, all distances are transformed by the same factor and all angle sizes are preserved. Thus, two rectangles are similar if, and only if, they have the same ratio of length to width. That is, $\dfrac{L}{W} = \dfrac{L'}{W'}$. A possible source of confusion is that the word "similar" in mathematical usage requires that very precise conditions are satisfied about shape and size. For example, in ordinary usage we might say two people look similar to one another, but this is not so in the strict mathematical sense of the word.

Copyright © 2015 Pearson Education, Inc.

29. Since each of the turns preserves orientation, the combined transformation must preserve orientation, so it is either a translation or a rotation. But there is an overall turn of 180°, so it remains only to determine the center of this rotation. On a 1 cm square grid, the two 90° rotations take O_1 to O_1' (and O_2 to O_2') as shown. Since the rotation center of a half turn is half-way between any point and its image, it must be the point O. The motion is equivalent to a 180° rotation (half turn) about the point O.

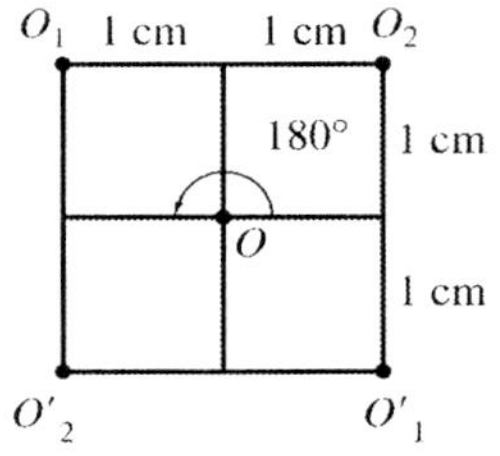

31. (a), (b) Reflection of $\triangle ABC$ across m_1 gives $\triangle A_1B_1C_1$. Reflection of $\triangle A_1B_1C_1$ across m_2 gives $\triangle A_2B_2C_2$. Finally, reflection $\triangle A_2B_2C_2$ across m_3 gives $\triangle A'B'C'$. For part (b), l is the line that contains the midpoints of $\overline{AA'}$, $\overline{BB'}$, and $\overline{CC'}$. (It is also the perpendicular bisector of any one of these.) See figure below.

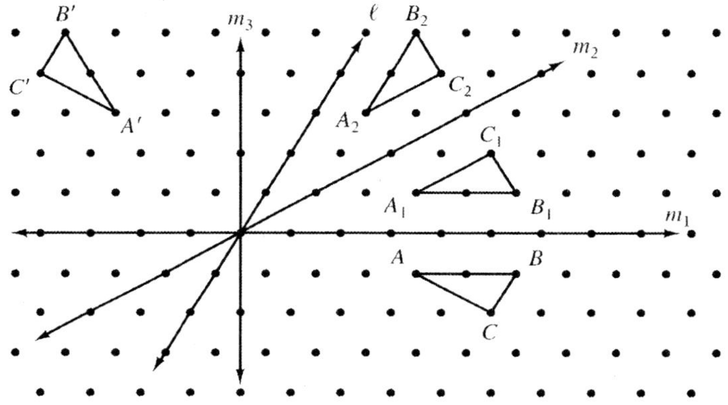

33. (a), (b) m_1 is the perpendicular bisector of $\overline{AA'}$. m_2 is the perpendicular bisector of $\overline{BB'}$. The image of C_1 across m_2 is C'.

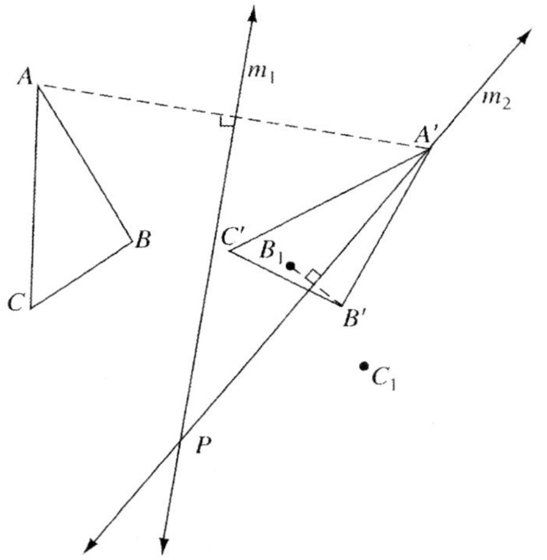

(c) The rigid motion that takes $\triangle ABC$ to $\triangle A'B'C'$ is a reflection across m_1 followed by a reflection across m_2. Since two consecutive reflections across intersecting lines are equivalent to a rotation, the basic rigid motion is a rotation about the point P of intersection of m_1 and m_2, through an angle twice the measure x of the directed angle from line m_1 toward line m_2.

35. (a) A translation: Six reflections give an orientation preserving rigid motion, so it is either a rotation or a translation. Since a rotation has a fixed point (namely the rotation center), the motion is a translation.

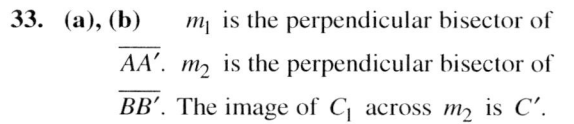

114 *Chapter 11 Transformations, Symmetries, and Tilings*

(b) A reflection: Eleven reflections give an orientation reversing transformation, so it is either a reflection or a glide-reflection. A glide-reflection leaves no point unmoved, so the motion is a reflection.

(c) A rotation: Two glide-reflections give an orientation-preserving rigid motion, so it is either a translation or a rotation. Since some point is unmoved, the motion is a rotation. It is possible the motion is the identity motion, which leaves all points unmoved.

37. (a) The line $\overleftrightarrow{PP'}$ passes through O, so O is the intersection of $\overleftrightarrow{PP'}$ and line l.

(b) Since $\overline{PQ}$ and $\overline{P'Q'}$ are parallel, Q' lies on the line through P' that is parallel to $\overline{PQ}$. Since Q' also lies on $\overrightarrow{OQ}$ (which is part of line l), Q' is the intersection of line l and the line through P' that is parallel to $\overline{PQ}$.

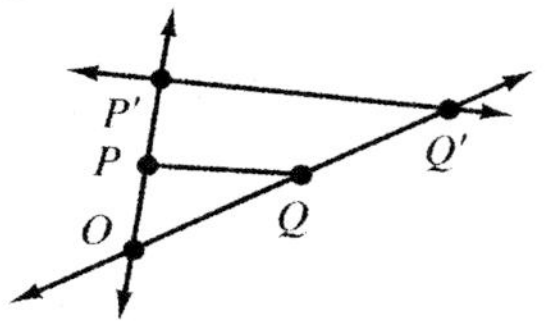

39. (a) Yes, this simple system will cycle the mattress through all four positions it can be put on the bed frame. If the letter A and B are put on the upper-facing side of the head and foot of the mattress as shown, and A' and B' on the lower-side, A is replaced by B', then B, then A', and then returns to A.

(b) Yes, this system cycles the mattress through all eight positions. To see why, letter the upper right-facing side of the mattress A, B, C, D in clockwise order, with A', B', C', and D' on the lower-facing side. The letter showing at the head of the bed is then successively A, C', B, B', C, A', D, D', and then A again to restart the cycle.

(c) Note that $PS' = PR + RS' = PR + RS$ and $QS' = QS$. Substituting in the inequality $PQ + QS' > PS'$ gives $PQ + QS > PR + RS$.

41. (a) Construct the image P' of P across m and the image S' of S across l. Then draw $\overline{P'S'}$ to find Q and R.

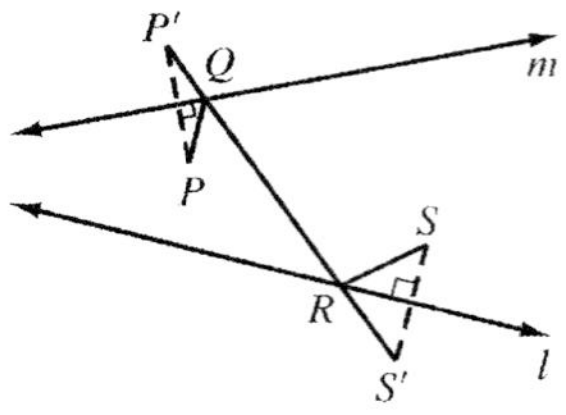

(b) Construct the image P'' of P across l and the image S'' of S across m. Then draw $\overline{P''S''}$ to find A and B.

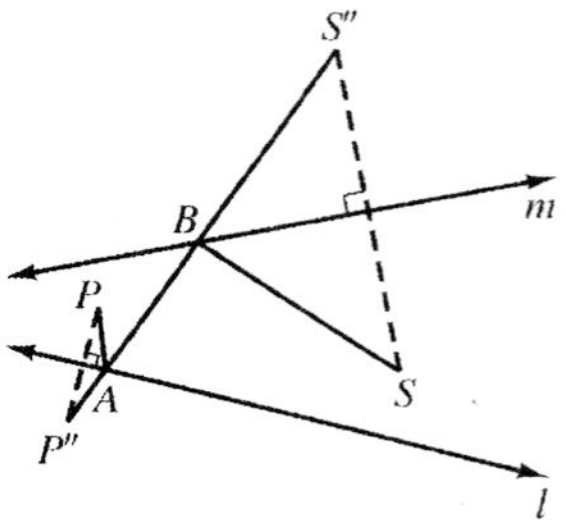

43. Draw $\overline{PP'''}$. Then "fold" the copies of the table back on top of the table to give the billiard path.

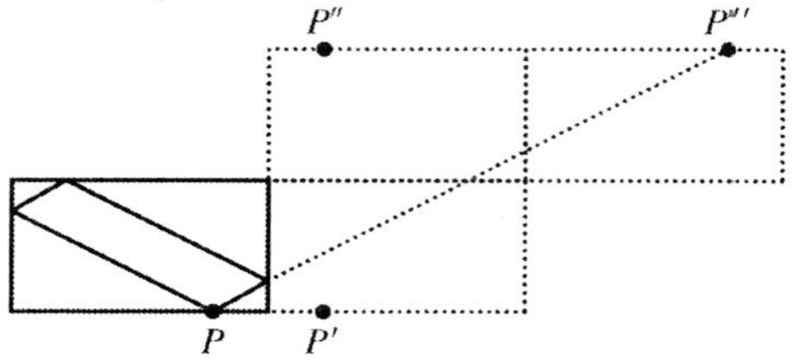

45. (a)

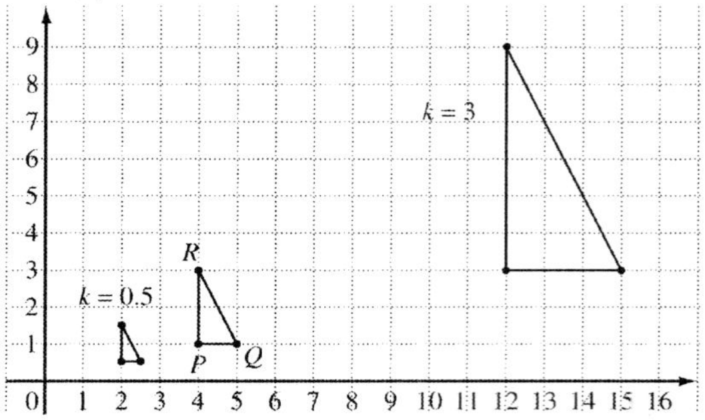

(b) After the size transformation with scale factor 2 that transformed $\triangle PQR$ into $\triangle P'Q'R'$, subtract 7 from the x-coordinates of the points to translate $\triangle P'Q'R'$ horizontally to the left by seven units.

Copyright © 2015 Pearson Education, Inc.

Solutions to Problem Set 11.2 **115**

(c) $\triangle PQR$ is first dilated with scale factor $k = 2$ to $\triangle P'Q'R'$, and then replacing all coordinates with their negatives gives a half-turn rotation about the origin that moves $\triangle P'Q'R'$ to $\triangle ABC$. Altogether, the *x*- and *y*-coordinates in the plane are multiplied by -2.

47. D. The coordinates of *T* are $(1, -3)$. Translating this point three units to the left and five units up gives $T'(-2, 2)$.

49. D.   **51.** D.

Section 11.2
Patterns and Symmetries

Problem Set 11.2

1. (a) 6 lines of symmetry   **(b)**   1 line of symmetry

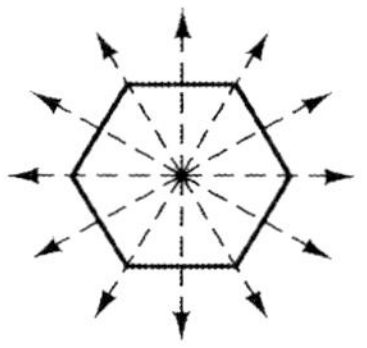

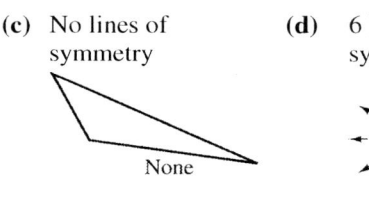

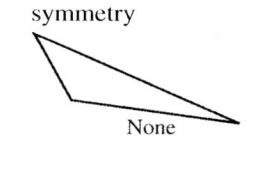

(c) No lines of symmetry   **(d)**   6 lines of symmetry

None

(e) 1 line of symmetry

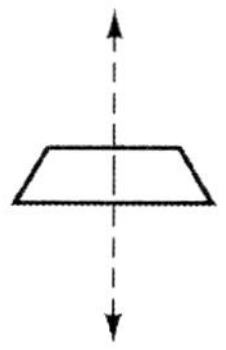

3. (a) Answers will vary. One possibility is an isosceles triangle.

(b) Answers will vary. For example, most parallelograms have no reflection symmetry, but they do have 180° rotation symmetry.

(c) Not possible. If the rotation symmetry is by an angle other than 180°, then the original line of symmetry can be rotated by this angle to produce another line of symmetry. If the rotation symmetry is half-turn symmetry, then the figure is the same when rotated 180° as when reflected over the original line of symmetry, implying that there is a second line of symmetry which is perpendicular to the first. Therefore, it is not possible for the figure to have rotational symmetry and exactly one line of symmetry.

5. (a)

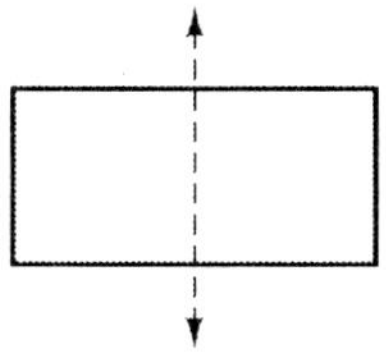

(b)

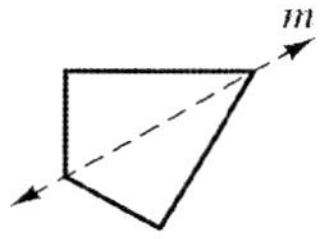

(c)

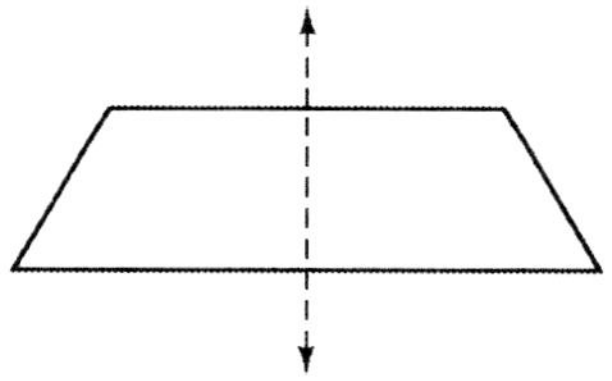

(d)

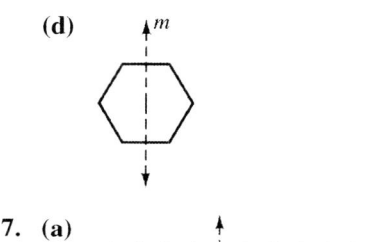

7. (a)

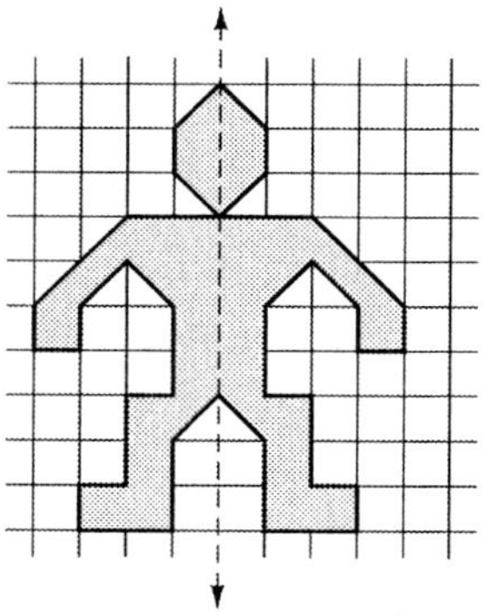

Copyright © 2015 Pearson Education, Inc.

116 *Chapter 11 Transformations, Symmetries, and Tilings*

(b)

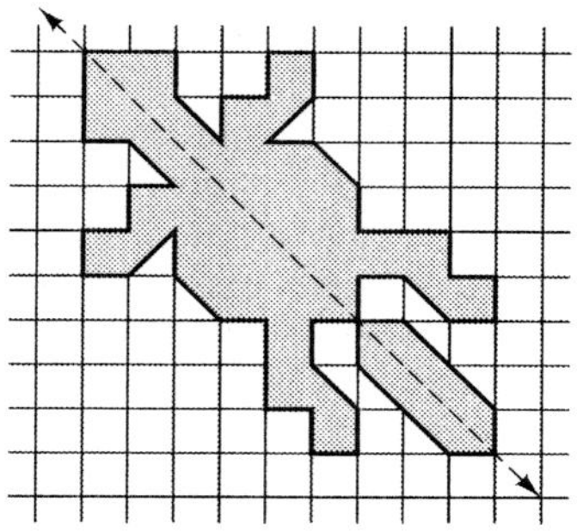

9. **(a)** One vertical line of symmetry

(b) Three lines of symmetry and 120° rotational symmetry

(c) 120° rotational symmetry and no line of symmetry

(d) Point symmetry about the center (provided that the ® symbol is ignored)

(e) Point symmetry about the center

11. **(a)**

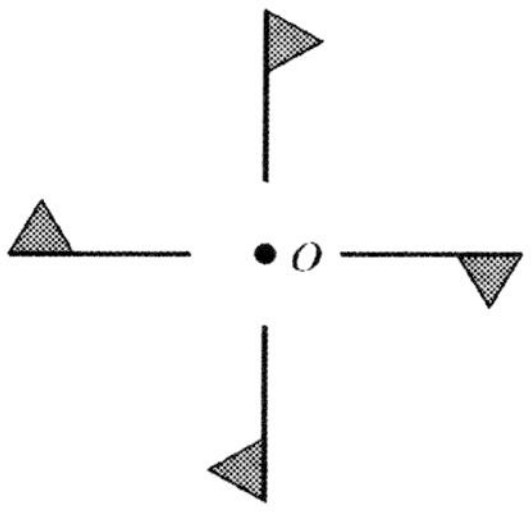

(b)

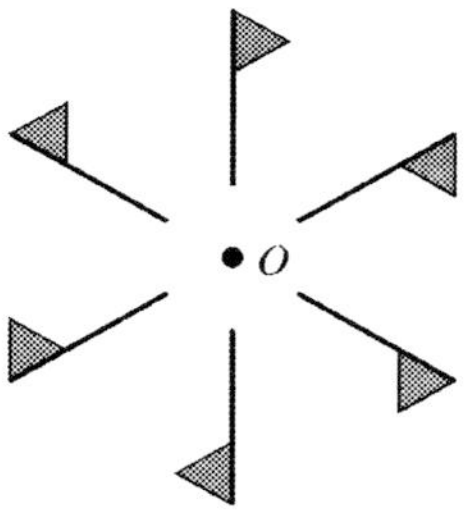

13. **(a)** 0, 8   **(b)** 0, 3, 8

(c) 0, 8   **(d)** 0, 8

15. **(a)** i, l, o, t, u, v, w, x

(b) c, l, o, x

(c) l, o, x

(d) l, o, s, x, z

17. **(a)** Translation symmetry, vertical line symmetry, horizontal line symmetry, glide-reflection symmetry, and half-turn symmetry: *mm*

(b) Translation symmetry and vertical line symmetry: *m*1
(Note that an M is not the same as a horizontally reflected W.)

(c) Translation symmetry and half-turn symmetry: 12

19. **(a)** **(i)** *mm*   **(ii)** 1*g*

(iii) 1*m*   **(iv)** 12

(b) Remaining patterns are *mg*, *m*1, and 11. Patterns created may vary.

mg ... ↑↓↑↓↑ ...
*m*1 ... ↗↖ ↗↖ ↗ ...
11 ... ↗↗↗↗↗ ...

21. Answers will vary.

23. **(a)** The lines of symmetry of the non-square rectangle are correct. The isosceles trapezoid has a vertical line of symmetry, but not a horizontal line of symmetry.

(b) Use a Mira to compare the reflections over the vertical and horizontal lines, explaining to Stacey why the vertical line is a line of symmetry and why the horizontal line is not.

25. Lynn can always place a penny at the position that is the image of Kelly's last move under a half-turn centered at the center of the table. Since Lynn can always make a move, Kelly will be the first player unable to find space for an additional penny on the table.

27. **(a)** The sequence of letters is the same backwards as forwards.

(b) The sequence of words is the same backwards as forwards.

(c) The sequence of letters and spaces is the same backwards as forwards.

29. **(a)** The pattern has no vertical line of symmetry and, disregarding the color scheme, it has horizontal glide symmetry. Its type is 1*g*.

Copyright © 2015 Pearson Education, Inc.

(b) The pattern has a horizontal line of symmetry but no vertical line of symmetry. Its type is 1*m*.

(c) The pattern has vertical line symmetry and horizontal glide symmetry. Its type is *mg*.

(d) The pattern has vertical and horizontal line symmetry. Its type is *mm*.

31. (a) The motion is a translation. The pattern does not have vertical line symmetry, horizontal glide-reflection, or half-turn symmetry. Its type is 11.

p	p	p	p	p	p

(b) The motion is a reflection across a vertical line. The pattern has vertical line symmetry but does not have horizontal glide-reflection or half-turn symmetry. Its type is *m*1.

p	q	p	q	p	q

(c) The motion is a half-turn. The pattern has half-turn symmetry only. Its type is 12.

p	d	p	d	p	d

(d) The motion is a glide-reflection. The pattern has glide-reflection symmetry only. Its type is 1*g*.

p	b	p	b	p	b

33. (a) As left-handed people know well, not all scissors are symmetric.

(b) A tee shirt has bilateral symmetry

(c) A dress shirt is not symmetric due to buttons and pockets.

(d) Most golf clubs are either left-handed or right-handed, so they are not symmetric. Some putters, designed to be used by both left-handed and right-handed people, have bilateral symmetry.

(e) Tennis rackets have two planes of bilateral symmetry (ignoring such details as how they are strung and wrapping around the handle).

(f) The blacked-out squares usually form a symmetric pattern which may include reflection symmetry and/or 90° rotation symmetry.

35. Note that we do not include translation symmetry or glide-reflection symmetry, because these patterns do not extend outside the square shown.
(a) one diagonal line of symmetry;
(b) two diagonal lines of symmetry and half-turn symmetry;
(c) two diagonal lines of symmetry, vertical and horizontal lines of symmetry through the center, and 90° rotation symmetry
(d) two lines of diagonal symmetry and half-turn symmetry;
(e) vertical and horizontal lines of symmetry through the center and half-turn symmetry;
(f) two diagonal lines of symmetry and half-turn symmetry.

37. Answers will vary.

39. D

Section 11.3
Tilings and Escher-like Designs

Problem Set 11.3

1. Many different tilings can be formed. For example: The pattern shown consists of squares within squares—all formed from identical triangles.

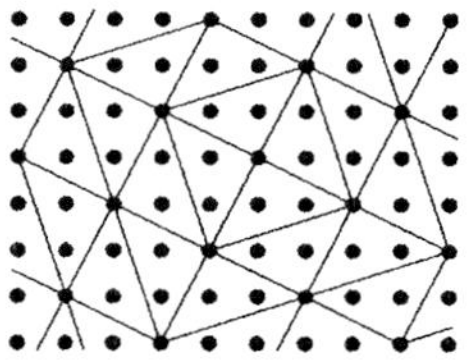

3. As shown in Example 11.12, any quadrilateral can tile the plane.

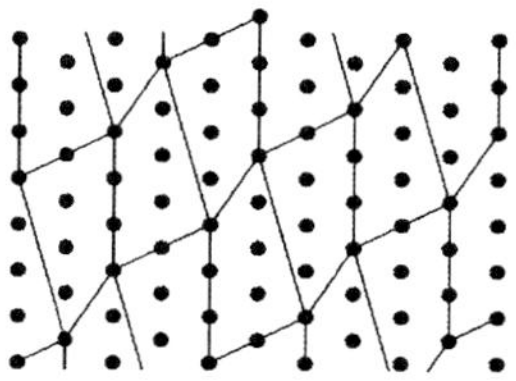

118 *Chapter 11 Transformations, Symmetries, and Tilings*

5.

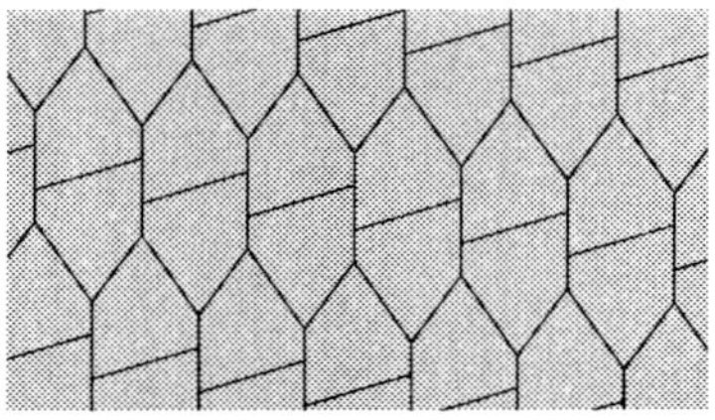

7. The hexagon will tile using translations only. No rotation is necessary.

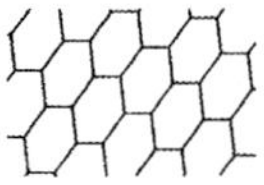

9. (a) The measure of an interior angle of a regular hexagon is $\dfrac{180°(6-2)}{6} = 120°$, while the measure of each angle in an equilateral triangle is 60°. Thus, the sum of the angles at the vertex V is $120° + 60° + 60° + 120° = 360°$.

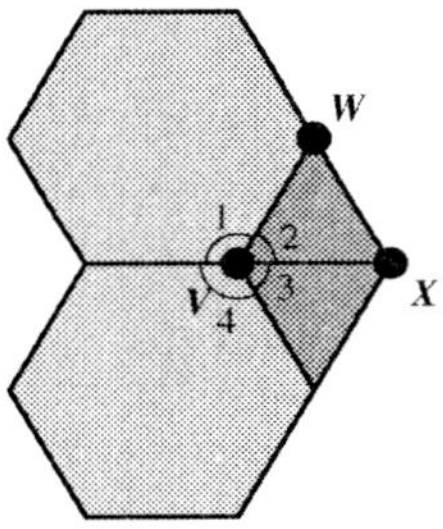

(b) If the vertex figure at V is repeated at W, the vertex figure at X has 3 triangles, making it different from the figure at V. (There are other answers as well. For example, repeating the vertex figure at X forces the two hexagons at W to be separated by a triangle.)

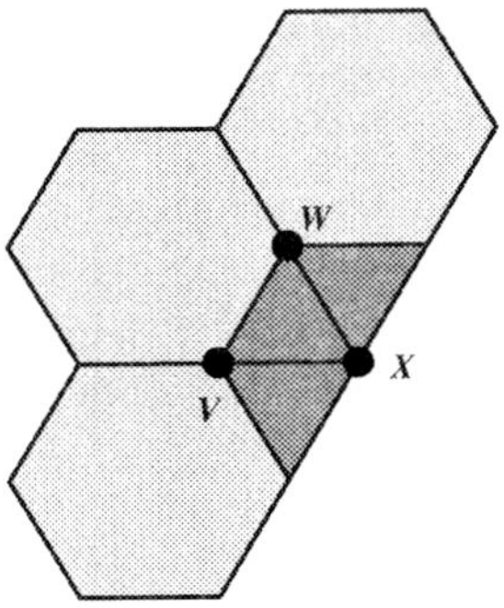

11. The interior angle of a square is 90° and for a regular pentagon is $\dfrac{(5-2)(180°)}{5} = 108°$ and for a regular 20-gon is $\dfrac{(20-2)(180°)}{20} = 162°$. Then, $90° + 108° + 162° = 360°$.

13. (a)

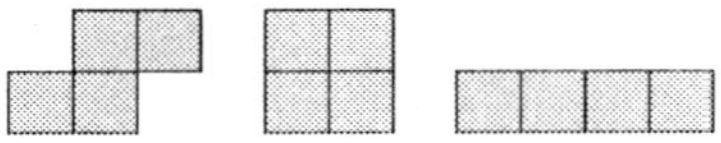

(b) Each tetromino tiles the plane.

(c) Yes. An example is shown below.

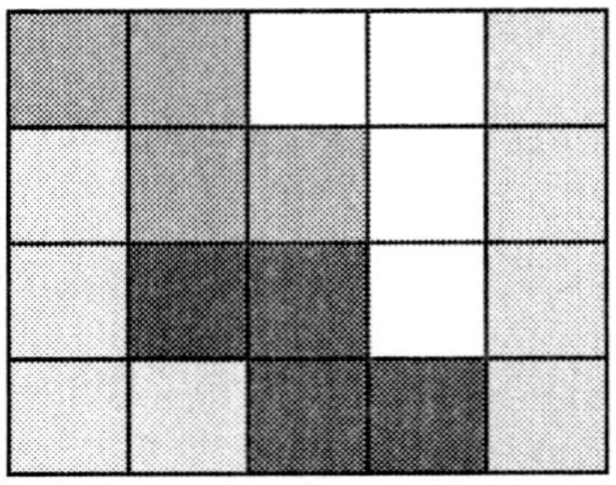

15. Answers will vary.

17. Miguel needs to be reminded that the tiles must meet edge to edge, so that a vertex of any polygon can only be placed coincident with vertices of other polygons.

19. (a) Wailea is correct that the four polygons form a vertex figure since the sum of the angle measures of the four polygons is $60° + 60° + 90° + 150° = 360°$.

(b) Remind Wailea that every vertex figure must be identical, so at vertex B there must be an equilateral triangle outward from edge $\overline{BC}$, but this means three equilateral triangles surround the vertex at C.

21. (a) Squares, pentagons, hexagons, heptagons, octagons

(b) Actually, it is a "real" tiling by "fake" regular polygons. Many vertex figures that appear in the tiling cannot correspond to regular polygons. For example, a pentagon, hexagon, and octagon meet at each of many vertices. If these were regular figures, they would have interior angles of 108°, 120°, and 135°, respectively. But $108° + 120° + 135° = 363° \neq 360°$, so these cannot be all regular polygons.

23. Designs will vary.

25. Designs will vary.

Copyright © 2015 Pearson Education, Inc.

27. (a) Any four identical triangles can be arranged to form a new triangle that has the same shape, but is twice as large.

(b)   **(c)**

(d)

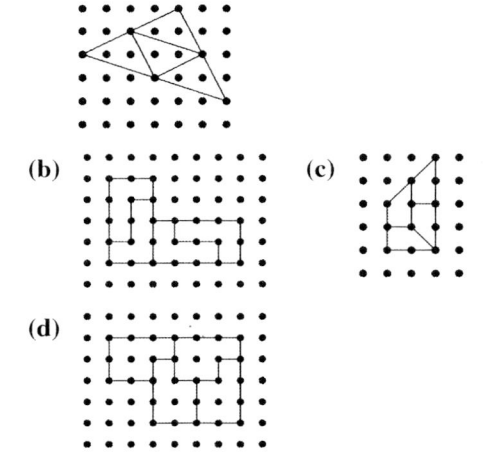

29. (a) Four Sphinxes can be arranged to create a larger sphinx.

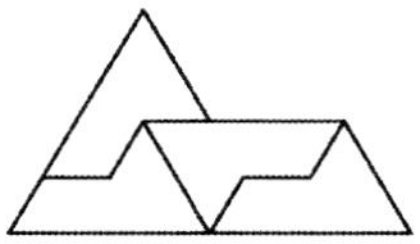

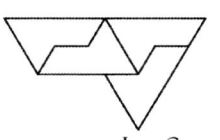

(b) A third-generation Sphinx requires 16 copies of the original Sphinx. It consists of four second generation Sphinxes.

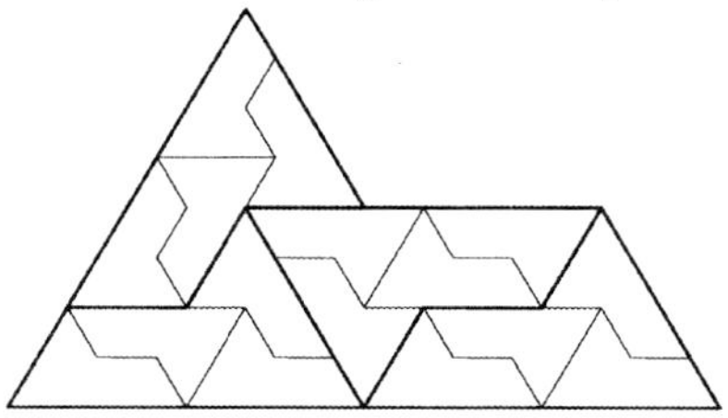

(c) There is no end to the number of generations that can be formed.

31. A

33. A, C. The criterion of using "only a clockwise rotation" is not clear. Clearly, the rotation must be performed three times, but it is not clear if the rotation must always be applied with the same center, or if the rotation must be applied to the original figure only or to a previously rotated shape. Although the C tessellation can be obtained with a fixed rotation center using clockwise rotations of 90°, 180°, and 270°, some students may also claim with some justification that the A tessellation is also a correct answer.

Chapter 11 Review Exercises

1. Move each point 3 units right and 1 unit down.

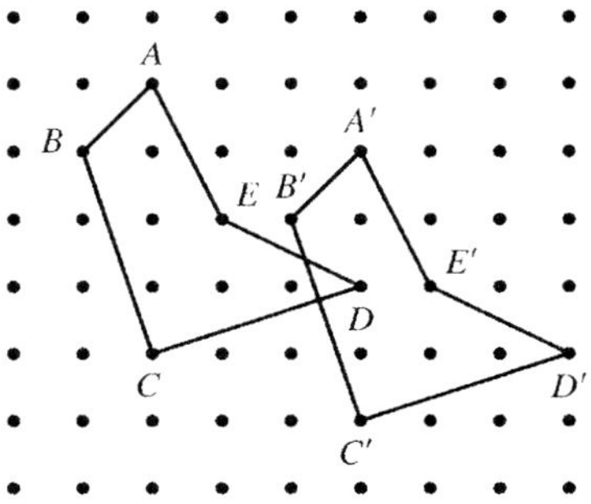

3. Reflection across line l where line l is determined by any two of the midpoints of $\overline{AA'}$, $\overline{BB'}$, and $\overline{CC'}$. It is also the perpendicular bisector of each of these segments.

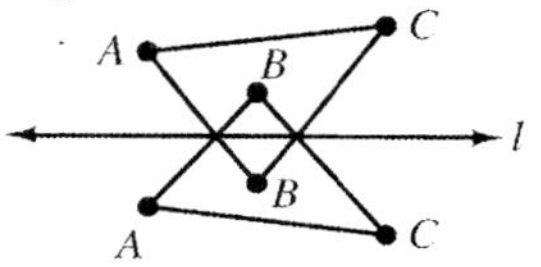

5.

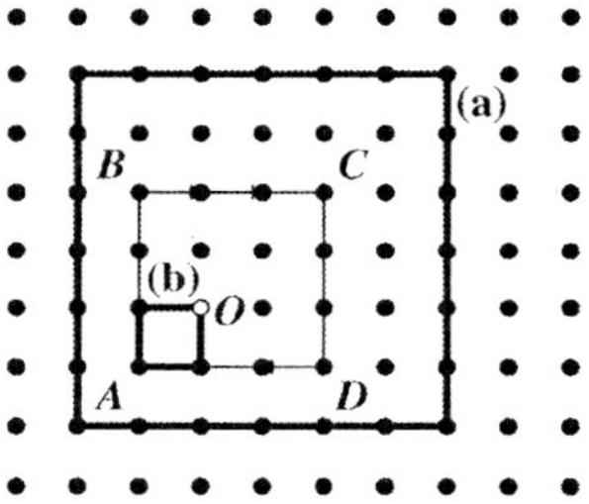

7. (a) Glide-reflection, since the first two reflections accomplish a translation in a direction parallel to the third line of reflection.

(b) Reflection, since the 3 lines of reflection are parallel.

(c) Reflection, since the 3 lines of reflection are concurrent.

(d) Glide-reflection, since the 3 lines of reflection are neither parallel nor concurrent.

9. The letters in the upper row have either rotational or mirror symmetry, unlike those in the second row.

Copyright © 2015 Pearson Education, Inc.

120 *Chapter 11 Transformations, Symmetries, and Tilings*

11. **(a)** equilateral triangle

    (b) parallelogram

    (c) a regular 9-gon (enneagon, or nonagon)

13. **(a)** Translation symmetry and vertical line symmetry (type $m1$)

    (b) Translation symmetry and horizontal line symmetry (type $1m$)

    (c) Translation symmetry and glide-reflection symmetry (type $1g$)

    (d) Translation symmetry, vertical line symmetry, horizontal reflection (and glide-reflection) symmetry, and half-turn symmetry (type mm)

    (e) Translation symmetry and half-turn symmetry (type 12)

(f) Translation symmetry, vertical line symmetry, horizontal glide-reflection symmetry, and half-turn symmetry (type mg)

15. Answers will vary. Two examples are shown.

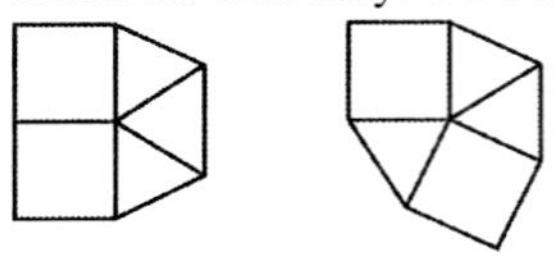

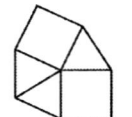

17. (a), (b), (c), (d), and (e) will tile the plane because any triangle, quadrilateral, pentagon with two parallel sides, hexagon with one pair or three pairs of opposite parallel sides of the same length will tile the plane. The plane cannot be tiled with any convex polygon having seven or more sides.

Copyright © 2015 Pearson Education, Inc.

Chapter 12 Congruence, Constructions, and Similarity

Section 12.1
Congruent Triangles

Problem Set 12.1

1. (a) $L \leftrightarrow K$, $H \leftrightarrow W$, $S \leftrightarrow T$

   (b) $\overline{LH} \leftrightarrow \overline{KW}$, $\overline{HS} \leftrightarrow \overline{WT}$, $\overline{SL} \leftrightarrow \overline{TK}$

   (c) $\angle L \leftrightarrow \angle K$, $\angle H \leftrightarrow \angle W$, $\angle S \leftrightarrow \angle T$

   (d) $\triangle LHS \equiv \triangle KWT$

3. (a) $\overline{KL} \leftrightarrow \overline{BC}$, so $KL = BC$. Thus $KL = 4$.

   (b) $\overline{LJ} \leftrightarrow \overline{CA}$, so $LJ = CA$. Thus $LJ = 6$.

   (c) $\angle L \leftrightarrow \angle C$, so $m(\angle L) = m(\angle C)$. Thus $m(\angle L) = 56°$.

   (d) $\angle J \leftrightarrow \angle A$, so $m(\angle J) = m(\angle A)$. Thus $m(\angle J) = 41°$.

5. Use Construction 1 to construct a segment $\overline{AB}$ of length x. Next, construct the circle of radius y centered at B, and the circle of radius z centered at A. Finally, let C be one of the points of intersection of the two circles of radii y and z.

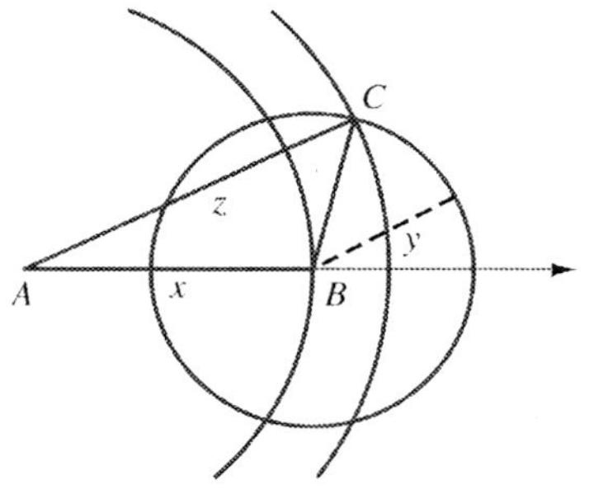

7. Answers will vary, but should follow Construction 2 described in the text.

9. (a) Use the ruler to draw segment $\overline{AB}$ of length 5 cm. Use the protractor to construct $\angle A$ with measure 28°. Along the terminal side, use the ruler to locate a point C that is 5 cm away from A. Draw segment $\overline{BC}$ to complete the triangle.

   (b) Use the ruler to draw segment $\overline{AB}$ of length 6 cm. Set the compass to 6 cm. Use the compass to draw intersecting arcs centered at A and B. Label the intersection point C. Draw segments $\overline{AC}$ and $\overline{BC}$ to complete the triangle.

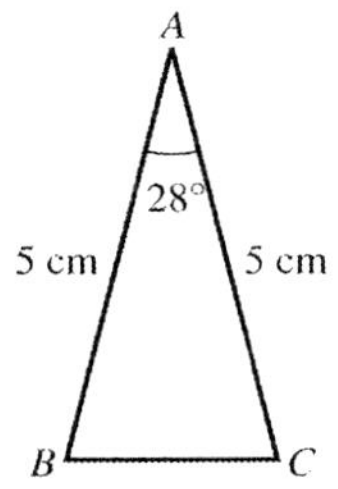

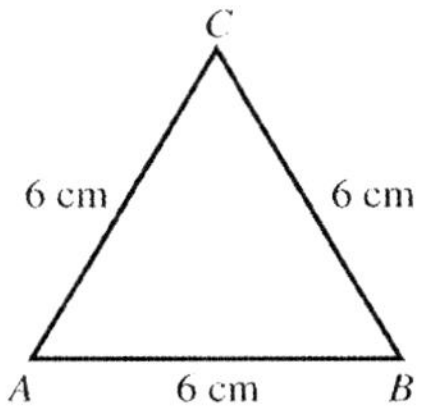

   There is only one possible triangle.

   (c) This is impossible by the triangle inequality since $2 + 5 < 8$.

11. (a) $\overline{BD} \equiv \overline{BD}$ and $\overline{AD} \equiv \overline{CD}$. $\angle BDA \equiv \angle BDC$ since they are right angles. $\triangle ABD \equiv \triangle CBD$ by the SAS property.

    (b) $\angle A \equiv \angle L$, $\angle B \equiv \angle E$, and $\overline{AB} \equiv \overline{FE}$. $\triangle ABC \equiv \triangle FED$ by the ASA property.

    (c) $\overline{AB} \equiv \overline{EF}$, $\overline{BC} \equiv \overline{FD}$, and $\overline{AC} \equiv \overline{ED}$. $\triangle ABC \equiv \triangle EFD$ by the SSS property.

    (d) $\angle ACB \equiv \angle DCE$, $\angle B \equiv \angle E$, and $\angle A \equiv \angle D$, but no conclusion is possible because no lengths are given.

13. (a) $\overline{AC} \equiv \overline{AC}$, $\angle ACD \equiv \angle ACB$, and $\angle D \equiv \angle B$. $\triangle ACD \equiv \triangle ACB$ by the AAS congruence property.

    (b) $\overline{AB} \equiv \overline{AD}$, $\overline{AC} \equiv \overline{AC}$, and $\angle BAC \equiv \angle DAC$. $\triangle ABC \equiv \triangle ADC$ by the SAS congruence property.

Copyright © 2015 Pearson Education, Inc.   121

122 Chapter 12 Congruence, Constructions, and Similarity

(c) $m(\angle ADC) = m(\angle ADB) + m(\angle BDC)$
$= m(\angle BCA) + m(\angle ACD) = m(\angle BCD)$, so
$\angle ADC \cong \angle BCD$, $\angle ACD \cong \angle BDC$ and
$DC \cong CD$. $\triangle ADC \cong \triangle BCD$ by the ASA congruence property.

15. Let $\triangle ABC$ be equilateral. Since $AB = BC$, it follows from the isosceles triangle theorem that $\angle A \cong \angle C$. In the same way, since $BC = CA$ it follows that $\angle B \cong \angle A$. Thus all three angles are congruent.

17. (a) Both $\angle 1$ and $\angle 2$ should measure close to half that of $\angle BAC$.

(b) $\triangle AFD \cong \triangle AFE$ by the SSS property.

(c) $\angle 1$ and $\angle 2$ are corresponding angles in the triangles shown congruent in part (b), so $\angle 1 \cong \angle 2$.

19. (a) Since alternate interior angles between parallel lines are congruent, $\angle ABD \cong \angle CDB$ and $\angle ADB \cong \angle CBD$. $DB \cong BD$, thus, $\triangle ABD \cong \triangle CDB$ by the ASA property.

(b) Since $\triangle ABD \cong \triangle CDB$, $AD \cong CB$ and $AB \cong CD$. Thus $AD = CB$ and $AB = CD$.

(c) Since $\triangle ABD \cong \triangle CDB$, $\angle BAD \cong \angle DCB$ or $m(\angle BAD) = m(\angle DCB)$.
$m(\angle ADC) = m(\angle ADB) + m(\angle BDC)$
$= m(\angle CBD) + m(\angle DBA) = m(\angle CBA)$.

21. (a) By definition of a rhombus, $AB = BC = CD = DA$. By problem 20, M is the bisector of each diagonal of the rhombus, so $AM = MC$ and $BM = MD$. Therefore, the four triangles ABM, CBM, CDM, and ADM are congruent by the SSS property.

(b) The four angles at M are congruent and have measures summing to $360°$. Thus each angle measures $90°$.

23. (a) The sum of the three sides is 14 cm, so the fourth side must be less than 14 cm: $0 < s < 14$ cm, where s is the length of the fourth side.

(b) If A, B, and C are not collinear, then ABC forms a triangle, so $AC < AB + BC$ by the triangle inequality. If A, B, and C are collinear, then $AC = AB + BC$. Hence, $AC \le AB + BC$. Similarly, if A, C, and D are not collinear, then ACD forms a triangle, so $AD < AC + CD$ by the triangle inequality. If A, C, and D are collinear, then $AD = AC + CD$. Hence, $AD \le AC + CD$. Combine the two inequalities to get $AD \le AB + BC + CD$.

25. Dane is mistaken. Have him make a hinged quadrilateral, as described in the Into the Classroom found in section 12.2. The shape can be changed even though the lengths of the edges remain the same throughout a motion.

27. Following the suggestion give the figure shown. Since the diagonals of a rectangle intersect at a point the same distance from the vertices of the rectangle (see problem 20(b)), $OA = OB = OC = OD$. Thus, it is clear that the circle at the midpoint of the hypotenuse through point C passes through all of the vertices of the rectangle, which include points A, B, and C.

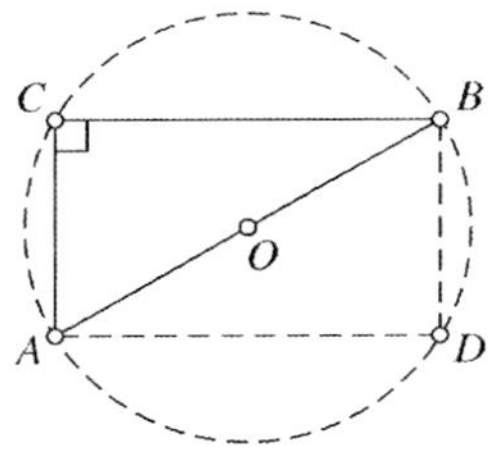

29. Construct a ray $\overrightarrow{QS}$, and use your compass to construct the segment $\overline{QR} \cong \overline{BC}$. At both Q and R, construct parallel rays $\overrightarrow{QT}$ and $\overrightarrow{RU}$ for which $\angle TQR \cong \angle URS \cong \angle B$. Finally construct the ray $\overrightarrow{RV}$ for which $\angle URV \cong \angle A$. Label P the intersection of $\overrightarrow{QT}$ and $\overrightarrow{RV}$. Thus $\triangle PQR \cong \triangle ABC$.

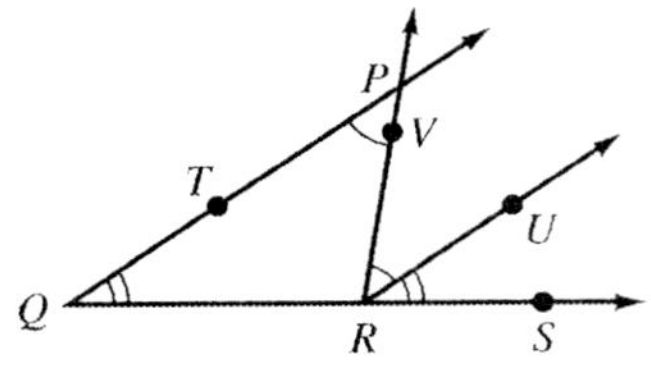

31. If $AB = CD = a$, $BC = AD = b$, and $AC = BD = c$, then each face of the tetrahedron is a triangle with sides of length a, b, and c. By the SSS property, the faces of the tetrahedron are congruent to one another.

Copyright © 2015 Pearson Education, Inc.

Solutions to Problem Set 12.2   123

33. The best location for P is the center of the regular hexagon, where it would lie on the intersection of the diagonals $\overline{AD}$, $\overline{BE}$, and $\overline{CF}$. By the triangle inequality,
$AD < AQ + DQ$, $BE < BQ + EQ$, and $CF < CQ + FQ$. Also, $AP + DP = AD$, $BP + EP = BE$, and $CP + FP = CF$.
Combining the inequalities and equalities,
$AP + BP + CP + DP + EP + FP <$
$AQ + BQ + CQ + DQ + EQ + FQ$.

35. **(a)** The framework forms a parallelogram, but not necessarily a rectangle.

   (b) Let $ABCD$ be a parallelogram. By problem 19(b), $AB = CD$ and $AD = BC$. $AC = BD$ is given. Then $\triangle ABC \cong \triangle BAD$ $\equiv \triangle CDA \equiv \triangle DCB$ by the SSS property. Thus, the corner angles are congruent. That is, $\angle A \equiv \angle B \equiv \angle C \equiv \angle D$. Since the sum of their measures is 360°, each corner angle is a right angle, so $ABCD$ is a rectangle.

   (c) The diagonal brace is nailed across in order to fix the wall so it remains a rectangle. The brace forms two triangles with fixed side lengths. The two triangles are rigid by the SSS property.

37. $\triangle POA$ is isosceles since $OP = OA$. Therefore, the base angles have the same measure x whose sum $2x$ is the measure of the opposite exterior angle. That is,
$m\angle AOQ = 2x = 2m\angle OPA$.
Similarly, $m\angle BOQ = 2y = 2m\angle OPB$.
Therefore,
$m\angle AOB = m\angle AOQ + m\angle BOQ$
$= 2m\angle OPA + 2m\angle OPB$
$= 2m\angle APB$.

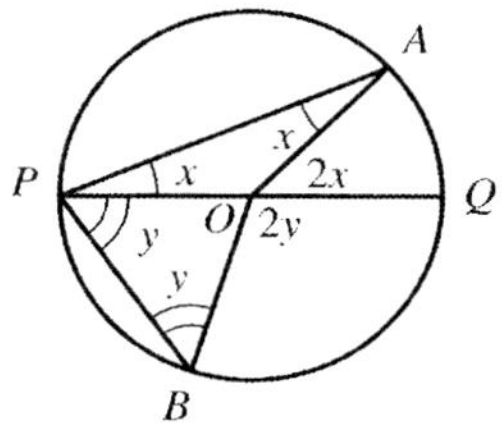

39. Construct the four radii from the center O, so each side $\overline{AB}, \overline{BC}, \overline{CD}, \overline{DA}$ is the base of an isosceles triangle. Let w, x, y, z be the measure of the base angles of the respective triangles, as shown in the diagram below. The sum of the interior angles of the quadrilateral is 360°, so $(w + x) + (x + y) + (y + z) + (z + w) = 360°$.
That is, $w + x + y + z = 180°$. But then
$m\angle A + m\angle C = (w + x) + (y + z) = 180°$.
Similarly,
$m\angle B + m\angle D = (x + y) + (z + w) = 180°$.

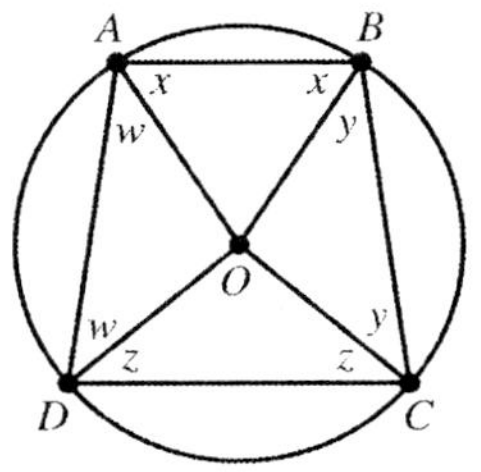

40. B.

41. Answers will vary, but should point out clearly that all six triangles have the same size and shape. The triangles are in different orientations, but all six triangles are right triangles with legs of length half that of a side of the equilateral triangle, and a hypotenuse that is $\sqrt{3}/2$ times the length of a side of the equilateral triangle.

43. C.

Section 12.2
Constructing Geometric Figures

Problem Set 12.2

1. Use Construction 4, since the point C is the point of reflection.

3. Step 2. Use Construction 4 to construct the lines k and l perpendicular to line m that pass through A and B.
   Step 3. Construct the circles of radius AB centered at A and B. These circles intersect k and l at the points C and D so that $ABCD$ is the required square.

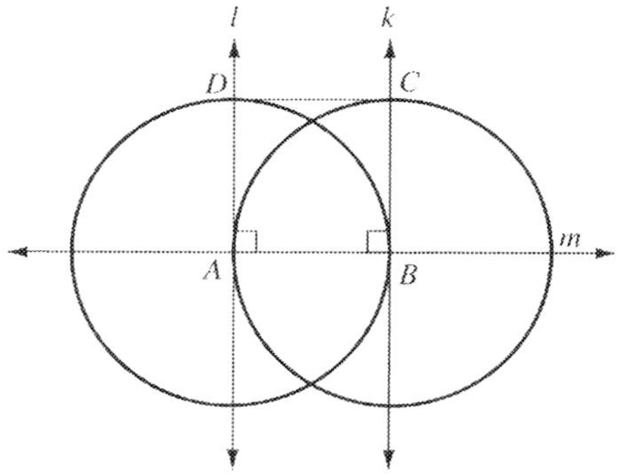

Copyright © 2015 Pearson Education, Inc.

124 Chapter 12 Congruence, Constructions, and Similarity

5. Step 1. Construct a circular arc centered at A that passes through point B; let D be the intersection of the arc with the ray $\overrightarrow{AP}$.

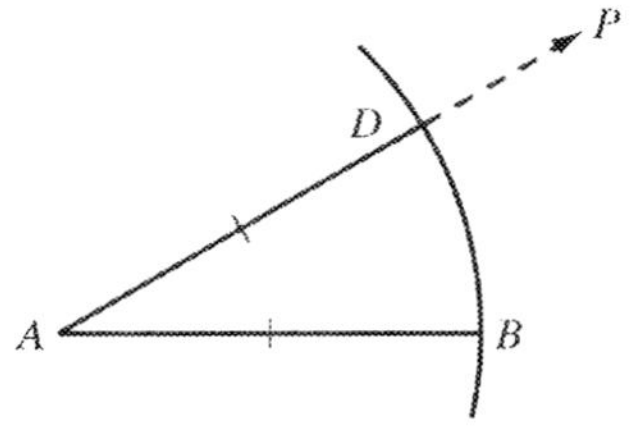

Step 2. Construct a circular arc of radius AB centered at B and D; let C be the point of intersection of these two arcs. Then $ABCD$ is the required rhombus.

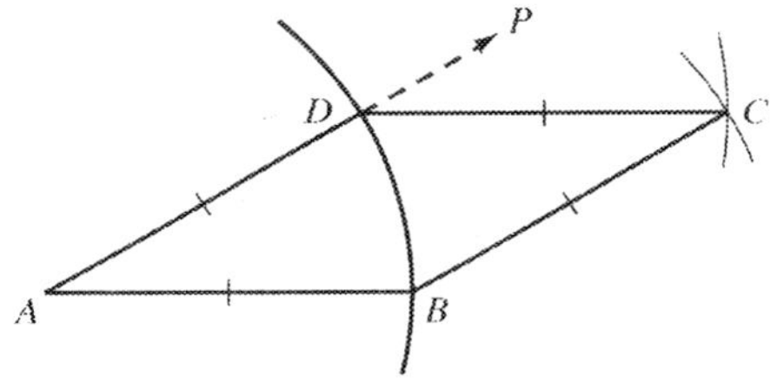

7. Step 1. Use Constructions 3 and 4 to construct lines k and m through A that are respectively perpendicular and parallel to line L. Let k and l intersect at B.

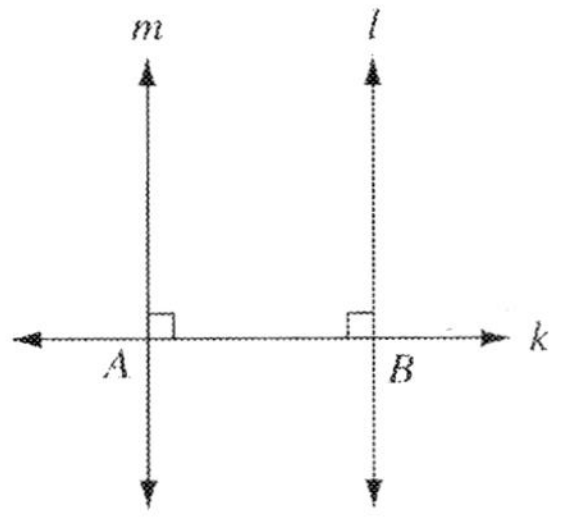

Step 2. Construct the circles of radius AB centered at A and B, and let them intersect l and m at C and D. Then $ABCD$ is the required square.

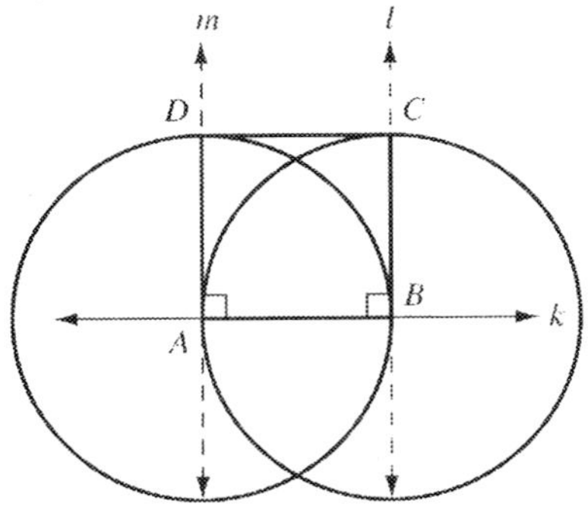

9. Pivot the Mira about P until l coincides with its reflection to construct the line m through P that is perpendicular to l.

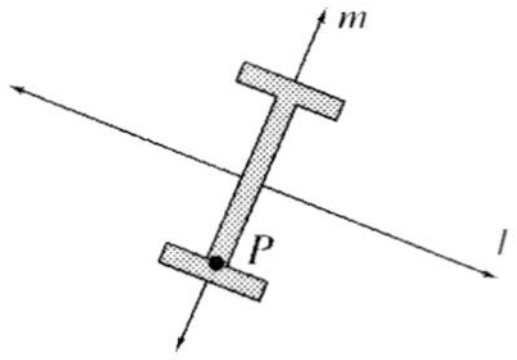

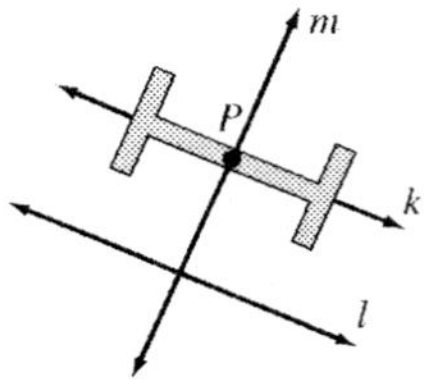

11. (a) The corresponding angles property guarantees that m is parallel to l.

(b) Align the ruler with the line; slide the drafting triangle, with one leg of the right triangle on the ruler, until the second leg of the triangle meets point P.

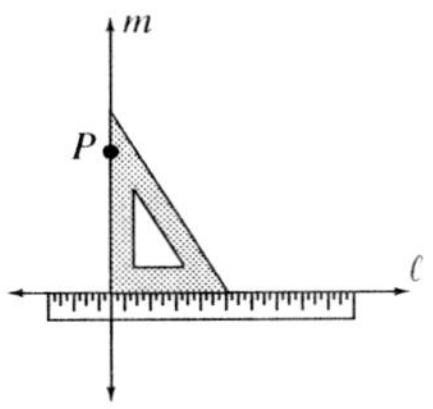

Pivot the Mira about P until m coincides with its reflection to construct the line k through P that is perpendicular to m. Thus k is parallel to l.

13. Construct the perpendicular to side $\overrightarrow{BA}$ at T (construction 5) and the angle bisector (construction 7). Let O be their point of intersection. The circle centered at O passing through T is the desired circle.

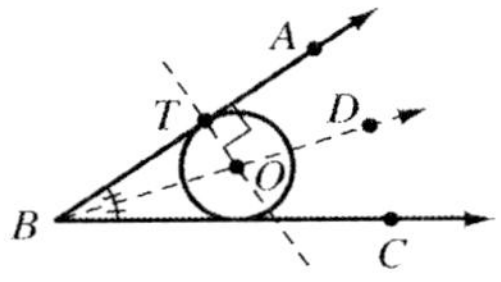

Copyright © 2015 Pearson Education, Inc.

Solutions to Problem Set 12.2   125

15. The circumcenter is the intersection of the perpendicular bisectors of the sides of the triangle.

   (a) The circumcenter will be inside the acute triangle.

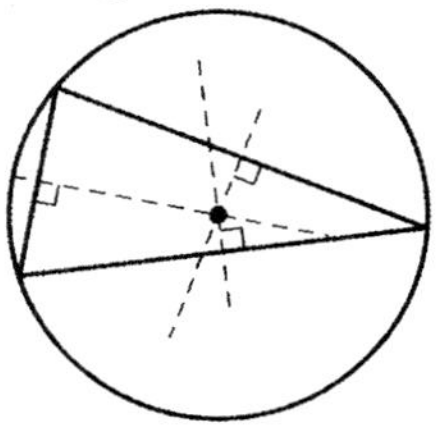

   (b) The circumcenter will be at the midpoint of the hypotenuse of the right triangle.

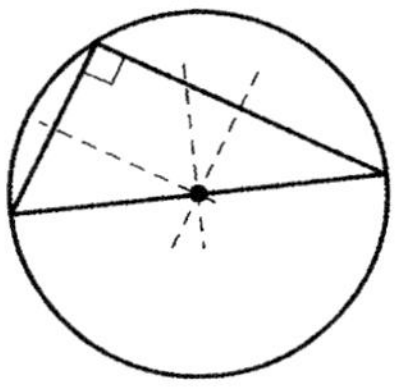

   (c) The circumcenter will be outside the obtuse triangle.

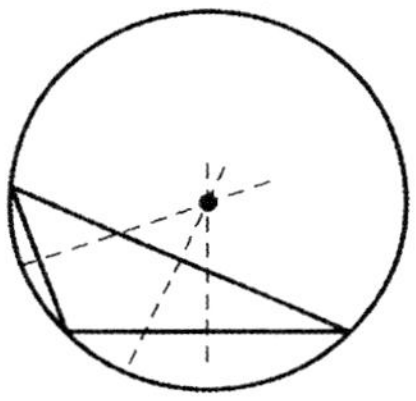

   (d) The circumcenter is inside, on, or outside a triangle if, and only if, the triangle is acute, right, or obtuse, respectively.

17. Since $\triangle PQS$ is inscribed in a circle with diameter $\overline{PQ}$, it has a right angle at point S by Thale's theorem. Thus $\overline{PS} \perp \overline{SQ}$. Similarly, $\triangle PQT$ is inscribed in a semi-circle with diameter $\overline{PQ}$, so $\overline{PT} \perp \overline{TQ}$. Hence $\overline{PS}$ and $\overline{PT}$ are tangent to the circle at Q.

19. By Thale's theorem, $\angle ADB$ is a right angle. $\triangle ODB$ is an equilateral triangle since all sides have the length of the radius of the circles, so $\triangle ODB$ is equiangular by problem 15 of Problem Set 12.1. Thus, $m(\angle ODB) = 60°$. Moreover, $ODBE$ is a rhombus, so $\overline{DE}$ is a bisector of $\angle ODB$, so
$$m(\angle ADE) = m(\angle ADB) - m(\angle EDB)$$
$$= 90° - 30° = 60°.$$

Similarly, $m(\angle AED) = 60°$. Therefore, all angles of $\triangle ADE$ measure $60°$, so $\triangle ADE$ is equilateral.

21. Constructions may vary.

   (a) Extend $\overline{AB}$. Construct perpendicular rays to $\overline{AB}$ at both A and B, to the same side of $\overline{AB}$. Let the bisect the right angle at A. Let the bisector's intersection with the ray at B determine C. Construct the line through C perpendicular to $\overline{BC}$. Let D be the intersection with the ray constructed at A. Thus $ABCD$ is a square with the given side $\overline{AB}$.

   (b) Construct the equilateral triangle $\triangle ABO$ with the given side $\overline{AB}$. The reflection of A across $\overline{OB}$ determines the point C. Repeat reflections to determine the remaining vertices D, E, and F to construct the regular hexagon $ABCDEF$.

23. (a) Suppose the perpendicular bisector of chord $\overline{AB}$ intersects the circle at a point C. Then the circle is the circumscribing circle of $\triangle ABC$. The center of the circumscribing circle is the point of concurrence of the perpendicular bisectors of all three sides of $\triangle ABC$. In particular, it is on the perpendicular bisector of side $\overline{AB}$ contains the center of the circle.

   (b) Construct two non-parallel chords, and their respective perpendicular bisectors. By part (a), each bisector contains the center of the circle. Hence, the bisectors intersect at the center.

   (c) The dashed lines are the perpendicular bisectors of chords $\overline{AB}$ and $\overline{BC}$. By part (b), these bisectors intersect at the center of the larger circle.

25. (a) Waun seems to have constructed an equilateral pentagon, since all five sides have the same length. However, not all of the angles are congruent to each other, so Waun's pentagon is not regular.

   (b) Waun needs to check that all sides are congruent and all angles are congruent.

Copyright © 2015 Pearson Education, Inc.

126 *Chapter 12 Congruence, Constructions, and Similarity*

27. Trisecting an angle with compass and straightedge is impossible. In the case of a 150° angle, trisecting the related chord gives angles close to 25°, 100°, and 25°.

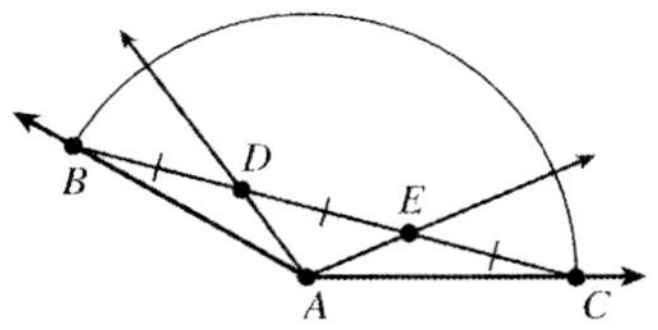

29. (a) There are five points on *C* on line *l* for which triangle *ABC* is isosceles. As shown below, they are constructed by the circle at *A* through *B*, the circle at *B* through *A*, and the perpendicular bisector of $\overline{AB}$.

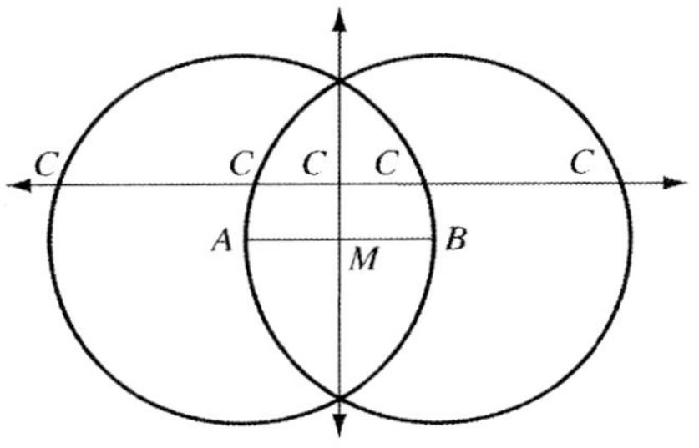

(b) There are four points *D* on line *l* for which triangle *ABD* is a right triangle. As shown below, they are constructed by the circle at *M*, the midpoint of $\overline{AB}$, through *A*, and the perpendiculars to $\overline{AB}$ at *A* and *B*.

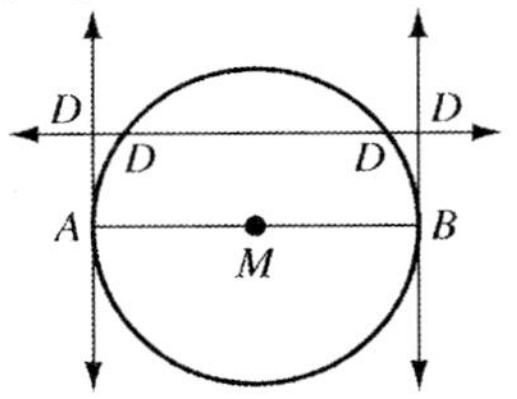

31. Draw △*ABC* and construct the angle bisectors of ∠*A* and ∠*B*. Call their point of intersection *P*. Using the equidistance property of the angle bisector, the distance from *P* to $\overline{AB}$, *PE*, is the same as the distance from *P* to $\overline{BC}$, *PG*, because *P* is on the angle bisector of ∠*B*. Likewise, the distance from *P* to $\overline{AB}$, *PE*, is the same as the distance from *P* to $\overline{AC}$, *PF* because *P* is on the angle bisector of ∠*A*. Thus the distance from *P* to $\overline{BC}$, *PG* is the same as the distance from *P* to $\overline{AC}$, *PF*, so *P* lies on the angle bisector of ∠*C*. Therefore, the three angle bisectors are concurrent.

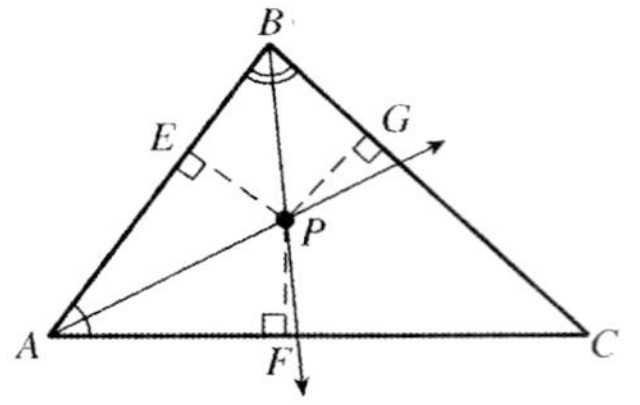

33. (a) Construction should match the illustration in the text.

(b) A ruler and protractor will verify that *PENTA* is regular.

35. (a) Construction should match the illustration in the text.

(b) *BFGHIJK* is a close approximation since 7 is not a Fermat prime.

37. By the Gauss-Wantzel Constructibility Theorem, the constructible regular polygons have *n* sides where $n = 2^r \cdot 4$ (where *r* is a whole number), $2^r \cdot p$ (where *p* is a fermat prime and *r* is a whole number), or $n = 2^r \cdot p_1 \cdot \ldots p_m$ (where $p_1, \ldots p_m$ are distinct fermat primes and *r* is a whole number). The only Fermat primes we need be concerned with are 3, 5, and 17, since the remaining Fermat primes are greater than 100. Therefore, the polygon is constructible for the following values of *n*:

$2^r \cdot 4$: 4, 8, 16, 32, 64, 128, ...

$2^r \cdot 3$: 3, 6, 12, 24, 48, 96, 192, ...

$2^r \cdot 5$: 5, 10, 20, 40, 80, 160, ...

$2^r \cdot 17$: 17, 34, 68, 136, ...

$2^r \cdot 3 \cdot 5$: 15, 30, 60, 120, ...

$2^r \cdot 3 \cdot 17$: 51, 102, ...

$2^r \cdot 5 \cdot 17$: 85, 170, ...

$2^r \cdot 3 \cdot 5 \cdot 17$: 255, 510, ...

In summary we have the following list.
Constructible: 3, 4, 5, 6, 8, 10, 12, 15, 16, 17, 20, 24, 30, 32, 34, 40, 48, 51, 60, 64, 68, 80, 85, 96
Nonconstructible: 7, 9, 11, 13, 14, 18, 19, 21, 22, 23, 25, 26, 27, 28, 29, 31, 33, 35, 36, 37, 38, 39, 41, 42, 43, 44, 45, 46, 47, 49, 50, 52, 53, 54, 55, 56, 57, 58, 59, 61, 62, 63, 65, 66, 67, 69, 70, 71, 72, 73, 74, 75, 76, 77, 78, 79, 81, 82, 83, 84, 86, 87, 88, 89 90, 91, 92, 93, 94, 95, 97, 98, 99, 100

Copyright © 2015 Pearson Education, Inc.

Solutions to Problem Set 12.3   **127**

39. Since the 70° measure of the exterior angle is the sum of the measures of the opposite interior angles, one of which is 35°, the interior angle at *C* also has measure 35°. Therefore, $\triangle ABC$ is isosceles, and $AB = BC$, making the lighthouse at *C* four miles from *B*. In general, "doubling the angle" will give a distance that is equal to the distance moved from the initial point where the angel is first observed.

41. B.

Section 12.3
Similar Triangles

Problem Set 12.3

1. (a) $m(\angle A) = 180° - m(\angle B) - m(\angle C)$
$= 180° - 60° - 90° = 30°$.
$\triangle ABC \sim \triangle PNO$ by the AA similarity property. The scale factor from $\triangle ABC$ to $\triangle PNO$ is $\dfrac{PN}{AB} = \dfrac{12}{8} = \dfrac{3}{2}$.

(b) $\dfrac{QS}{EF} = \dfrac{SR}{FD} = \dfrac{RQ}{DE} = 2$, so $\triangle DEF \sim \triangle RQS$ by the SSS similarity property. The scale factor is 2.

3. (a) Yes; since all angles measure 60°, equilateral triangles are similar by the AA similarity property.

(b) No; there are many possible examples. For example, let the first triangle have angles with measures 50°, 50°, and 80°, and let the second triangle have angles with measures 70°, 70°, and 40°.

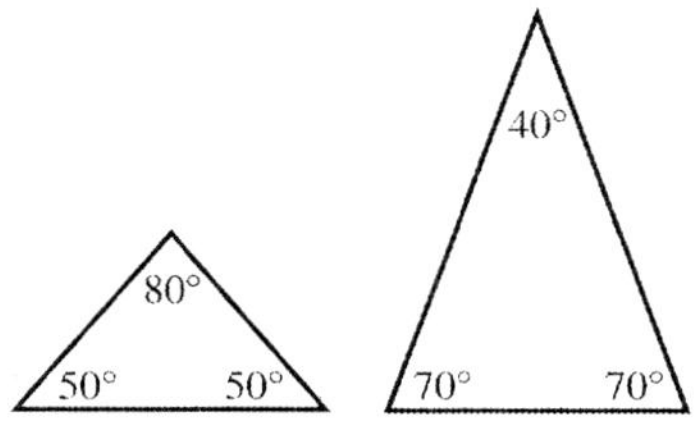

(c) Yes; since any two such triangles have angles of measure 36° and 90°, they are similar by the AA similarity property.

5. (a) $\dfrac{15}{12} = \dfrac{a}{8}$; $a = \dfrac{15}{12} \cdot 8 = 10$

(b) $\dfrac{8}{6} = \dfrac{12}{b}$; $b = 12 \cdot \dfrac{6}{8} = 9$

7. (a) By the Pythagorean theorem, the hypotenuse of triangle *XYZ* is $\sqrt{5^2 + 12^2} = \sqrt{169} = 13$, so its perimeter is 5 + 12 + 13 = 30. Since the perimeter of triangle *ABC* is 3000 feet, the scale factor is 100. Thus, triangle *ABC* is a right triangle with legs of length 500 and 1200, and a hypotenuse of length 1300.

(b) The area of triangle *XYZ* is $\dfrac{5 \times 12}{2} = 30 \text{ ft}^2$ and the area of triangle *DEF* is 100 times as large. Therefore, the scale factor is $\sqrt{100} = 10$ and we see that triangle *DEF* is a right triangle with legs of length 50 feet and 120 feet, and a hypotenuse of length 130 feet.

9. No; for example, a square and a nonsquare rectangle are convex quadrilaterals with congruent angles, yet they are not similar.

11. There are two choices for *F*.

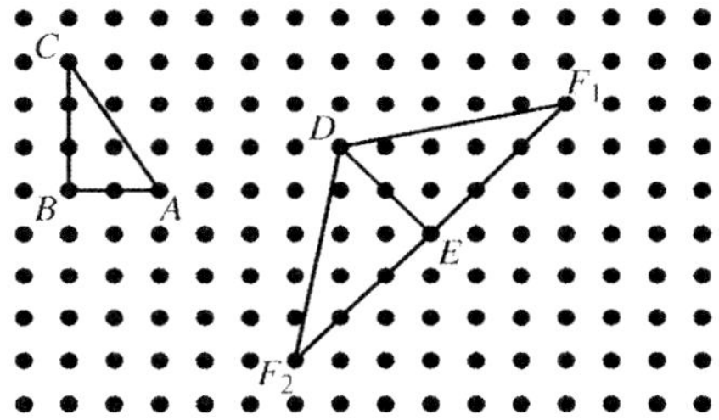

13. $\angle AEB \cong \angle DEC$ since they are vertical angles. Also, since $\overline{AB}$ and $\overline{CD}$ are parallel, $\angle BAE \cong \angle CDE$ by the alternate interior angles theorem. By the AA similarity property, $\triangle ABE \sim \triangle DCE$, so $\dfrac{AE}{DE} = \dfrac{BE}{CE}$.
Therefore, $\dfrac{a}{b} = \dfrac{x}{y}$ or $a \cdot y = x \cdot b$.

15. (a) $m(\angle A) + m(\angle B) + m(\angle C) = 180°$, and $m(\angle B) = m(\angle C)$ since the triangle is isosceles. Therefore, $m(C) = 72°$. In $\triangle BCD$, $\angle C \cong \angle CDB$, so $m(\angle CBD) = 180° - 72° - 72° = 36°$. Thus $\triangle BCD$ is a golden triangle.

Copyright © 2015 Pearson Education, Inc.

128 *Chapter 12 Congruence, Constructions, and Similarity*

(b) Answers will vary. Two possible answers are:

17. All three angles are congruent, so triangle ABC is equilateral.

19. Answers will vary.

21. This would be a good opportunity to use a Venn diagram to show the attributes of similar and congruent shapes. Provide examples for Jackson to sort, and have him justify his choices using the relationships common to similar or congruent figures. In particular, all congruent triangles are also similar triangles with a scale factor of 1. However, similar triangles are not congruent if their scale factor is other than 1.

23. Reef's reasoning is correct, and explains why addition won't generally give similar triangles (the only exception is an equilateral triangle). Miley needs to understand that multiplication is the key: when each side is be multiplied by the same number, namely the scale factor, the two triangles are similar.

25. (a) $\angle ACB \cong \angle CDA$ since they are right angles. $\angle BAC \cong \angle CAD$ since they are the same angle. $\triangle ABC \sim \triangle ACD$ by the AA similarity property.
$\angle BCA \cong \angle BDC$ since they are right angles. $\angle ABC \cong \angle CBD$ since they are the same angle. $\triangle ABC \sim \triangle CBD$ by the AA similarity property.

(b) Using $\triangle ACD \sim \triangle ABC$, then $\dfrac{AD}{AC} = \dfrac{AC}{AB}$

or $\dfrac{x}{b} = \dfrac{b}{c}$. Similarly, $\triangle CBD \sim \triangle ABC$

gives $\dfrac{BD}{BC} = \dfrac{CB}{AB}$ or $\dfrac{y}{a} = \dfrac{a}{c}$.

(c) $x = \dfrac{b^2}{c}$ and $y = \dfrac{a^2}{c}$. Also, $x + y = c$, so

$$c = \frac{b^2}{c} + \frac{a^2}{c}, \text{ or } c^2 = a^2 + b^2.$$

27. (a) $DJ = \sqrt{1^2 + \left(\dfrac{1}{2}\right)^2} = \sqrt{\dfrac{5}{4}} = \dfrac{1}{2}\sqrt{5}$

(b) By similarity, $\dfrac{PS}{AD} = \dfrac{TS}{JD}$. Therefore, using part (a),

$$PS = AD \cdot \frac{TS}{JD} = 1 \cdot \frac{\frac{1}{2}}{\frac{\sqrt{5}}{2}} = \frac{1}{\sqrt{5}} = \frac{\sqrt{5}}{5}.$$

(c) Area $(PQRS) = (PS)^2 = \dfrac{1}{5}$. That is, the inner small square has 20% of the area of the large square $ABCD$.

29. The lines perpendicular to the sides $\overline{AB}$ and $\overline{AC}$ cross at an angle congruent to $\angle A$. That is, $\angle P \cong \angle A$. Similarly, $\angle Q \cong \angle B$, so $\triangle PQR \sim \triangle ABC$ by the AA similarity property.

31. As corresponding angles to parallel lines, $\angle EPD \cong \angle ABD$. Since $\angle EDP = \angle ADB$ it follows from the AA similarity property that $\triangle EPD \sim \triangle ABD$. Thus $\dfrac{x}{a} = \dfrac{p}{p+q}$. A similar argument shows that $\triangle AEP \sim \triangle ADC$, so that $\dfrac{x}{b} = \dfrac{q}{p+q}$. Therefore, adding these proportions gives $\dfrac{x}{a} + \dfrac{x}{b} = \dfrac{p+q}{p+q} = 1$, so $x = \dfrac{1}{\frac{1}{a} + \frac{1}{b}} = \dfrac{ab}{a+b}$. The same procedures show that $y = \dfrac{ab}{a+b}$.

Copyright © 2015 Pearson Education, Inc.

Solutions to Chapter 12 Review Exercises **129**

33. $\angle APC \cong \angle BPD$ since they are vertical angles. $\angle PDB \cong \angle PCA$ since they are right angles. $\triangle ACP \sim \triangle BDP$ by the AA similarity property. Since $\dfrac{AC}{BD} = \dfrac{4}{2} = 2$, the scale factor is 2. Therefore, $CP = 2DP$ and $AP = 2BP$. Since $CD = 4$ and $CD = CP + DP = 3DP$, $DP = \dfrac{4}{3}$, and $CP = \dfrac{8}{3}$. Using the Pythagorean theorem,

$$BP = \sqrt{2^2 + \left(\dfrac{4}{3}\right)^2} = \sqrt{4 + \dfrac{16}{9}} = \sqrt{\dfrac{52}{9}} = \dfrac{2}{3}\sqrt{13}$$

and $AP = 2BP = \dfrac{4}{3}\sqrt{13}$.

35. If Mingxi and the tree are both vertical on flat (not necessarily level) ground, the two triangles are similar by the AA property. Therefore, the height h of the tree satisfies $\dfrac{h}{5.75'} = \dfrac{(75+7)'}{7'}$. Solving, we have

$$h = \dfrac{(5.75')82'}{7'} \approx 67.36' \approx 67'4''. \qquad \textbf{37.}$$

By similar triangles $\dfrac{h}{3.5''} = \dfrac{2640'}{2'}$. Thus the cliff has height

$$h = 3.5'' \cdot \dfrac{2640'}{2'} = 4620'' = 385'.$$

39. In the figure shown, G is the general's eye, and his staff hangs vertically downward from point A. Then, $\triangle GAB \sim \triangle FDB$ and $\triangle GAC \sim \triangle EDC$. Since $AG = 2$ ft, $AB = 1$ ft, and $DB = 44$ ft, it follows that $DF = 44 \cdot 2 = 88$ ft.
In the same way, since $AG = 2$ ft, $AC = 3$ ft, and $DC = 42$ ft, it follows that

$DE = 42 \cdot \dfrac{2}{3} = 28$ ft. Therefore, the river is

$DF - DE = 88 - 28 = 60$ ft wide.

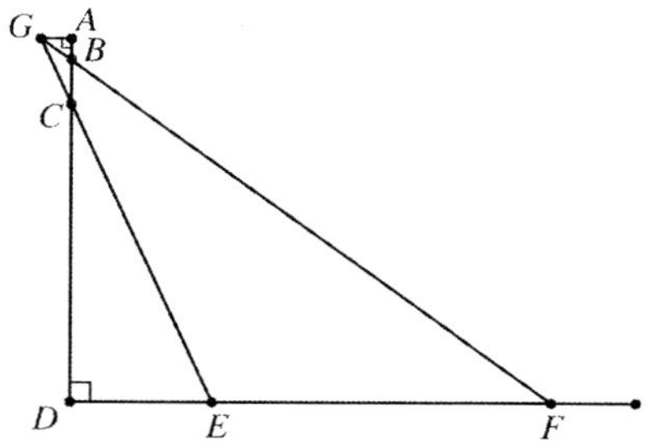

41. C.

43. C.

$$\dfrac{XZ}{PR} = \dfrac{YZ}{QR} \Rightarrow \dfrac{XZ}{6} = \dfrac{12}{4} \Rightarrow$$
$$XZ = 6\left(\dfrac{12}{4}\right) = 18 \text{ cm}$$

Chapter 12 Review Exercises

1. **(a)** True. $\angle A \cong \angle D$, $\angle B \cong \angle E$, $\angle C \cong \angle F$, $\overline{AB} \cong \overline{DF}$, and $\overline{BC} \cong \overline{DE}$.

   (b) False. After pairing up the congruent angles, the sides with equal lengths are not corresponding sides in the triangles, so the two triangles are not congruent.

3. $m(\angle B) = m(\angle G) = 62º$, $AB = FG = 2.1$ cm, and $BC = GH = 3.2$ cm, so $\triangle ABC \cong \triangle FGH$ by SAS.

   (a) $AC = FH = 2.9$ cm

   (b) $m(\angle H) = 180º - m(\angle F) - m(\angle G)$
   $= 180º - 78º - 62º = 40º$

   (c) $m(\angle A) = m(\angle F) = 78º$

   (d) $m(\angle C) = m(\angle H) = 40º$

5. $\angle B \cong \angle C$ by the isosceles triangle theorem. By construction, $BF = CD$ and $BD = CE$. Therefore, $\triangle BDF \cong \triangle CED$ by SAS, so $DE = DF$.

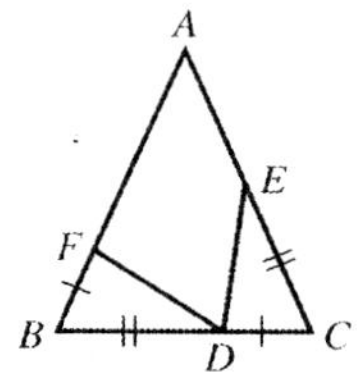

7. **(a)** See construction 7 of Section 12.2.

   (b) See construction 4 of Section 12.2

   (c) See construction 6 of Section 12.2.

   (d) Draw a circle at any point A on line m, and let it intersect line l at B and C. Construct $\overline{AB}$ and $\overline{BC}$. Draw circles of the same radius at B and C to determine the respective midpoints M and N of $\overline{AB}$ and $\overline{AC}$. (See construction 6 of Section 12.2) Then $k = \overleftrightarrow{MN}$ is the desired line.

   (continued on next page)

Copyright © 2015 Pearson Education, Inc.

130 *Chapter 12 Congruence, Constructions, and Similarity*

(*continued*)

Alternatively, construct a line perpendicular to l at a point on l. This determines a perpendicular segment between l and m. The perpendicular bisector of the segment is the desired line k.

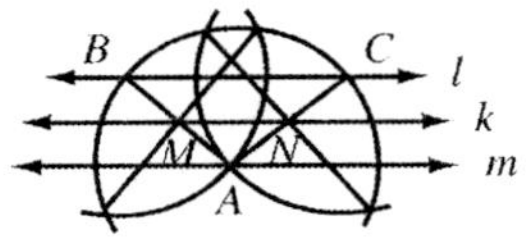

9. **(a)** Construct $\angle A$, mark off length AB, and draw a circle at B of radius BC. The circle intersects the other ray from A at two points, C_1 and C_2, giving two triangles $\triangle ABC_1$ and $\triangle ABC_2$.

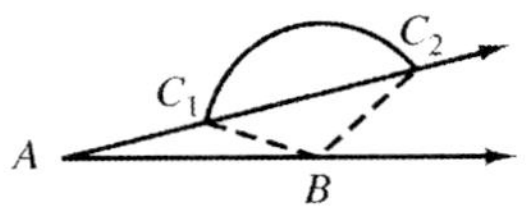

(b) Only $\triangle ABC_1$ has $\angle C = \angle C_1$ obtuse.

11. **(a)** Yes; measure the side lengths and see if the SSS similarity property applies.

(b) Yes; measure the angles and see if the AA similarity property applies.

13. **(a)** $\angle A \cong \angle Q$ and $\dfrac{QP}{AB} = \dfrac{QR}{AC} = \dfrac{3}{2}$, so $\triangle ABC \sim \triangle QPR$ by the SAS similarity property. The scale factor is $\dfrac{3}{2}$.

(b) $\dfrac{XZ}{CB} = \dfrac{XY}{CA} = \dfrac{YZ}{AB} = 3$, so $\triangle ABC \sim \triangle YZX$ by the SSS similarity property. The scale factor is 3.

(c) $m(\angle F) = m(\angle H) = 70°$ by using the isosceles triangle theorem. $\triangle ABC \sim \triangle HGF$ by the AA similarity property. The scale factor is $\dfrac{HG}{AB} = \dfrac{4}{6} = \dfrac{2}{3}$.

(d) $\dfrac{BD}{AB} = \dfrac{DC}{BD} = \dfrac{BC}{AD} = \dfrac{1}{2}$. $\triangle ABD \sim \triangle BDC$ by the SSS similarity property. The scale factor is $\dfrac{1}{2}$.

15. Let K be the top of the stick and Y the top of the pyramid. Then $\triangle KS_1S_2 \sim \triangle YP_1P_2$ with scale factor $\dfrac{P_1P_2}{S_1S_2} = \dfrac{270}{2} = 135$. Therefore the height of the pyramid is 135×3 feet $= 405$ ft.

Copyright © 2015 Pearson Education, Inc.

Chapter 13   Statistics: The Interpretation of Data

Section 13.1
Organizing and Representing Data

Problem Set 13.1

1. **(a)** Any score in the 70's could be considered typical or average. However, it is hard to tell with the data all mixed up.

   (b)

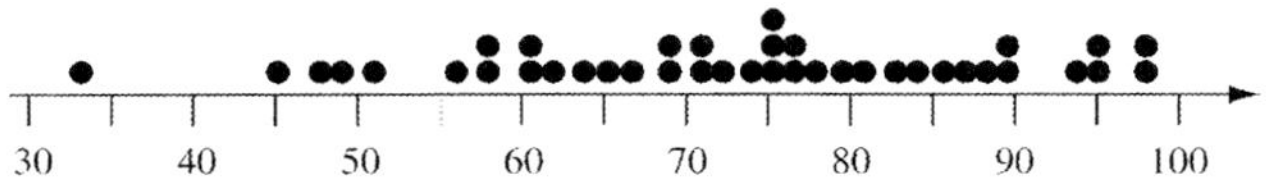

   (c) 75 is a typical score.

   (d) The score of 33 is more than 10 points below the next lowest score and therefore seems to be quite atypical.

   (e) Most of the scores are fairly evenly distributed between 55 and 89 with a small peak at 75. Those scoring above 90 did exceptionally well and should receive an A. Those scoring below 60 should receive an F.

3. Follow the example given in Figure 13.2. Let the tens digits of the scores be the stems and let the units digits be the leaves.

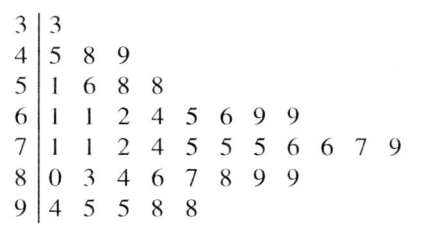

   ```
   3 | 3
   4 | 5 8 9
   5 | 1 6 8 8
   6 | 1 1 2 4 5 6 9 9
   7 | 1 1 2 4 5 5 5 6 6 7 9
   8 | 0 3 4 6 7 8 9 9
   9 | 4 5 5 8 8
   ```

5.

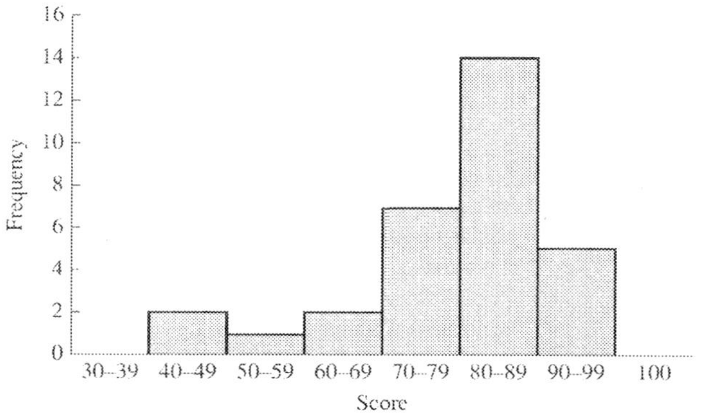

7. **(a)**

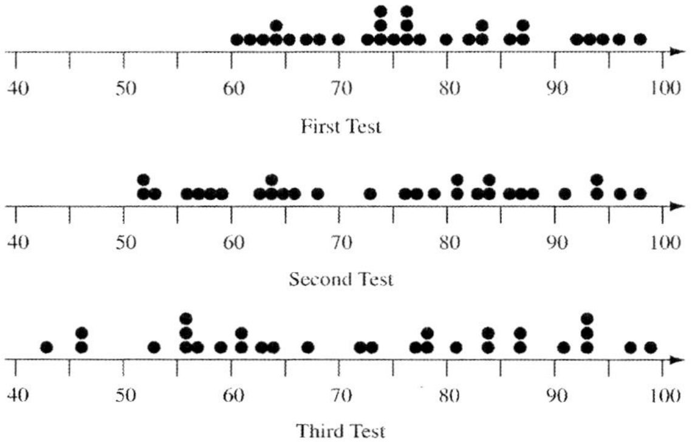

   First Test

   Second Test

   Third Test

   (b) All of the students did at least reasonably well on the first test–probably in part because not too much background information was required. However, the test scores indicate that the poorer students did increasingly less well as the term progressed, creating an increasingly bimodal distribution of grades. This is not surprising since mathematics is a sequential study where current success depends heavily on understanding earlier material.

9. **(a)** 30% of 360° is 108°.

   (b) 36% of 360° is 129.6°.

11. **(a)** 22% of 360° is approximately 79°.

    (b) 142° is approximately 39% of 360°.

Copyright © 2015 Pearson Education, Inc.   131

132 *Chapter 13 Statistics: The Interpretation of Data*

(c) Since $\dfrac{12\%}{39\%} \approx \dfrac{1}{3}$, the city's expenditure

for maintenance is about $\dfrac{1}{3}$ as much as its expenditure for police.

(d) The city spends about the same amount for Administration and Fire protection, since the sections of the pie are approximately equal.

13. (a)–(c) Answers will vary. Possible answer: 40 white, 100 black, 17 red, 29 gray, 57 green, 257 other. The bar graph should represent individual data. Based on this

data, one would expect 8% $\left(\dfrac{40}{500} = 0.08 \right)$

or 200(0.08) = 16 to be white.

(d) This is not possible to answer. If the street corner is close to the original one and there is a lot of traffic on the street, the logic of part (c) might be ok. However, if you are in a different city, if the street is not very busy, if the corner is in a different area from the original street corner, or if you are at the same corner but at a different time of day, there most likely would be no correlation at all.

15. (a)

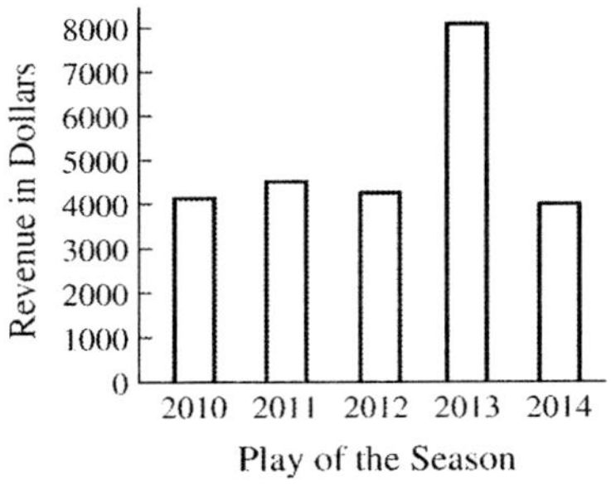

Play of the Season

(b) Since student interest may not correlate with revenue, the answer is "no." For example, we can't determine the number of students who bought tickets nor whether the price of a ticket changed in different years.

17.

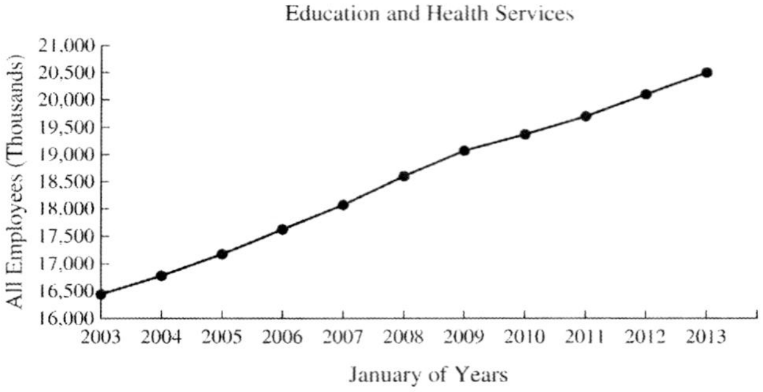

January of Years

19. (a) Answers will vary.

(b) Bar graphs should represent individual data.

(c) Answers will vary.

(d) Answers will vary.

21. (a) Since there are 25 pine trees, and

$\dfrac{25}{10} = 2.5$, there should be two and one half icons.

(b)

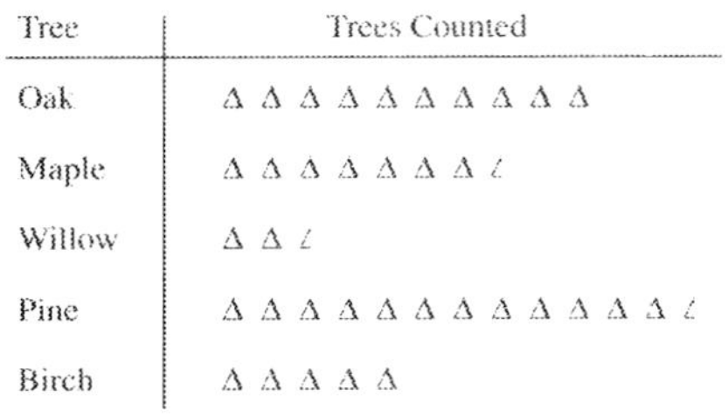

Tree	Trees Counted
Oak	⩜ ⩜ ⩜ ⩜ ⩜ ⩜ ⩜ ⩜ ⩜
Maple	⩜ ⩜ ⩜ ⩜ ⩜ ⩜ ⟋
Willow	⩜ ⩜ ⟋
Pine	⩜ ⩜ ⩜ ⩜ ⩜ ⩜ ⩜ ⩜ ⩜ ⩜ ⩜ ⟋
Birch	⩜ ⩜ ⩜ ⩜ ⩜

23. (a) Sarita is not using the key on the pictograph. She is assigning a value of one book to each picture instead of the value of two books. She arrived at the answer that Fred has read four more books than James by finding the person who had four more squares than James. For the second question, Sarita is correct that Fred read the most books. However, she assigned a value of one to each picture instead of using the key to interpret the graph.

(b) For question 1, Kristen has read exactly four more books than James. For question 2, Fred read the most books, ten.

Copyright © 2015 Pearson Education, Inc.

(c) Answers will vary. One possible answer is to have Sarita write the number 2 inside each square so she remembers the value of each picture. Another strategy is for her to focus on the key and the write down the total for each person before ever interpreting the question.

(b) Jeremy is correct in realizing that Sunday and Friday have the same amount, 15. However, the question asks what two days together equal Friday. He needs to be shown how to guess and check each pair of days until he finds the combination that sums to 15 (Monday and Tuesday). He also needs to be coached to read the questions more carefully.

25. A possible answer is below. There could be others.

Comparing Histograms and Bar Graphs

HISTOGRAMS	BAR GRAPHS
Organizes and summarizes data	Organizes and summarizes data
Data represented in intervals of numeric data	Usually displays categorical data
Vertical axis indicates frequency	Vertical axis indicates frequency
Usually "continuous data" adjacent to each other (bars must touch)	Can be bars that are separated (do not abut)

27. (a) The pictographs are misleading since the figure for 2012 appears to be more than twice as large as the one for 2003. One would guess the assets had more than doubled.

(b) The volume for 2012 is eight times $(2 \times 2 \times 2)$ the volume for 2003, not double. This is not an accurate impression for the assets.

(c) Yes.

29. (a) BT's net income might be about $4 \cdot 63,000,000 = \$252,000,000$.

(b)

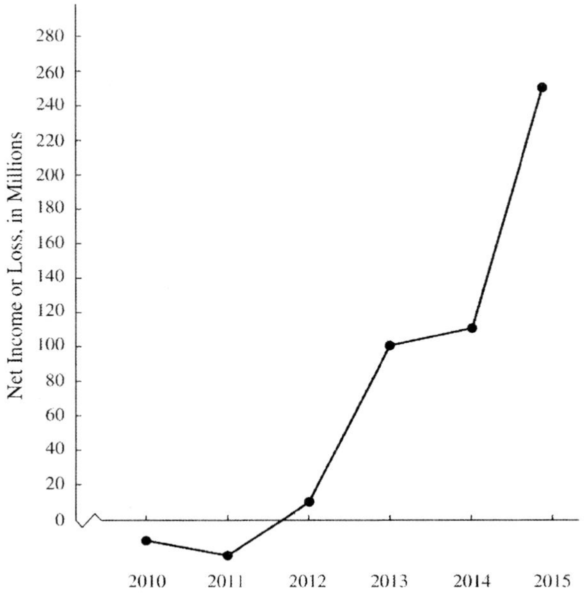

(c) About $\dfrac{105 - 5}{5} = 20 = 2000\%$

(d) About $\dfrac{252 - 100}{100} = 1.52 = 152\%$

31. (a)

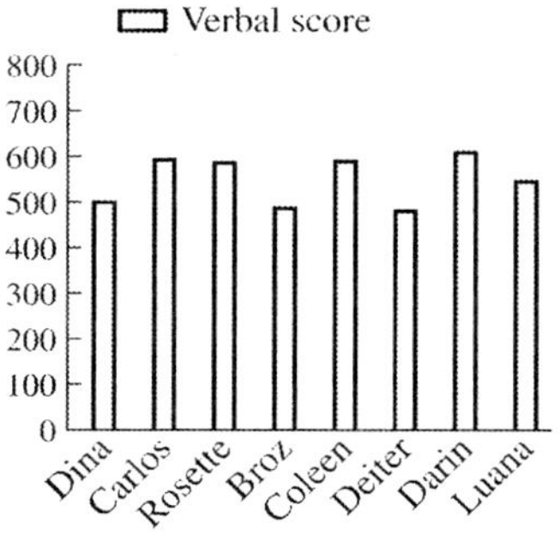

Copyright © 2015 Pearson Education, Inc.

134 *Chapter 13 Statistics: The Interpretation of Data*

(b)

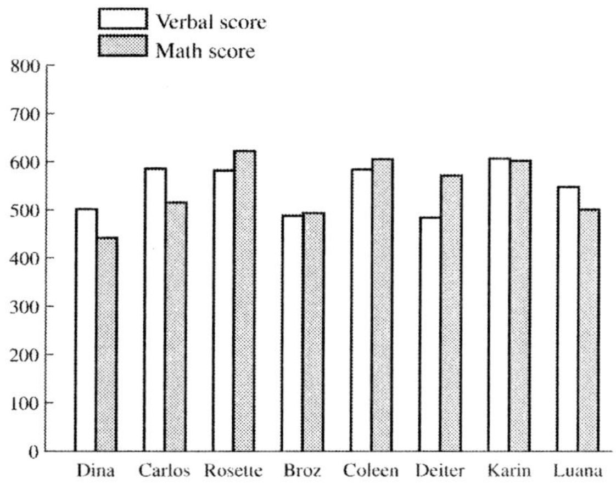

Verbal score
Math score

Bar chart with students: Dina, Carlos, Rosette, Broz, Coleen, Deiter, Karin, Luana

33.

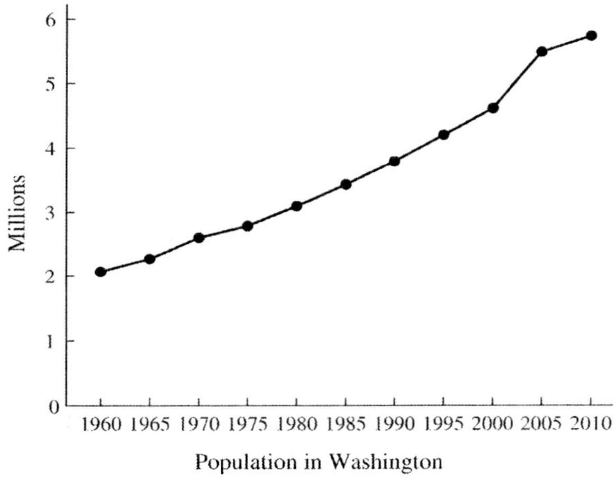

Line graph — Millions (y-axis) vs. years 1960–2010 (x-axis)

Population in Washington

35. B. $14 \div 4 = 3.5$

37. D. Recall that the mode is the value that appears most often in a set of data.

Section 13.2
Measuring the Center and Variation of Data

Problem Set 13.2

1. mean:

$$\overline{x} = \frac{\displaystyle\sum_{i=1}^{n} x_i}{n} = \frac{\displaystyle\sum_{i=1}^{16} x_i}{16}$$

$$= \frac{18 + 27 + \cdots + 27 + 30}{16} = \frac{329}{16}$$

$$= 20.5625 \doteq 20.6$$

median: arrange the values in order from smallest to largest
14 15 17 17 18 18 18 19 19 21 22 23 24 27 27 30

The median is the average of the two middle values. $\hat{x} = \frac{19 + 19}{2} = 19$

mode: The mode is the value that occurs most often. In this case, the mode is 18.

3. (a) List the values from Problem 1 in order:
14, 15, 17, 17, 18, 18, 18, 19, 19, 21, 22, 23, 24, 27, 27, 30

Q_L is the median for the first 8 values:

$$Q_L = \frac{17 + 18}{2} = 17.5$$

$\hat{x}$ is the median of the entire set: $\hat{x} = 19$.
Q_U is the median for the last 8 values:

$$Q_U = \frac{23 + 24}{2} = 23.5.$$

(b) The 5-number summary is: the smallest value $- Q_L - \hat{x} - Q_U -$ the largest value; i.e., $14 - 17.5 - 19 - 23.5 - 30$.

(c)

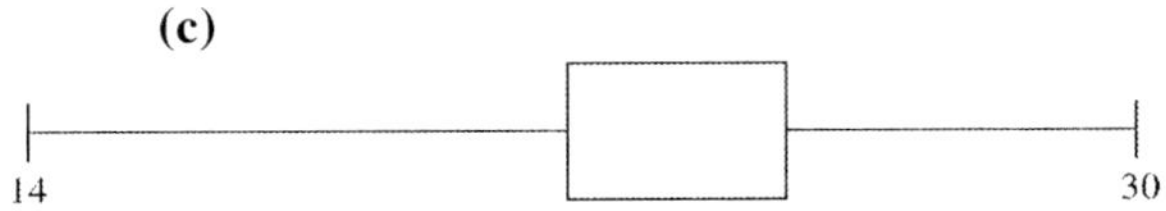

Box plot from 14 to 30

(d) $\text{IQR} = Q_U - Q_L = 23.5 - 17.5 = 6$

(e) $Q_L - 1.5 \cdot \text{IQR} = 17.5 - (1.5)6 = 8.5$
$Q_U + 1.5 \cdot \text{IQR} = 23.5 + (1.5)6 = 32.5$
Since none of the values are less than 14 or greater than 30, there are no outliers.

5. The lowest data point is 14 and the highest data point is 30, so the midrange is

$$\frac{14 + 30}{2} = 22.$$

Copyright © 2015 Pearson Education, Inc.

Solutions to Problem Set 13.2   **135**

7. (a) Class A: $63 - 68.5 - 75 - 85 - 96$
Class B: $47 - 65 - 70.5 - 88 - 95$

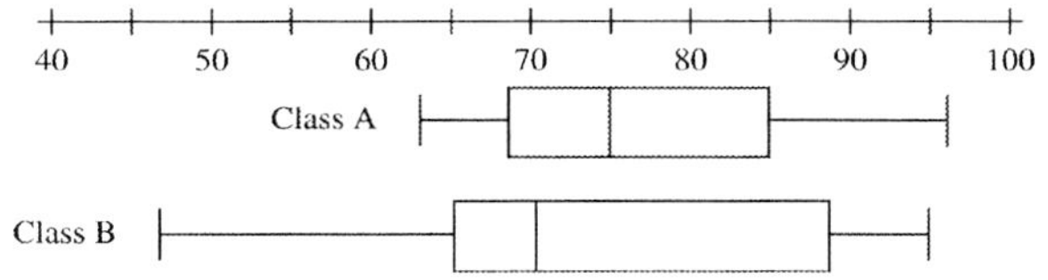

(b) Class B had a much larger range, the difference between the upper quartile and the lower quartile is larger, and the median is lower. As a whole, class A had higher final grades than class B.

(c) For class A, $\text{IQR} = Q_U - Q_L = 85 - 68.5 = 16.5$.
For class B, $\text{IQR} = 88 - 65 = 23$.

(d) For class A, $Q_L - 1.5 \cdot \text{IQR} = 68.5 - (1.5)(16.5) = 43.75$, and $Q_U + 1.5 \cdot \text{IQR} = 85 + (1.5)(16.5) = 109.75$.
Since no data values in class A are less than 43.75 and none are greater than 109.75, there are no outliers in class A. For class B, $Q_L - 1.5 \cdot \text{IQR} = 65 - (1.5)(23) = 30.5$, and $Q_U + 1.5 \cdot \text{IQR} = 88 + (1.5)(23) = 122.5$.
Since no data values in B are less than 30.5 and none are greater than 122.5, there are no outliers in class B.

9. (a) To be within one standard deviation of the mean, a data value must be in the interval $(20.56 - 4.44, 20.56 + 4.44) = (16.12, 25)$.

There are 11 out of 16 values between 16.12 and 25. $\frac{11}{16} \approx 0.69 = 69\%$

(b) To be within two standard deviations of the mean, a data value must be in the interval $(20.56 - (2) \cdot (4.44), 20.56 + (2) \cdot (4.44)) = (11.68, 29.44)$

There are 15 out of 16 values between 11.68 and 29.44. $\frac{15}{16} \approx 0.94 = 94\%$

(c) To be within three standard deviations of the mean, a data value must be in the interval $(20.56 - (3) \cdot (4.44), 20.56 + (3) \cdot (4.44)) = (7.24, 33.88)$.

All 16 values are between 7.24 and 33.88: $\frac{16}{16} = 1 = 100\%$

11. (a) The sum of the data points is $2(1) + 2(2) + 2(3) + 2(4) + 2(5) = 30$, so the mean is $30 \div 10 = 3$. Since every integer comes up the same number of times, every integer between 1 and 5 inclusive is a mode. The median is 3.

(b) The sum of the data points is $4(1) + 4(2) + 4(3) + 4(4) + 4(5) = 60$, so the mean is $60 \div 20 = 3$. Since every integer comes up the same number of times, every integer between 1 and 5 inclusive is a mode. The median is 3.

(c) The sum of the data points is $6(1) + 6(2) + 6(3) + 6(4) + 6(5) = 90$, so the mean is $90 \div 30 = 3$. Since every integer comes up the same number of times, every integer between 1 and 5 inclusive is a mode. The median is 3.

(d) Each of the data sets has the same mean, median, and modes.

13. (a) Yes, there will be five blocks in each stack.

(b) $\bar{x} = \dfrac{5 + 1 + 4 + 7 + 6 + 7}{6} = 5$

(c) The height of each of the stacks is the mean of the data set.

Copyright © 2015 Pearson Education, Inc.

136 *Chapter 13 Statistics: The Interpretation of Data*

(d) First, have the students consider stacks of equal height. Second, rearrange the stacks to be of unequal heights. In each case, discuss the notion of typical, or average, height.

(e) The 5, 1, 4, 7, 11, and 7 blocks cannot be arranged into six stacks with an equal number of blocks in each stack. The best that can be done is to have five stacks with 6 blocks each and one stack with five blocks. This suggests that the average is somewhat less than 6. Here it might be better to think of small pizzas. These could be divided equally among six students by giving each student five and five sixths pizzas and 5 5/6 = for the data 5, 1, 4, 7, 11, and 7.

15. Answers will vary. When we did the activity, we concluded that the female "cubit" was about 17.5 inches and the male "cubit" was about 19.5 inches.

17. (a) Instead of finding the median for the number of students, Joseph found the median for the number of books reported. When calculating the mode, Joseph looked at the most number of books possible to report, not the most number of students who had read, 8. The mode is 1, and the median is 2.5.

(b) Explain the difference between the label (number of books reported) and the outcome (number of students who read that many books).

19. Presumably, Yugi is crossing off numbers alternately from each side. In that case, Yugi's approach works exactly when there are an odd number of data points. If there is an even number of data points, he must take the average of the last two numbers. Showing him an example using a data set of 4 points would clarify the situation.

21. Marilyn is making a mistake talking about "the mode." There can be more than one mode in a data set. For example, the data set {1, 1, 2, 2, 5} has two modes. She should be using the expression "a mode" unless she is talking about a specific data set.

23. (a) First calculate the mean of the data values. Do this by finding the sum of all the data values, and then dividing the sum by the total number of data values. Thus, the mean is

$$\frac{22.2 + 23.5 + 22.5 + 22.6 + 23.0 + 22.8 + 22.4 + 22.2 + 23.0 + 23.3 + 23.9 + 22.7}{12} \approx 22.8.$$

Therefore, the sum of the absolute values of the differences of the data values from the mean is

$$|22.2 - 22.8| + |23.5 - 22.8| + \cdots + |22.7 - 22.8| \approx 4.9.$$

(b) We compute the mean or average value of the sum of the absolute values by dividing the result of part (a) by the number of data values, or 12. By doing so, we obtain $\frac{4.9}{12} \approx 0.41$. This number grows with greater variability, so could be used as a measure of variability.

(c) Compliment Leona on a good idea that avoids the canceling out of the effects of the various terms and note that her approach has an effect that is similar to the effect produced by squaring the terms in computing the standard deviation since squares are always positive. Here the standard deviation is 0.53, roughly the same as the 0.4 of part (b). The real reason for using the standard deviation is that it possesses useful theoretical properties that are too involved to discuss here while, at the same time, giving a reasonable measure of variability.

25. (a) mean: $\bar{x} = \dfrac{\sum\limits_{i=1}^{n} x_i}{n} = \dfrac{28 + 34 + 41 + 19 + 17 + 23}{6} = \dfrac{162}{6} = 27$

standard deviation $s = \sqrt{\dfrac{1 + 49 + 196 + 64 + 100 + 16}{6}} = \sqrt{\dfrac{426}{6}} \approx 8.4$

Copyright © 2015 Pearson Education, Inc.

Solutions to Problem Set 13.2   137

(b) mean:
$$\bar{x} = \frac{33 + 39 + 46 + 24 + 22 + 28}{6} = \frac{192}{6} = 32$$

standard deviation:
$$s = \sqrt{\frac{1 + 49 + 196 + 64 + 100 + 16}{6}} = \sqrt{\frac{426}{6}} \approx 8.4$$

(c) Adding (or subtracting) the same number from each one of a set of data leaves standard deviation unchanged, and the mean is increased (or decreased) by that number.

27. $\bar{x} = 37.4$, $\tilde{x} = 42$, mode = 16

```
  • •   • • • • •                              •
├──┼──┼──┼──┼──┼──┼──┼──┼──┤
15 20 25 30 35 40 45 50
```

There are 2 values below and 7 values above the mean so the mean of 37.4 is not a typical value. The mode of 16 is far below most of the data values so it is not a typical value. The median of 42 is centered in the group of values between 38 and 48 so it is a likely choice for the typical value for this set of data.

29. Answers will vary.

(a) Consider the condition: mean = median. Suppose $\bar{x} = \tilde{x} = 10$. Now choose 2 values below $\bar{x}$, say 5 and 7. But we must also choose 2 values above $\bar{x}$ so that $\frac{\sum x_i}{n} = \bar{x} = 10$, say 13 and 15. The condition mean = median has been met with the data: 5, 7, 10, 13, 15.
Consider the condition: mean < median < mode.
Using the data above, there must be 2 equal values larger than 10 to meet this condition. If 13 and 15 are replaced with their average of 14, then mean = median < mode for the data: 5, 7, 10, 14, 14. For these values, $\bar{x} = 10$, $\tilde{x} = 10$, and the mode is 14.

(b) 0, 10, 10, 11, 12, 13, 14; $\bar{x} = 10$, $\tilde{x} = 11$,
mode = 10

(c) 8, 9, 10, 11, 20;
$\bar{x} = 11\frac{1}{3}$, $\tilde{x} = 10$, mode = 10

31. (a) Given $n = 10$ and $\bar{x} = 3$, then $\sum x = n \cdot \bar{x} = 10 \cdot 3 = 30$. Since $\sum x = 30$ for 10 values of ones, twos, and threes, the 10 values must all be 3s.

(b)

1s	2s	3s
5	0	5
4	2	4
3	4	3	Possibilities
2	6	2
1	8	1
0	10	0

(c) Given $n = 10$ and $\bar{x} = 1$, then $\sum x = n \cdot \bar{x} = 10 \cdot 1 = 10$. Since $\sum x = 10$ for 10 values of ones, twos, and threes, the 10 values must all be 1s.

(d) No, for $\bar{x} = 1$, all 10 values must be 1 so $s = 0$.

33. (a) S has the smaller standard deviation. $\sum_{i=1}^{n}(45 - x_i)^2$ is the same for sets R and S but for set R, $n = 7$ and for set S, $n = 9$.

(b) No, $\sum_{i=1}^{9}(45 - x_i)^2$ is greater for set T than for set S because the data values of 45 and 45 have been replaced by 80 and 10, which are farther from the mean. Therefore, the standard deviation for set S is smaller.

35. The total of data values in A is $30 \cdot 45 = 1350$.
The total of data values in B is $40 \cdot 65 = 2600$.
For the combined data,
$$\bar{x} = \frac{1350 + 2600}{30 + 40} \approx 56.4.$$

37. B. There are 15 data points. Since the middle data point is at 15, the median length is 15.

138 Chapter 13 Statistics: The Interpretation of Data

39. B.
$$102,210(45.75) \doteq 100,000(45)$$
$$= 4,500,000$$
The actual answer is
$$102,210(45.75) = 4,676,107.50$$

41. A   **43.** A.

Section 13.3
Statistical Inference

Problem Set 13.3

1. (a) All freshmen in U.S. colleges and universities in 2010
   (b) All football players in U.S. colleges and universities in 2010
   (c) All people in the U.S.
   (d) All people in Los Angeles
   (e) It would be inappropriate to include known criminals and individuals in mental institutions in the population of parts (c) and (d). Likely young children and perhaps even teenagers should not be included in the population, although a judgment call—depending on the point of the study—would have to be made at this point.

3. Probably not. Only those who are listed in the phone book can be selected for an interview, so people without a phone and people with unlisted numbers are not included. Also, face-to-face interviews can create problems such as nonresponse, lying, and interviewer bias.

5. (a) This is surely a poor sampling procedure. The sample is clearly not random. The selection of the colleges or universities could easily reflect biases of the investigators. The choices of the faculty to be included in the study almost surely also reflects the bias of the administrators of the chosen schools.
   (b) Presumably the population is all college and university faculty. But the opinions of faculty at large research universities are surely vastly different from those of their colleagues at small liberal arts colleges. Indeed, there are almost surely four distinct populations here.

7. A random sample taken from a list of all voters in his district would be representative of his constituents. If the voters are numbered, he can use a random-number generator. Otherwise, since voter lists are essentially random, he might choose every nth person on the list of his constituents, where n is chosen large enough to guarantee a sample of reasonable size.

9. Some possible answers are as follows:
   (a) Using a numbered list of all registered students, select 20 valid numbers from a random number table that corresponds to the list of students.
   (b) The mean of the answers obtained should give a reasonable estimate.
   (c) The mean of all the means obtained should give a better estimate than that of part (b).

11.
    ```
       |-s-|-s-|-s-|-s-|-s-|-s-|
      16.4 19.1 21.8 24.5 27.2 29.9 32.6
                    x̄
    ```
    (a) $\bar{x} \pm \sigma$
       The limit 2.7 units less than 24.5 is 21.8.
       The limit 2.7 units more than 24.5 is 27.2.
    (b) $\bar{x} \pm 2\sigma$; between 19.1 and 29.9
    (c) $\bar{x} \pm 3\sigma$; between 16.4 and 32.6

13. Yes. Just continue taking samples until one finally shows up with eight out of the ten in the sample preferring White toothpaste. As long as the population of dentists contains at least 8 dentists who prefer White Toothpaste, eventually a sample will contain those 8 out of 10.

15. (a) $-1.6 + 0.9 + 0.4 + (-0.6) + 1.4 + (-0.6)$
       $= -0.1$
    (b) $-1.5 + 0.9 + (-1.0) + 0.4 + (-0.1) + 1.3$
       $= 0$

17. (a) A z-score of -1.75 corresponds to an area of $0.0401 = 0.04 = 4\%$ to the nearest hundredth.
    (b) A z-score of 0.26 corresponds to an area of $0.6026 = 0.60 = 60\%$ to the nearest hundredth.

Copyright © 2015 Pearson Education, Inc.

Solutions to Chapter 13 Review Exercises   139

19. Praise her. For a random sample each possible outcome must have an equal chance of occurring and this is certainly so when tossing a single die.

21. Rebecca is right if none of the two-digit numbers is to contain 0, 7, 8, or 9 as a digit. Otherwise, she is wrong. Let the red die generate the first digit of a two-digit number and let the green die generate the second digit. Since what occurs on the red die has no effect on what occurs on the green die and vice versa, each of the two-digit numbers 11, 12, 13, 14, 15, 16, 21, 22, 23, 24, 25, 26, 31, 32, 33, 34, 35, 36, 41, 42, 43, 44, 45, 46, 51, 52, 53, 54, 55, 56, 61, 62, 63, 64, 65, 66 has an equal chance of occurring. Of course, these are the only numbers that can appear in the generated sequence.

23. Yes, since all sides of the die are equally likely to come up, all sequences of 0s and 1s are equally likely to appear.

25. No. The statistic is based on voluntary responses which may or may not represent the opinion of all high school biology teachers. Those mailing back the survey are those who feel strongly about the issue—most likely, those with strong religious beliefs who believe in creationism, and those with strongly anti-religious views. Those with moderate views would probably be underrepresented.

27. (a) Since the z score indicates the location of a data point in a data set, in this case they are likely to be the same as in problem 14(a).

(b) The z scores, $z = \dfrac{x - \overline{x}}{s}$, are the same: $-1.6, 0.9, 0.4, -0.6, 1.4, -0.6$

29. (a) Probably twice what they were in problem 14(a).

(b) The z-scores, $z = \dfrac{x - \overline{x}}{s}$, are the same: $-1.6, 0.9, 0.4, -0.6, 1.4, -0.6$

(c) The data are all twice as big, but s is as well. Since the z scores are scaled by s, it is not surprising that they remain the same.

31. Compliment Mai Ling on good thinking and suggest that there must be a better way to generate the digits. One can use the random number generating software on a computer or the program RANDOM on a graphing calculator.

Chapter 13 Review Exercises

1. (a)

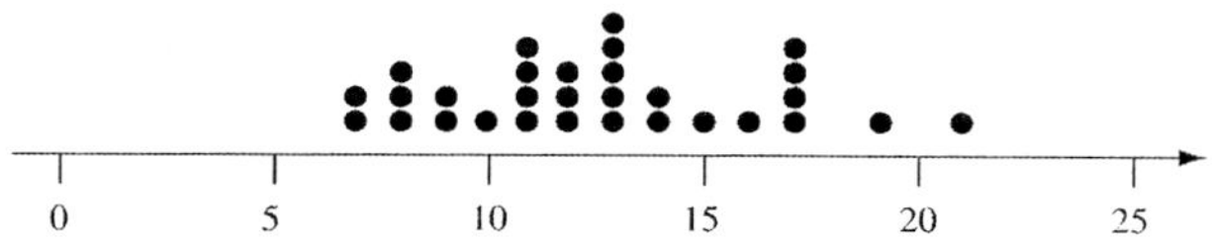

(b) About 13 hours per week.

3. Use a bar for each number of hours.

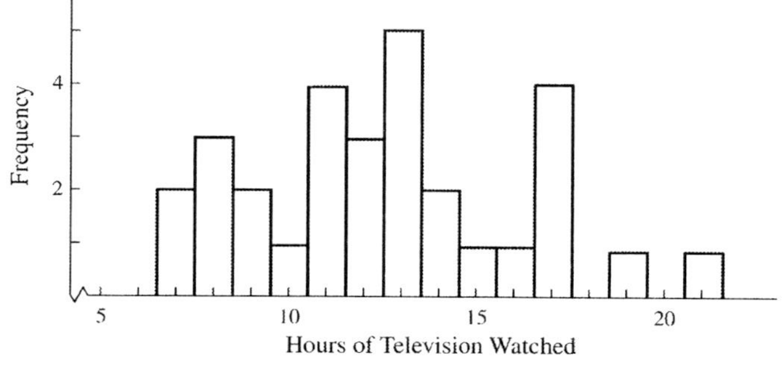

140 *Chapter 13 Statistics: The Interpretation of Data*

5. (a)

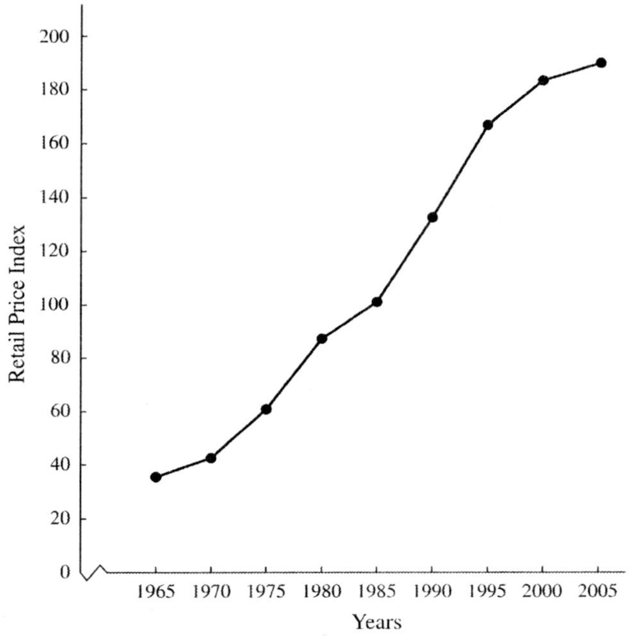

(b) Draw a line straight up from the year 1972 on the horizontal axis. Now draw a horizontal line from the point on the line graph above 1972 to the vertical axis. The retail price index for farm products in 1972 was about 48.

(c) Assuming the retail prices index continues the same upward trend as that from 2000 to 2005, the retail price index for farm products in 2010 will be about 200.

7. Find the central angle for each section of the budget.
Administration: 12% of 360° = 43.2°
New construction: 36% of 360° = 129.6°
Repairs: 48% of 360° = 172.8°
Miscellaneous: 4% of 360° = 14.4°

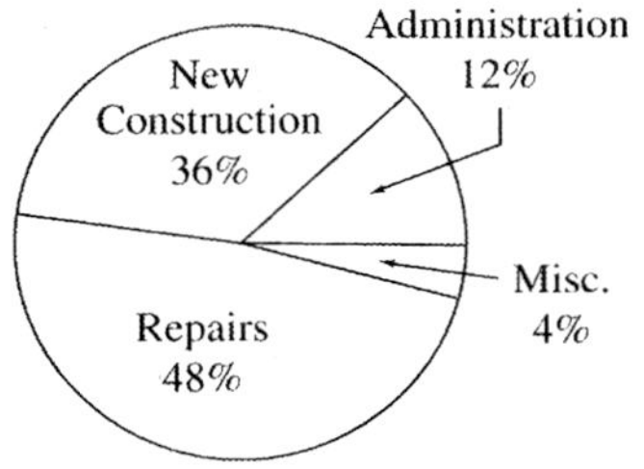

9. How is a "medical doctor" defined? Does this include all specialists? osteopaths? naturopaths? chiropractors? acupuncturists? How was the sampling done to determine the stated average?

11. (a) Q_L is the median of the first 15 ordered values, the 8th value.
$Q_L = 10$
Q_U is the median of the last 15 ordered values, the 23rd value.
$Q_U = 15$

(b) 5-number summary;
minimum $- Q_L - \hat{x} - Q_U -$ maximum
For Problem 1 data:
7 – 10 – 12.5 – 15 – 21

(c) $IQR = Q_U - Q_L = 15 - 10 = 5$
$Q_L - 1.5\ IQR = 10 - 1.5(5) = 2.5$
$Q_U + 1.5\ IQR = 15 + 1.5(5) = 22.5$
Since there are no data points less than 2.5 or greater than 22.5, there are no outliers.

(d) Q_L is the median for the first 12 ordered values, the average of the 6th and 7th values.
$$Q_L = \frac{8+8}{2} = 8$$
Median is the average of the 12th and 13th ordered values.
$$\hat{x} = \frac{9+9}{2} = 9$$
Q_U is the median for the last 12 ordered values, the average of the 18th and 19th values.
$$Q_U = \frac{11+11}{2} = 11$$

Copyright © 2015 Pearson Education, Inc.

(e) 5-number summary for the Problem 4
data: $6 - 8 - 9 - 11 - 13$

(f) $\text{IQR} = Q_U - Q_L = 11 - 8 = 3$

$Q_L - 1.5\ \text{IQR} = 8 - 1.5(3) = 3.5$

$Q_U + 1.5\ \text{IQR} = 11 + 1.5(3) = 15.5$

Since there are no data points less than
3.5 or greater than 15.5, there are no
outliers.

(g)

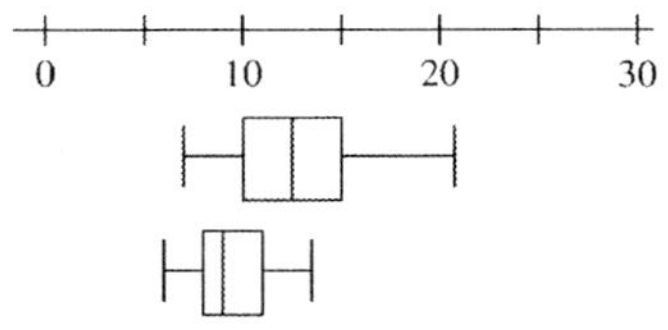

13. There are $21 \cdot 77 = 1617$ points for the 21
students. So there are $1617 + 69 + 62 + 91 = 1839$ points for all 24 students. Thus, the
average is $1839 \div 24 \approx 76.6$.

15. No. The sample only represents the population
of students at State University, not university
students nationwide.

17. Voluntary responses to mailed questionnaires
tend to come primarily from those who feel
strongly (either positively or negatively) about
an issue or who represent narrow special
interest groups. They are rarely representative
of the population as a whole.

19. Since 12 is the 6th largest number in the data
set, and $\dfrac{6}{7} = 0.8571$ to the nearest ten-
thousandth, 12 is the 85.71 percentile to the
nearest hundredth.

21. From Table 13.4, the entry for -0.9 is 0.1841
and the entry for 0.9 is 0.8159. Therefore, the
desired percentage is $81.59\% - 18.41\% = 63.18\%$.

Chapter 14   Probability

Section 14.1
The Basics of Probability

Problem Set 14.1

1. **(a)** There are two possible outcomes for each coin: Head (H) or Tail (T). The list of all outcomes for the experiment in order of penny, nickel, dime, and quarter are:

   HHHH THHH TTHH TTTH TTTT
          HTHH THTH TTHT
          HHTH THHT THTT
          HHHT HTTH HTTT
               HTHT
               HHTT

   (b) $P(\text{HHTT}) = \dfrac{1}{16} = 0.0625$

   (c) There are 16 possible outcomes for the experiment and six of the outcomes have two heads and two tails (center column of outcomes from (a)). Therefore,

   $$P(\text{2 heads and 2 tails}) = \frac{6}{16} = \frac{3}{8} = 0.375 .$$

   (d) $P(\text{at least one head}) = 1 - P(\text{no heads})$
   $$= 1 - P(\text{TTTT})$$
   $$= 1 - \frac{1}{16} = \frac{15}{16}$$

3. **(a, b)**

   (c) *A* and *B* are mutually exclusive since no outcomes belong to both events. Similarly, *A* and *C* are mutually exclusive. *B* and *C* are not mutually exclusive, since the outcomes (4, 5), (5, 5) and (6, 5) belong to both events.

5. There are four possibilities for the two children, BB, BG, GB, and GG. We know that the family has at least one boy. Therefore, the possibilities are BB, BG, and GB. Thus, the probability that the other child is a girl is $\dfrac{2}{3}$.

7. $S = \{B1, B2, B3, B4, B5, W1, W2, W3, W4, W5, W6, W7\}$
   There are 12 equally likely balls to randomly choose from.

   (a) Of the 12 balls to select, four are numbered 1 or 2 (B1, W1, B2, W2).
   $$P(1 \text{ or } 2) = P(1) + P(2)$$
   $$= \frac{2}{12} + \frac{2}{12} = \frac{4}{12} \approx 0.33$$

   (b) $P(5 \text{ or white})$
   $$= P(5) + P(W) - P(5 \text{ and white})$$
   $$= \frac{2}{12} + \frac{7}{12} - \frac{1}{12} = \frac{8}{12} \approx 0.67$$

   (c) $P(5 \mid \text{white}) = \dfrac{1}{7} \approx 0.14$

9. $S = \left\{ 00, 01, 02, \cdots, 98, 99 \right\}$. There are 100 equally likely 2-digit numbers possible.

   (a) The set of numbers greater than 80 is $\left\{ 81, 82, \cdots, 98, 99 \right\}$ and there are 19 such numbers. Therefore,
   $$P(\text{a number greater than } 80) = \frac{19}{100} = 0.19 .$$

   (b) $P(\text{a number less than } 10) = \dfrac{10}{100} = 0.1$

   (c) The set of numbers that are a multiple of 3 is $\left\{ 00, 3, 6, 9, \cdots, 96, 99 \right\}$ and there are $\dfrac{99}{3} = 33 + 1$ such numbers. Therefore,
   $$P(\text{a number is a multiple of } 3) = \frac{34}{100}$$
   $$= 0.34$$

   (d) $P(\text{even or} < 50)$
   $$= P(\text{even}) + P(< 50) - P(\text{even and} < 50)$$
   $$= \frac{50}{100} + \frac{50}{100} - \frac{25}{100} = \frac{75}{100} = 0.75$$

(e) The set of even numbers less than 50 is $\{00, 02, \cdots, 46, 48\}$ and there are $\dfrac{50}{2} = 25$ such numbers. Therefore,

$$P(\text{an even number less than 50}) = \dfrac{25}{100}$$
$$= 0.25$$

(f) $P(\text{even} \mid < 50) = \dfrac{25}{50} = 0.5$

11. You must select one heart from the remaining 9 hearts and 38 other cards.

Therefore, $P(\text{flush}) = \dfrac{9}{47} \approx 0.19$.

13. (a) $n(S) = 6 \times 12 = 72$

(b) Obtain 2 by rolling a 1 on each die. Obtain 18 by rolling 6 on the first die and 12 on the other. Thus,

$$P(2 \text{ or } 18) = \dfrac{2}{72}.$$

(c) There are six ways to obtain each outcome. For example 7 can be obtained as follows:

Die 1	Die 2
1	6
2	5
3	4
4	3
5	2
6	1

Thus, there are $6 \times 7 = 42$ ways to obtain the elements of the event. Therefore, the probability of rolling a number in the event $\{7, 8, 9, 10, 11, 12, 13\}$ is $\dfrac{42}{72} = \dfrac{7}{12}$.

(d) There is one way to obtain 2; there are two ways to obtain 3 and three ways to obtain 4. Thus, the probability of rolling a number less than or equal to 4 is $\dfrac{6}{72} = \dfrac{1}{12}$. Therefore, the probability of rolling a number greater than 4 is $1 - \dfrac{1}{12} = \dfrac{11}{12}$.

15. (a)

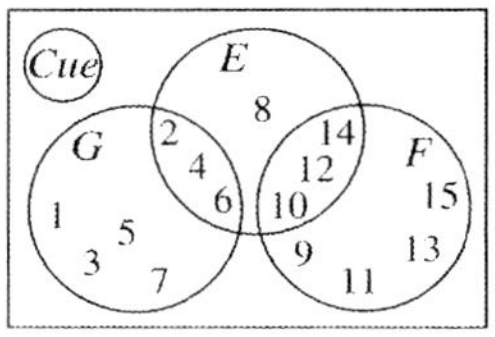

(b) $P(\text{solid or stripe}) = \dfrac{14}{16} = \dfrac{7}{8} = 0.875$

(c) $P(\text{odd}) = \dfrac{8}{16} = \dfrac{1}{2} = 0.5$

(d) $P(\text{solid or even})$
$= P(\text{solid}) + P(\text{even}) - P(\text{solid and even})$
$= \dfrac{7}{16} + \dfrac{7}{16} - \dfrac{3}{16} = \dfrac{11}{16} = 0.6875$

17. The sections 1–10 have equal areas so the central angle for each sector is $\dfrac{360°}{10} = 36°$. Sections a–e have equal areas so the central angle for each sector is $\dfrac{360°}{5} = 72°$. The probability for each of the following solutions are determined by ratios of angular measures of appropriate regions.

(a) Three of the 10 regions are shaded so

$$P(\text{shaded}) = \dfrac{3 \cdot 36°}{360°} = \dfrac{108°}{360°} = 0.3.$$

(b) $P(\text{region 1 or region 2})$
$= P(\text{region 1}) + P(\text{region 2})$
$= \dfrac{36°}{360°} + \dfrac{36°}{360°} = \dfrac{72°}{360°} = 0.2$

(c) Regions 10 and 6 represent two of the 10 regions so
$P(\text{region 10 or region 6})$
$= P(\text{region 10}) + P(\text{region 6})$
$= \dfrac{36°}{360°} + \dfrac{36°}{360°} = \dfrac{72°}{360°} = 0.2$

(d) $P(\text{region e}) = \dfrac{72°}{360°} = 0.2$

(e) Region 8 is one of three shaded areas so

$$P(\text{region 8} \mid \text{shaded area}) = \dfrac{36°}{3 \cdot 36°} = \dfrac{36°}{108°}$$
$$= \dfrac{1}{3} \approx 0.33$$

Copyright © 2015 Pearson Education, Inc.

144 *Chapter 14 Probability*

(f) $P(\text{a vowel} \mid \text{odd numbered region}) = \dfrac{2 \cdot 36°}{5 \cdot 36°} = \dfrac{72°}{180°} = 0.4$.

(g) The vowels a and e cover 4 regions: 7, 8, 9, and 10. The odd regions are 1, 3, 5, 7, and 9. Regions 7 and 9 are odd and contain a vowel. Therefore,

$$P(\text{vowel} \cup \text{odd-numbered region}) = P(\text{vowel}) + P(\text{odd-numbered region})$$
$$- P(\text{vowel and odd-numbered region})$$
$$= \frac{4 \cdot 36°}{360°} + \frac{5 \cdot 36°}{360°} - \frac{2 \cdot 36°}{360°} = \frac{144° + 180° - 72°}{360°} = \frac{252°}{360°} = 0.7$$

19. Answers will vary, but should indicate why the information is helpful. Suppose the marble drawn is white. If the initially placed marble is black, there is just one way to draw a white marble. But, if the initially placed marble is white, there are two ways to draw a white marble—either the initial or the added marble. That is, there is a 2/3 probability of drawing a white marble, in which case the initial marble in the sack was also white. Thus, if the marble drawn from the sack is black, it must be the initially placed marble with probability 1. If the drawn marble is white, the probability is 2/3 that the initial marble is white.

21. (a) Jose is incorrect, since there are 18 ways for Susan to get an odd number and only 9 ways for Bill to get a 4, 8, or 12.

(b) Answers will vary, but it will be helpful to refer to the table of the 36 equally likely outcomes.

23. The probability is slightly worse, since there is a small probability of drawing the same card twice.

25. $P(E \text{ and } F) = P(E) + P(F) - P(E \text{ or } F) = \dfrac{1}{2} + \dfrac{1}{4} - \dfrac{1}{4} = \dfrac{1}{2} = P(E)$

Therefore, $F \subseteq E$.

27.

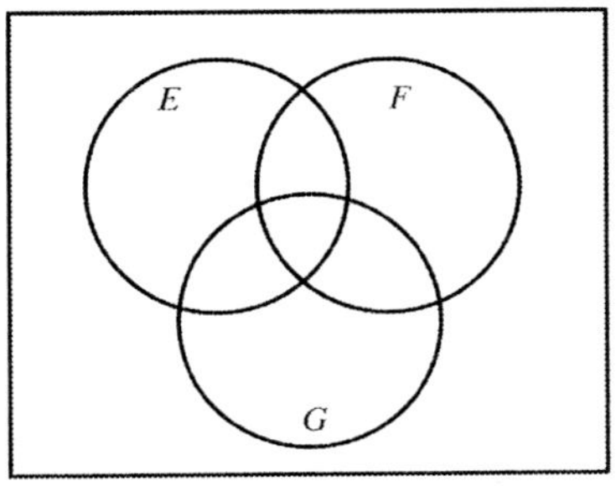

(a) If x belongs only to set E, the only non-zero term on the right side of the formula that counts x is $n(E)$. If y belongs to E and F, but not G, there is a 1 added by each of $n(E)$ and $n(F)$, and a 1 removed by $n(E \cap F)$, so y is added once since $1 + 1 - 1 = 1 = n(E)$. If z belongs to E, F, and G, there is a 1 in each term of the right side of the formula, so y is added once since $1 + 1 + 1 - 1 - 1 - 1 + 1 = 1$.

(b) From part (a), we have
$$n(E \cup F \cup G) = n(E) + n(F) + n(G) - n(E \cap F) - n(E \cap G) - n(F \cap G) + n(E \cap F \cap G)$$

If we divide each term by $n(S)$, we have
$$\frac{n(E \cup F \cup G)}{n(S)} = \frac{n(E)}{n(S)} + \frac{n(F)}{n(S)} + \frac{n(G)}{n(S)} - \frac{n(E \cap F)}{n(S)} - \frac{n(E \cap G)}{n(S)} - n\frac{(F \cap G)}{n(S)} + \frac{n(E \cap F \cap G)}{n(S)}.$$

Note that $\dfrac{n(E \cup F \cup G)}{n(S)} = P(E \text{ or } F \text{ or } G)$, $\dfrac{n(E)}{n(S)} = P(E)$, $\dfrac{n(F)}{n(S)} = P(F)$, etc. Therefore, we have

$$P(E \text{ or } F \text{ or } G) = P(E) + P(F) + P(G) - P(E \cap F) - P(E \cap G) - P(F \cap G) + P(E \cap F \cap G).$$

Copyright © 2015 Pearson Education, Inc.

29. (a)

	1	3	4	5	6	8
1	(1, 1)	(1, 3)	(1, 4)	(1, 5)	(1, 6)	(1, 8)
2	(2, 1)	(2, 3)	(2, 4)	(2, 5)	(2, 6)	(2, 8)
2	(2, 1)	(2, 3)	(2, 4)	(2, 5)	(2, 6)	(2, 8)
3	(3, 1)	(3, 3)	(3, 4)	(3, 5)	(3, 6)	(3, 8)
3	(3, 1)	(3, 3)	(3, 4)	(3, 5)	(3, 6)	(3, 8)
4	(4, 1)	(4, 3)	(4, 4)	(4, 5)	(4, 6)	(4, 8)

(b) A five is the event $\{(1, 4), (2, 3), (2, 3), (4, 1)\}$. The probability is $4/36 = 1/9$.

(c) The most likely roll is a seven, with the favorable outcomes $(1, 6), (2, 5), (2, 5), (3, 4), (3, 4), (4, 3)$. The probability is $6/36 = 1/6$.

(d) The probability of any roll is the same with either pair of dice, so the game is even.

31. (a) $0.10 + 0.05 - 0.02 = 0.13$

(b) $0.05 - 0.02 = 0.03$

(c) $1 - 0.13 = 0.87$

33. D.

Color	Number of spins resulting in this color	%
Blue	33	33%
Red	34	34%
Yellow	17	17%
Green	16	16%

Since the spinner has 6 equal sections, the probability of landing on a particular section is 1/6, or about 17%. Since the probability of getting Yellows is about the same as the probability of landing on a particular section, there is 1 White section. Similarly there is 1 Green section. Since the probability of getting Blue is about twice the probability of landing on a particular section, there are 2 Blue sections. Similarly, there are 2 Red sections. The spinner contains 2 Blue sections, 2 Red sections, 1 Yellow section, and 1 Green section. The only spinner with this combination is choice D.

35. E. In this case, it is probably best for young students to list all the possibilities and then compute the desired probability. The possible choices are ($1, $5), ($1, $10), and ($5, $10). The only possibility that works is ($5, $10).

Thus, the desired probability is $\dfrac{1}{3}$.

37. D. There are twelve faces, of which six are even and six are odd. There are seven numbers greater than 5 and eight numbers less than 9.

Section 14.2
Applications of Counting Principles to Probability

Problem Set 14.2

1. There are 3 ways to roll a 4:
$4 = 1 + 3 = 2 + 2 = 3 + 1$. This is represented by the event $\{(1, 3), (2, 2), (3, 1)\}$.
There are 5 ways to roll a 6:
$6 = 1 + 5 = 2 + 4 = 3 + 3 = 4 + 2 = 5 + 1$. This is represented by the event $\{(1, 5), (2, 4), (3, 3), (4, 2), (5, 1)\}$. The events are mutually exclusive in the uniform sample space of 36 outcomes, so the probability of rolling a 4 or a 6 is $\dfrac{3}{36} + \dfrac{5}{36} = \dfrac{8}{36} = \dfrac{2}{9}$.

3. The $2 \cdot 6 = 12$ possible outcomes are:
H1 H2 H3 H4 H5 H6 T1 T2 T3 T4 T5 T6

(a) {H2, H4, H6} is the event of a head and an even number, so its probability is
$\dfrac{3}{12} = \dfrac{1}{4}$.

(b) {H1, H2, H3, H4, H5, H6, T2, T4, T6} is the event of a head or an even number, so its probability is $\dfrac{9}{12} = \dfrac{3}{4}$.

5. Let x students speak both languages, so $24 = 11 + 9 + x$. Therefore, $x = 4$, so the probability that a student speaks both languages is $\dfrac{4}{24} = \dfrac{1}{6}$.

146 *Chapter 14 Probability*

7. Let R be the set of red face cards.
{jack of diamonds, queen of diamonds, king of diamonds, jack of hearts, queen of hearts, king of hearts}
Let A be the set of black aces.
{ace of spades, ace of clubs}
Notice that $R \cap A = \varnothing$ since there are no elements that are common to both set R and set A. $n(R) = 6$, so $P(R) = \dfrac{6}{52}$. $n(A) = 2$, so

$P(A) = \dfrac{2}{52}$. The events are mutually exclusive,

so
$$P(R \cap A) = P(R) + P(A)$$
$$= \frac{6}{52} + \frac{2}{52} = \frac{8}{52} = \frac{2}{13}.$$

9. Using the same reasoning as in problem 8, we obtain these results.

   (a) $5 \cdot 4 \cdot 3 = 60$ **(b)** $2 \cdot 4 \cdot 3 = 24$

   (c) $2 \cdot 4 \cdot 3 = 24$

   (d) The first digit can be chosen in two ways, either 2 or 4; the last digit in three ways, 1, 3, or 5; and the middle digit in three ways from the remaining digits not yet selected. This gives $2 \cdot 3 \cdot 3 = 18$ numbers in the event, with a probability of
   $$\frac{18}{60} = \frac{3}{10}.$$

11. The digits 1, 3, 5, 7, and 9 are odd and the digits 0, 2, 4, 6, and 8 are even.

    (a) There are 5 choices for the first odd digit, 5 choices of the second odd digit, 5 choices for the third odd digit, 5 choices for the fourth even digit, and 5 choices for the last even digit. Therefore, there are
    $5 \cdot 5 \cdot 5 \cdot 5 \cdot 5 = 5^5 = 3125$ five-digit numbers with the first three digits odd and the last two digits even when repetition of digits is allowed.

    (b) $5 \cdot 4 \cdot 3 \cdot 5 \cdot 4 = 1200$

13. **(a)** To draw the possibility tree, begin by showing the 3 possible letters for the first letter and label these 3 limbs "*a*", "*b*", and "*c*". Now there are two possible letters remaining for the second letter so draw 2 limbs from each of the first letters and label each limb with the remaining possible letters. There is only one choice

for the third letter after the first two have been selected so draw a limb and label the remaining letter.

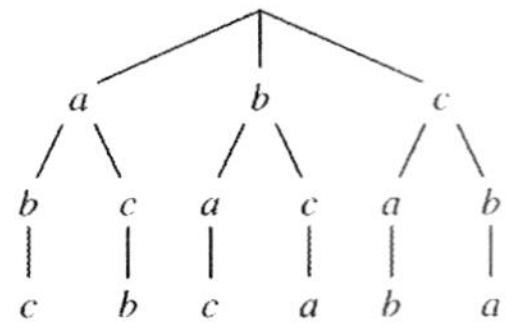

Following the limbs of the possibility tree, the code words are: *abc, acb, bac, bca, cab, cba.*

(b) There are three of the 3! = 6 permutations with *a* preceding *b*, namely *abc, acb,* and *cab.* The probability is $\dfrac{3}{6} = \dfrac{1}{2}$.

15. $u = 1 - \dfrac{2}{3} = \dfrac{1}{3}$, $v = 1 - \dfrac{2}{5} = \dfrac{3}{5}$, $w = \dfrac{2}{3} \cdot \dfrac{3}{5} = \dfrac{2}{5}$

$x = \dfrac{1}{4} \div \dfrac{1}{3} = \dfrac{3}{4}$, $y = 1 - \dfrac{3}{4} = \dfrac{1}{4}$, $z = \dfrac{1}{3} \cdot \dfrac{1}{4} = \dfrac{1}{12}$

17. **(a)** There are $9 \cdot 10 \cdot 10 \cdot 10 \cdot 10 = 90,000$ different five-digit numbers.

    (b) There are $9 \cdot 10 \cdot 10 \cdot 10 \cdot 5 = 45,000$ different five-digit even numbers.

    (c) Since the leading digit cannot be a zero, there are $4 \cdot 9^4 = 26,244$ five-digit numbers that have exactly one zero.

    (d) There are $9^4 = 6561$ five-digit numbers that begin with a seven followed by four digits, none of which is a 7. There are $8 \cdot 4 \cdot 9^3 = 23,328$ five digit numbers with the digit 7 in exactly one of the four places that follows the leading digit which is neither a 7 nor a 0. So, there are $6561 + 23,328 = 29,889$ five digit numbers that have exactly one 7.

19. Answers will vary, but all should emphasize that Alejandro is double-counting the numbers that are divisible by both 4 and 5. That is, the fifty numbers 20, 40, …, 1000 that are divisible by 20 are counted twice by simply summing 200 + 250. The correct answer is $450 - 50 = 400$.

21. **(a)** Answers will vary but should make it clear to Ernie that the multiplication principle gives the answer 8. Ernie has added 2 + 4 instead of multiplying.

Copyright © 2015 Pearson Education, Inc.

(b) Answers will vary but should remind Donna that there are still two crust types available, so the correct answer is 6.

23. (a) Let FP, JP, ..., FT, and JT denote the events of a slate with a freshman president, junior president, ..., freshman treasurer, and junior treasurer. Then, the possibility tree on the next page shows there are $10 + 15 + 10 + 6 + 15 + 6 = 62$ possible slates where the offices are not held by individual from the same class.

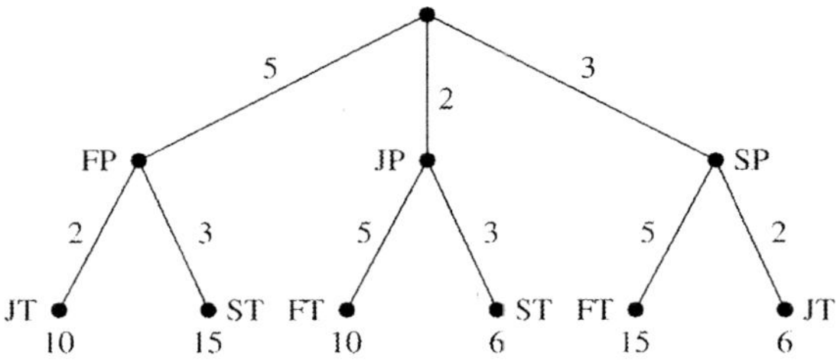

```
                    •
            5   /   2   \   3
          /        |        \
      FP •       JP •        • SP
    2 /  \ 3   5 /  \ 3    5 /  \ 2
  JT •    • ST  FT •   • ST  FT •   • JT
   10     15    10     6     15     6
```

(b) Let $\overline{F}$, $\overline{J}$, and $\overline{S}$ denote the events that no freshman, no junior, and no senior holds an office. To determine $n\left(\overline{F}\right)$, notice that there are 2 choices of a junior and 3 choices of a senior, and 2 ways to assign them to the offices of president and treasurer. Therefore, $n\left(\overline{F}\right) = 2 \cdot 3 \cdot 2 = 12$.

Similar reasoning shows that
$n\left(\overline{J}\right) = 5 \cdot 3 \cdot 2 = 30$ and
$n\left(\overline{S}\right) = 5 \cdot 2 \cdot 2 = 20$.

Since the events are mutually exclusive, there are $12 + 30 + 20 = 62$ different slates of officers in the chess club.

25. (a) $P = \dfrac{12}{52} \cdot \dfrac{12}{52} = \dfrac{9}{169}$

(b) $P = \dfrac{12}{52} \cdot \dfrac{11}{52} = \dfrac{33}{676}$

27. There are $10 \cdot 7 = 70$ ways to give a baseball card and a basketball card. There are $10 \cdot 9 = 90$ ways to give a baseball card and a football card, and there are $7 \cdot 9 = 63$ ways to give a basketball card and a football card. So there are $70 + 90 + 63 = 223$ ways for Marc to give away two cards from different sports. Thus, the probability that the cards are from different sports is $\dfrac{223}{26 \cdot 25} = \dfrac{223}{650}$

29. (a) Since one of the nine keys is the correct one, $p_1 = \dfrac{1}{9}$. The first key tried is incorrect, with probability $\dfrac{8}{9}$, and the second key is correct with probability $\dfrac{1}{8}$. So, by conditional probabilities, $p_2 = \dfrac{8}{9} \cdot \dfrac{1}{8} = \dfrac{1}{9}$. Similarly, $p_3 = \dfrac{8}{9} \cdot \dfrac{7}{8} \cdot \dfrac{1}{7} = \dfrac{1}{9}$, and so on.

(b) For n keys, the reasoning followed in part (a) shows that
$$p_1 = \frac{1}{n}, \quad p_2 = \frac{n-1}{n} \cdot \frac{1}{n-1} = \frac{1}{n},$$
$$p_3 = \frac{n-1}{n} \cdot \frac{n-2}{n-1} \cdot \frac{1}{n-2} = \frac{1}{n}, \cdots.$$
The expected number of keys tried is then
$$e_n = 1 \cdot \frac{1}{n} + 2 \cdot \frac{1}{n} + \cdots + n \cdot \frac{1}{n}$$
$$= \frac{1}{n}\left(1 + 2 + \cdots + n\right)$$
$$= \frac{1}{n} \cdot \frac{n \cdot (n+1)}{2} = \frac{n+1}{2}$$

31. (a)

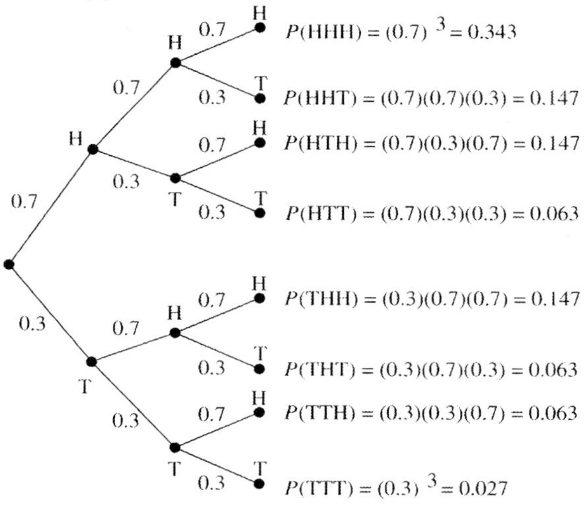

```
                                    0.7   H
                               H  •——— • P(HHH) = (0.7)³ = 0.343
                          0.7 /
                            /      0.3   T
                     H   •    0.3  •——— • P(HHT) = (0.7)(0.7)(0.3) = 0.147
                0.7 / •             0.7   H
                  /   \        •——— • P(HTH) = (0.7)(0.3)(0.7) = 0.147
                 /     0.3  T
          0.7 /         •    0.3   T
           •                •——— • P(HTT) = (0.7)(0.3)(0.3) = 0.063
            \
             \                0.7   H
              \          H  •——— • P(THH) = (0.3)(0.7)(0.7) = 0.147
          0.3 \      0.7 /
               \       /      0.3   T
                •   •    0.3  •——— • P(THT) = (0.3)(0.7)(0.3) = 0.063
                T        0.7   H
                     •——— • P(TTH) = (0.3)(0.3)(0.7) = 0.063
                 0.3  T
                     •    0.3   T
                          •——— • P(TTT) = (0.3)³ = 0.027
```

(b) $P(HHH) = 0.343$

(c) $P(HHT \text{ or } HTH \text{ or } THH)$
$= 0.147 + 0.147 + 0.147 = 0.441$

(d) $P(H \text{ appears}) = 1 - P(TTT)$
$= 1 - 0.027 = 0.973$

148 *Chapter 14 Probability*

(e) P(H and T appear)

$= 1 - P$(HHH) $- P$(TTT)

$= 1 - 0.343 - 0.027 = 0.63$

33. Fix the positions of the red, white, and black wires. Then the number of possible connections is just the number of ways you can order the yellow, blue, and green wires; i.e., $3! = 6$. Thus, the probability that the connection is correct is $\dfrac{1}{6}$.

35. (a) 2^4, or 16 words.

(b) 2^6, or 64 words.

(c) At least ten bits, since $2^9 = 512$ but $2^{10} = 1024$.

(d) The five code words 0000, 0001, 0011, 0111, and 1111 have no 0 following a 1, so the probability is $\dfrac{5}{16}$.

37. (a) The population of the U.S. reached 300 million in 2006. There are $26^6 = 308,915,776$ possible codes, so every person in the U.S. can be given a distinct six-letter identification.

(b) Yes, since $52^5 = 380,204,032$

(c) Seven letters suffice because $26^7 = 8,031,810,176$ but $26^6 = 308,915,776$.

39. D. There are two cup sizes, five flavors, and two types of sprinkles or none at all. Thus, by the multiplicative principle for counting, there are $2 \cdot 5 \cdot 3 = 30$ ways to place an order.

41. C. The probability of picking a lime jelly bean on the first pick is $\dfrac{4}{16} = \dfrac{1}{4}$, and the probability of choosing a lime jelly bean on the second pick is $\dfrac{3}{15} = \dfrac{1}{5}$, so the probability is $\dfrac{1}{4} \cdot \dfrac{1}{5} = \dfrac{1}{20}$.

Section 14.3
Permutations and Combinations

Problem Set 14.3

1. (a) $P(7, 3) = 7 \cdot 6 \cdot 5 = 210$

(b) $P(7, 7) = 7 \cdot 6 \cdot 5 \cdot 4 \cdot 3 \cdot 2 \cdot 1 = 5040$

(c) $P(7, 4) = 7 \cdot 6 \cdot 5 \cdot 4 = 840$

(d) $C(7, 3) = \dfrac{7 \cdot 6 \cdot 5}{3 \cdot 2 \cdot 1} = 35$

(e) $C(7, 7) = \dfrac{7!}{7!} = 1$

(f) $C(7, 4) = \dfrac{7 \cdot 6 \cdot 5 \cdot 4}{4 \cdot 3 \cdot 2 \cdot 1} = 35$

3. (a) combination.
$C(10, 5) = \dfrac{10 \cdot 9 \cdot 8 \cdot 7 \cdot 6}{5 \cdot 4 \cdot 3 \cdot 2 \cdot 1} = 252$

(b) permutation. $P(10, 4) = 10 \cdot 9 \cdot 8 \cdot 7 = 5040$

(c) permutation:
$P(12, 5) = 12 \cdot 11 \cdot 10 \cdot 9 \cdot 8 = 95,040$

(d) combination: $C(15, 3) = \dfrac{15 \cdot 14 \cdot 13}{3 \cdot 2 \cdot 1} = 455$

(e) permutation:
$P(5, 5) = 5! = 5 \cdot 4 \cdot 3 \cdot 2 \cdot 1 = 120$

5. There are 7 ways to choose the unicyclist and $P(6, 2) = 6 \cdot 5 = 30$ ways to choose the front and rear riders for the tandem bike. So there are $7 \cdot 30 = 210$ ways that a group of seven people can travel.

7. (a) The club members can line up in $14! \approx 90$ billion ways.

(b) The girls can be arranged in $8! = 40,320$ ways, and the boys can be arranged in $6! = 720$ ways, so they can be arranged in $8! \cdot 6! = 29,030,400$ ways.

(c) A slate of three officers can be chosen in $P(14, 3) = 14 \cdot 13 \cdot 12 = 2184$ ways.

Copyright © 2015 Pearson Education, Inc.

(d) The two boys can be chosen in

$$C(6,2) = \frac{6 \cdot 5}{2 \cdot 1} = 15 \text{ ways. The two girls}$$

can be chosen in $C(8,2) = \frac{8 \cdot 7}{2 \cdot 1} = 28$. The

committee can be chosen in
$C(6,2) \cdot C(8,2) = 15 \cdot 28 = 420$ ways.

9. For $n = 6$, we have
$$C(6, 0) = 1$$
$$C(6, 1) = 6 = 5 + 1 = C(5, 1) + C(5, 0)$$
$$C(6, 2) = \frac{6 \cdot 5}{2 \cdot 1} = 15$$
$$= 10 + 5 = C(5, 2) + C(5, 1)$$
$$C(6, 3) = \frac{6 \cdot 5 \cdot 4}{3 \cdot 2 \cdot 1} = 20$$
$$= 10 + 10 = C(5, 3) + C(5, 2)$$
$$C(6, 4) = \frac{6 \cdot 5 \cdot 4 \cdot 3}{4 \cdot 3 \cdot 2 \cdot 1} = 15$$
$$= 5 + 10 = C(5, 4) + C(5, 3)$$
$$C(6, 5) = \frac{6 \cdot 5 \cdot 4 \cdot 3 \cdot 2}{5 \cdot 4 \cdot 3 \cdot 2 \cdot 1} = 6$$
$$= 1 + 5 = C(5, 5) + C(5, 4)$$
$$C(6, 6) = \frac{6 \cdot 5 \cdot 4 \cdot 3 \cdot 2 \cdot 1}{6 \cdot 5 \cdot 4 \cdot 3 \cdot 2 \cdot 1} = 1$$

Thus, row 6 of Pascal's triangle is
1   6   15   20   15   6   1.
Similarly, we find that row 7 is
1   7   21   35   35   21   7   1.

11. (a) There are $C(12, 3) = \frac{12 \cdot 11 \cdot 10}{3 \cdot 2 \cdot 1} = 220$

ways to draw three face cards and

$$C(52, 3) = \frac{52 \cdot 51 \cdot 50}{3 \cdot 2 \cdot 1} = 22{,}100 \text{ ways to}$$

draw three cards. Thus, the probability of drawing three face cards is

$$\frac{C(12, 3)}{C(52, 3)} = \frac{12 \cdot 11 \cdot 10}{52 \cdot 51 \cdot 50} = \frac{220}{22{,}100} \doteq 0.01.$$

(b) There are $4 \cdot 13^3$ ways to draw three cards of different suits and

$$C(52, 3) = \frac{52 \cdot 51 \cdot 50}{3 \cdot 2 \cdot 1} \text{ ways to draw three}$$

cards. Thus, the probability of drawing three cards of different suits is

$$\frac{4 \cdot 13^3}{C(52, 3)} = \frac{4 \cdot 13 \cdot 13^2}{\dfrac{52 \cdot 51 \cdot 50}{3 \cdot 2 \cdot 1}} = \frac{52 \cdot 13^2}{\dfrac{52 \cdot 51 \cdot 50}{3 \cdot 2 \cdot 1}}$$
$$\doteq 0.398.$$

13. The probability that no two girls are side-by-

side is $\dfrac{6!5!}{11!} = \dfrac{6!5!}{11 \cdot 10 \cdot 9 \cdot 8 \cdot 7 \cdot 6!} \doteq 0.002$

15. There are $C(26, 5)$ ways to have five red cards and $C(12, 5)$ ways to have five face cards. These events are not mutually exclusive since there are red face cards. There are $C(6, 5)$ ways to have five of the six red face cards. Thus, the probability that a randomly drawn hand of five cards contains all red cards or all face cards is

$$\frac{C(26, 5) + C(12, 5) - C(6, 5)}{C(52, 5)}$$
$$= \frac{65{,}780 + 792 - 6}{2{,}598{,}960} = \frac{66{,}566}{2{,}598{,}960} \doteq 0.026$$

17. (a) There are 2^{10} ways to answer the quiz. There is 1 way to answer all ten questions correctly, and there are 10 ways to answer nine of the ten questions correctly. Thus, there are 11 ways to score 90% or better. The probability that Igor will score 90%

or better is $\dfrac{11}{2^{10}} = \dfrac{11}{1024} \doteq 0.011$.

(b) There are $C(10, 2) = 45$ ways to score 80% (i.e., answer eight questions correctly), $C(10, 3) = 120$ ways to score 70% (i.e., answer seven questions correctly), $C(10, 4) = 210$ ways to score 60% (i.e., answer six questions correctly), and $C(10, 5) = 252$ ways to score 50% (i.e., answer five questions correctly). Thus, the probability the Igor will pass the quiz with a score of 50% or better is

$$\frac{1 + 10 + 45 + 120 + 210 + 252}{1024} = \frac{638}{1024}$$
$$\doteq 0.62.$$

150 Chapter 14 Probability

19. (a) There are $C(8, 4)$ ways to draw 4 red balls and $C(8 + 5 + 6, 4) = C(19, 4)$ ways to draw any 4 balls from the urn. Therefore,

$$P(\text{all four red}) = \frac{n(\text{ways to draw 4 red balls})}{n(S)}$$

$$= \frac{C(8, 4)}{C(19, 4)} = \frac{\frac{8 \cdot 7 \cdot 6 \cdot 5}{4 \cdot 3 \cdot 2 \cdot 1}}{\frac{19 \cdot 18 \cdot 17 \cdot 16}{4 \cdot 3 \cdot 2 \cdot 1}}$$

$$= \frac{1680}{93{,}024} \approx 0.02.$$

(b) $P(2 \text{ red and } 2 \text{ green}) = \frac{C(8, 2) \cdot C(6, 2)}{C(19, 4)}$

$$= \frac{28 \cdot 15}{3876} \approx 0.11$$

(c) $P(\text{exactly 2 are red}) = \frac{C(8, 2) \cdot C(11, 2)}{C(19, 4)}$

$$= \frac{\frac{8 \cdot 7}{1 \cdot 2} \cdot \frac{11 \cdot 10}{1 \cdot 2}}{\frac{19 \cdot 18 \cdot 17 \cdot 16}{1 \cdot 2 \cdot 3 \cdot 4}} \approx 0.40$$

$P(\text{exactly 2 are green}) = \frac{C(6, 2) \cdot C(13, 2)}{C(19, 4)}$

$$= \frac{\frac{6 \cdot 5}{1 \cdot 2} \cdot \frac{13 \cdot 12}{1 \cdot 2}}{\frac{19 \cdot 18 \cdot 17 \cdot 16}{1 \cdot 2 \cdot 3 \cdot 4}} \approx 0.30$$

Therefore, $P(\text{exactly two are red or exactly two are green}) = P(\text{exactly two are red}) + P(\text{exactly two are green}) - P(\text{exactly two are red and exactly two are green}) \approx 0.40 + 0.30 - 0.11 = 0.59$

21. Both Murial and Jisoo have given correct answers. Murial counted each of the mutually exclusive cases (four seniors, three seniors and one junior, or two seniors and two juniors) to get $C(4, 4) + C(4, 3)C(7, 1) + C(4, 2)C(7, 2) = 1 + 28 + 6 \cdot 21 = 155$. Jisoo correctly counted the number of complementary ways to get $C(11, 4) - C(7, 4) - 4C(7, 3) = 330 - 35 - 4 \cdot 35 = 155$.

Shelley's answer of $C(4, 2)C(9, 2) = 216$ is much too high since her method counts many of the committees more than once. To understand why, suppose the seniors are $A, B, C,$ and D, and the juniors are a, b, c, d, f, g. If A and B are place on the committee, it can be filled out by also including C and g, resulting in the committee $\{A, B, C, g\}$. Now suppose that A and C are placed on the committee and it is then filled out by including B and g. The same committee $\{A, B, C, g\}$ is formed, but Shelley has counted it more than once.

23. There are $P(26, 5) = 26 \cdot 25 \cdot 24 \cdot 23 \cdot 22 = 7{,}893{,}600$ five-letter code words without repetition of letters.

(a) If a code begins with the letter a, then there are 25 choices for the second letter, 24 choices for the third letter, 23 choices for the fourth letter, and 22 choices for the last letter. There are $P(25, 4) = 25 \cdot 24 \cdot 23 \cdot 22 = 303{,}600$ ways to arrange the last 4 letters. Therefore,

$P(\text{a code word begins with } a)$

$$= \frac{P(25, 4)}{P(26, 5)} = \frac{303{,}600}{7{,}893{,}600} \approx 0.04$$

(b) As shown above, there are $P(26, 5) = 7{,}893{,}610$ possible code words. If we think of cd as a single unit, there are $4 \cdot 24 \cdot 23 \cdot 22 = 48{,}576$ code words with c immediately followed by d. Therefore, $P(\text{a code word with } c \text{ followed by } d)$

$$= \frac{4 \cdot 24 \cdot 23 \cdot 22}{26 \cdot 25 \cdot 24 \cdot 23 \cdot 22} = \frac{4}{26 \cdot 25} \approx 0.006.$$

(c) Since there are 5 vowels and 21 consonants, there are 5 ways to choose the first letter and $24 \cdot 23 \cdot 22$ ways to choose the other three letters. Therefore, $P(\text{a code word starts with a vowel and ends with a consonant}) = \frac{5 \cdot 21 \cdot 24 \cdot 23 \cdot 22}{7{,}893{,}600} \approx 0.16.$

(d) By symmetry, the number of code words with d immediately followed by c is the same as the number with c immediately followed by d. Therefore, by part (a), the desired number is $2 \cdot 48{,}576 = 97{,}152.$

25. Since the order of selection of the three balls does not matter, one answer is given by the combination number $C(m + n, 3)$. Taking the respective cases where all three balls are red, all three are blue, one is red and two are blue, and one is blue and two are red gives the sum $C(m, 3) + C(n, 3) + mC(n, 2) + nC(m, 2)$.

27. (a) $C(8, 5)$

(b) $C(7, 5)$, since there are 7 non-red marbles from which to choose 5 marbles.

Copyright © 2015 Pearson Education, Inc.

Solutions to Problem Set 14.3   151

29. (a) Choose eight of the fifteen members in $C(15, 8) = 6435$ ways, and the choose the two co-captains from the eight delegates in $C(8, 2) = 28$ ways. Thus, there are $6435 \cdot 28 = 180,180$ ways to choose the delegation and co-captains.

(b) First choose the co-captains in $C(15, 2) = 105$ ways. Then fill out the delegation by choosing six of the remaining thirteen members in $C(13, 6) = 1716$ ways. This gives $105 \cdot 1716 = 180,180$ ways to choose the delegation and co-captains.

(c)
$$C(15, 8) \cdot C(8, 2) = \frac{15!}{8!7!} \cdot \frac{8!}{8!2!6!} = \frac{15!}{7!6!2!} = 180,180$$
$$C(15, 2) \cdot C(13, 6) = \frac{15!}{13!2!} \cdot \frac{13!}{6!7!} = \frac{15!}{2!7!6!} = 180,180$$

(d) From 19 members, send a delegation of seven, including three co-captains.

31. (a) There are 10 tiles altogether. There are $C(10, 2) = \binom{10}{2} = 45$ ways to choose where two red tiles are placed along the walkway. From the remaining eight places, there are $C(8, 3) = \binom{8}{3} = 56$ ways to place the blue tiles. There are now five empty positions for the five green tiles. These can be placed in $C(5, 5) = \binom{5}{5} = 1$ way. Thus, there are
$$\binom{10}{2}\binom{8}{3}\binom{5}{5} = 45 \cdot 56 \cdot 1 = 2520 \text{ ways to}$$
choose the tiles.

(b) $\binom{10}{2}\binom{8}{3}\binom{5}{5} = \frac{10!}{2! \, 8!} \cdot \frac{8!}{3! \, 5!} \cdot \frac{5!}{5!} = \frac{10!}{2! \, 3! \, 5!}$

(c) There are 10! ways to pave the walkway with ten different tiles. Each color pattern appears the same if the two red tiles are permuted in 2! ways, the three blue tiles are permuted in 3! ways, and the five green tiles are permuted in 5! ways. This means that each color pattern is formed in $2! \, 3! \, 5!$ ways, so there are $\frac{10!}{2! \, 3! \, 5!}$ different color patterns.

33. (a) There are 4 letters with two T's and two O's. The number of permutations of 4 things with 2 things alike and 2 other things alike is given by:
$$\frac{4!}{2! \, 2!} = \frac{4 \cdot 3 \cdot 2 \cdot 1}{2 \cdot 1 \cdot 2 \cdot 1} = \frac{4 \cdot 3}{2 \cdot 1} = 6.$$
There are 6 different arrangements of the letters in TOOT.

(b) $\frac{7!}{2! \, 2! \, 2! \, 1!} = 630$

35. (a) $C(6 + 4 - 1, 6) = C(9, 6) = C(9, 3)$
$$= \frac{7 \cdot 8 \cdot 9}{1 \cdot 2 \cdot 3} = 84$$

(b) $C(8 + 6 - 1, 8) = C(13, 8) = C(13, 5)$
$$= \frac{9 \cdot 10 \cdot 11 \cdot 12 \cdot 13}{1 \cdot 2 \cdot 3 \cdot 4 \cdot 5} = 1287$$

(c) $C(4 + 32 - 1, 4) = C(35, 4)$
$$= \frac{32 \cdot 33 \cdot 34 \cdot 35}{1 \cdot 2 \cdot 3 \cdot 4} = 52,360$$

Copyright © 2015 Pearson Education, Inc.

152 *Chapter 14 Probability*

(d) The number of combinations with repetition is equivalent to determining the number of ways n ice cream cones can be ordered from k flavors. That is, it is the number of lists on n check marks and $k - 1$ bars that separate one flavor from the next. Therefore, n objects can be chosen from k types, with repetition allowed, in $C(n + k - 1, n)$ ways.

37. (a) There are

$$C(13, 3) = \frac{13 \cdot 12 \cdot 11}{3 \cdot 2 \cdot 1} = \frac{1716}{6} = 286 \text{ ways}$$

to choose the three boys for the committee. There are

$$C(11, 3) = \frac{11 \cdot 10 \cdot 9}{3 \cdot 2 \cdot 1} = \frac{990}{6} = 165 \text{ ways to}$$

choose the three girls for the committee. That means there are $C(13, 3) \cdot C(11, 3) = 286 \cdot 165 = 47,190$ ways to select a committee of three boys and three girls.

(b) Either Lourdes is not on the committee and Andy is (so choose 3 girls from 10 girls and 2 boys from 12 boys), *or* Lourdes is on the committee and Andy is not, (so choose 2 girls from 10 girls and 3 boys from 12 boys), *or* neither is on the committee (so choose 3 girls from 10 girls and 3 boys from 12 boys). Thus, using the second and third principles of counting, there are $C(10, 3) \cdot C(12, 2) + C(10, 2) \cdot C(12, 3) + C(10, 3) \cdot C(12, 3)$

$$= \frac{10 \cdot 9 \cdot 8}{1 \cdot 2 \cdot 3} \cdot \frac{12 \cdot 11}{1 \cdot 2} + \frac{10 \cdot 9}{1 \cdot 2} \cdot \frac{12 \cdot 11 \cdot 10}{1 \cdot 2 \cdot 3}$$
$$+ \frac{10 \cdot 9 \cdot 8}{1 \cdot 2 \cdot 3} \cdot \frac{12 \cdot 11 \cdot 10}{1 \cdot 2 \cdot 3}$$

$= 44,220$ committees without Lourdes and Andy on the same committee.

39. There are $3! \, 4! \, 5! = 17,280$ different ballots.

41. (a) There are $C(80, 20)$ ways to draw 20 numbers from the barrel of 80. There are $C(8, 5)$ ways to choose the 5 numbers to match those marked as spots, and $C(72, 15)$ to choose the other 15 numbers from the 72 numbers not marked as spots. This gives the probability:

$$\frac{C(8,5) \cdot C(72,15)}{C(80,20)} \approx 0.0183$$

(b) $\dfrac{C(8,6) \cdot C(72,14)}{C(80,20)} \approx 0.00237$

(c) $\dfrac{C(8,7) \cdot C(72,13)}{C(80,20)} \approx 0.00016$

(d) $\dfrac{C(8,8) \cdot C(72,12)}{C(80,20)} \approx 0.0000043$

Section 14.4
Odds, Expected Values, Geometric Probability, and Simulations

Problem Set 14.4

1. (a) There are 5 ways to roll a six so there must be $36 - 5 = 31$ ways to not roll a six. If 6 represents the event of rolling a score of six, then the odds in favor of 6 are $n(6)$ to $n(\overline{6})$ or 5 to 31 written 5:31.

(b) $n(\overline{6})$ to $n(6)$ or 31:5

(c) $P(6) = \dfrac{5}{36} \approx 0.14$

3. Since A and C are mutually exclusive,

$$P(A \text{ or } C) = P(A) + P(C) = \frac{1}{2} + \frac{1}{6} = \frac{4}{6} = \frac{2}{3}.$$

Therefore, the odds in favor of A or C are given by $\dfrac{P(A \text{ or } C)}{1 - P(A \text{ or } C)} = \dfrac{\frac{2}{3}}{1 - \frac{2}{3}} = \dfrac{2}{3} \cdot \dfrac{3}{1} = \dfrac{2}{1}$ or 2:1.

5. Since $P(A) = 0.3$ and $P(B) = 0.5$, then $P(C) = 1 - (0.3 + 0.5) = 0.2 = \dfrac{1}{5}$. Thus, the odds that C occurs are 1:4.

Copyright © 2015 Pearson Education, Inc.

Solutions to Problem Set 14.4 **153**

7. To determine the probability for each roll, count the number of ways to roll the score and divide by 36.

roll	2	3	4	5	6	7	8	9	10	11	12
probability	$\dfrac{1}{36}$	$\dfrac{2}{36}$	$\dfrac{3}{36}$	$\dfrac{4}{36}$	$\dfrac{5}{36}$	$\dfrac{6}{36}$	$\dfrac{5}{36}$	$\dfrac{4}{36}$	$\dfrac{3}{36}$	$\dfrac{2}{36}$	$\dfrac{1}{36}$

Expected win

$$= 4\left(\frac{1}{36}\right) + 6\left(\frac{2}{36}\right) + 8\left(\frac{3}{36}\right) + 10\left(\frac{4}{36}\right) + 20\left(\frac{5}{36}\right) + 40\left(\frac{6}{36}\right) + 20\left(\frac{5}{36}\right) + 10\left(\frac{4}{36}\right) + 8\left(\frac{3}{36}\right) + 6\left(\frac{2}{36}\right) + 4\left(\frac{1}{36}\right)$$

$$= \frac{600}{36} \approx 16.67$$

You can expect to win about $16.67 per roll on this game.

9. (a) The net gain of a win is $100 − $1 = $99.

$$\$99 \cdot \frac{1}{100} + (-\$1) \cdot \frac{99}{100} = \$0$$

The expected value if 100 tickets are sold is $0.

(b) The net gain of a win is $100 − $1 = $99.

$$\$99 \cdot \frac{1}{200} + (-\$1) \cdot \frac{199}{200} = \$\frac{99 - 199}{200}$$
$$= -\$0.50$$

The expected value if 200 tickets are sold is −$0.50.

11. The expected value is

$$\left(\frac{36}{n} - 1\right)\left(\frac{n}{38}\right) + (-1)\left(\frac{38 - n}{38}\right)$$
$$= \left(\frac{36 - n}{n}\right)\left(\frac{n}{38}\right) + \frac{-38 + n}{38}$$
$$= \frac{36 - n - 38 + n}{38} = -\frac{2}{38} \doteq -0.05$$

Since the player can expect to lose about 5% of the amount of the bets placed, the game is not fair.

13. (a) $\dfrac{1}{4}$ **(b)** $\dfrac{1}{2}$

(c) $\dfrac{3}{4}$

15. (a) Each blue region can be paired with a congruent yellow region, so the blue and yellow areas are equal. $A \cong D$, $B \cong E$, and $C \cong F$. Therefore, the probability is $1/2$.

(b) Each region can be partitioned into two triangles of base $2''$ and altitude $3''$, so all six regions have the same area. Therefore, the probability of a point being in any chosen region is $1/6$.

17. If sufficiently many trials are conducted, it should be discovered that there is a match about 63% of the time. This means no student is given a correct key with a probability of about 0.37. The probability is nearly the same for any group of more than ten students.

19. Answers will vary.

21. Answers will vary.

23. Answers will vary.

25. (a) With a large number of players, there is an increased chance that more than one winner is declared and the prize must be shared.

(b) Nearly all of your dreams are more than met with $10 million! Having a better chance to win $10 million than $100 million may be very sensible.

27. $P(2) = P(GG) + P(BB) = \left(\frac{1}{2}\right)^2 + \left(\frac{1}{2}\right)^2 = \frac{1}{2}$

and $P(4) = 1 - \left(P(2) + P(3)\right) = 1 - \frac{1}{2} - \frac{1}{4} = \frac{1}{4}$.

Thus, the expected value is

$$2P(2) + 3P(3) + 4P(4) = 2 \cdot \frac{1}{2} + 3 \cdot \frac{1}{4} + 4 \cdot \frac{1}{4}$$
$$= \frac{11}{4} = 2.75$$

29. (a) The points V and X should each be reach about $1/3$ of the time.

Copyright © 2015 Pearson Education, Inc.

154 *Chapter 14 Probability*

(b) Points *T*, *U*, *Y*, and *Z* are rarely reached.

(c)

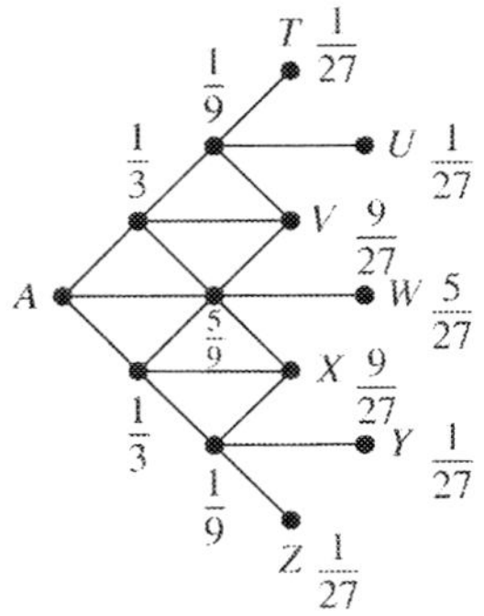

31. (a) There are 8 marbles altogether. Tracy can chose a marble in 1 of 5 ways and Nina can choose a marble in 1 of 3 ways, so there are $5 \cdot 3 = 15$ ways to choose the marbles. If the first choice is Tracy's and the second choice is Nina's, then we have the following possibilities: RR, GR, BR, WR, PR, RB, GB, BB, WB, PB, RW, GW, BW, WW, PW

(b) The probability that both marbles are the same color is $\dfrac{3}{15} = \dfrac{1}{5} = 0.2$.

(c) The probability that both marbles are the different colors is $1 - 0.2 = 0.8$.

Chapter 14 Review Exercises

1. (a) $S = \{(1, H), (1, T), (2, H), (2, T), (3, H), (3, T), (4, H), (4, T)\}$

(b)

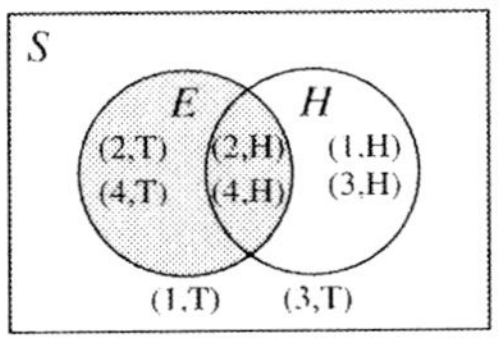

(c) $P(E) = \dfrac{4}{8} = \dfrac{1}{2}, \quad P(H) = \dfrac{4}{8} = \dfrac{1}{2}, \quad P(E \cup H) = \dfrac{6}{8} = \dfrac{3}{4}, \quad P(E \cap H) = \dfrac{2}{8} = \dfrac{1}{4}$

(d) No, *E* and *H* are not mutually exclusive since $E \cap H = \{(2, H), (4, H)\} \neq \varnothing$.

3.

Sum	2	3	4	5	6	7	8	9	10	11	12
Probability	$\dfrac{1}{36}$	$\dfrac{2}{36}$	$\dfrac{3}{36}$	$\dfrac{4}{36}$	$\dfrac{5}{36}$	$\dfrac{6}{36}$	$\dfrac{5}{36}$	$\dfrac{4}{36}$	$\dfrac{3}{36}$	$\dfrac{2}{36}$	$\dfrac{1}{36}$

Solution 1: To compute the probability of obtaining a sum of at most 11, add the probabilities associated with sums of 2 through 11.

$$P(\text{sum at most } 11) = \frac{1}{36} + \frac{2}{36} + \frac{3}{36} + \frac{4}{36} + \frac{5}{36} + \frac{6}{36} + \frac{5}{36} + \frac{4}{36} + \frac{3}{36} + \frac{2}{36} = \frac{35}{36} \approx 0.97$$

Solution 2: If you fail to obtain a sum of at most 11 you obtain a sum of 12.
Use $P(A) = 1 - P(\overline{A})$ where *A* is a sum of at most 11. Then,

$$P(\text{sum at most } 11) = 1 - P(\text{sum of } 12) = 1 - \frac{1}{36} = \frac{35}{36} \approx 0.97.$$

Copyright © 2015 Pearson Education, Inc.

Solutions to Chapter 14 Review Exercises **155**

5. **(a)** The probability that the card is a heart or a club is $\dfrac{13}{52}+\dfrac{13}{52}=\dfrac{26}{52}=\dfrac{1}{2}$.

(b) The probability that the card is a diamond or a king is $\dfrac{13}{52}+\dfrac{4}{52}-\dfrac{1}{52}=\dfrac{16}{52}=\dfrac{4}{13}$.

(c) The probability that the card is neither a diamond nor a king is $1-\dfrac{4}{13}=\dfrac{9}{13}$.

7. **(a)** Mischa can align the books on a shelf in $9! = 362{,}880$ ways.

(b) Mischa can align the books so that the math books are together and the physics books are together in $2\cdot4!\,5!=5760$ ways.

(c) The probability that no two of the physics books are together is $\dfrac{4!\,5!}{9!}=\dfrac{2880}{362{,}880}=\dfrac{1}{126}$.

9.

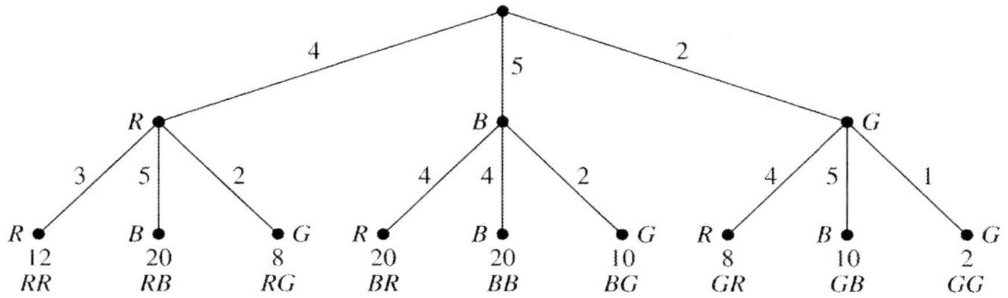

(a) $P(\text{both same color})=\dfrac{12+20+2}{(12+20+8)+(20+20+10)+(8+10+2)}=\dfrac{34}{110}=\dfrac{17}{55}\doteq0.31$

(b) $P(\text{at least one green})=\dfrac{8+10+8+10+2}{(12+20+8)+(20+20+10)+(8+10+2)}=\dfrac{38}{110}=\dfrac{19}{55}\doteq0.35$

11. There are 12 equal size outer sections, each with central angle $\dfrac{360°}{12}=30°$. There are 4 equal sized inner sections, each with central angle $\dfrac{360°}{4}=90°$. The probabilities are found using the appropriate ratios of angular measures.

(a) Sections b and 8 do not overlap so $b\cap8=\varnothing$. Therefore, $P(\text{b and }8)=0$.

(b) Since b and 8 do not overlap, they are mutually exclusive. Sum the probability for each event to find $P(\text{b or }8)$.

$P(\text{b or }8)=P(\text{b})+P(8)$
$=\dfrac{90°}{360°}+\dfrac{30°}{360°}=\dfrac{120°}{360°}$
$=\dfrac{1}{3}\approx0.33.$

(c) Given that the spinner lands in section 8, it cannot land in section b. Therefore, $P(b\,|\,8)=0$.

(d) Section 2 is a proper subset of section b. Therefore,

$P(\text{b and }2)=P(2)=\dfrac{30°}{360°}=\dfrac{1}{12}\approx0.08$

(e) Sections b and 2 are not mutually exclusive, therefore,

$P(\text{b or }2)=P(\text{b})+P(2)-P(\text{b and }2)$
$=\dfrac{90°}{360°}+\dfrac{30°}{360°}-\dfrac{30°}{360°}$
$=\dfrac{90°}{360°}=0.25.$

Copyright © 2015 Pearson Education, Inc.

156 *Chapter 14 Probability*

(f) Given that the spinner lands in section *b*, it can land on Sections 1, 2, or 3. Therefore, use (d) to find:

$$P\left(2|b\right) = \frac{P(2 \text{ and } b)}{P(b)} = \frac{\frac{30°}{360°}}{\frac{90°}{360°}}$$

$$= \frac{30°}{90°} = \frac{1}{3} \approx 0.33.$$

13. (a) The events are independent so multiply the probabilities.

$$P(\text{spade}) \cdot P(3 \text{ or } 5) \cdot P(H) = \frac{1}{4} \cdot \frac{2}{6} \cdot \frac{1}{2}$$

$$= \frac{1}{24}$$

(b) By complementary probabilities,

$$1 - P(\text{not spade}) \cdot P(1, 2, 4, \text{ or } 6) \cdot P(T)$$

$$= 1 - \frac{3}{4} \cdot \frac{4}{6} \cdot \frac{1}{2} = 1 - \frac{1}{4} = \frac{3}{4}$$

15. (a) $7! = 7 \cdot 6 \cdot 5 \cdot 4 \cdot 3 \cdot 2 \cdot 1 = 5040$

(b) $\dfrac{9!}{6!} = \dfrac{9 \cdot 8 \cdot 7 \cdot 6!}{6!} = 9 \cdot 8 \cdot 7 = 504$

(c) $\dfrac{8!}{(8-8)!} = \dfrac{8!}{0!} = \dfrac{8 \cdot 7 \cdot 6 \cdot 5 \cdot 4 \cdot 3 \cdot 2 \cdot 1}{1}$

$= 40,320$

(d) $7 \cdot 6! = 7! = 5040$

(e) $P(8, 5) = 8 \cdot 7 \cdot 6 \cdot 5 \cdot 4 = 6720$

(f) $P(8, 8) = 8 \cdot 7 \cdot 6 \cdot 5 \cdot 4 \cdot 3 \cdot 2 \cdot 1 = 8!$

$= 40,320$

(g) $C(9, 3) = \dfrac{9 \cdot 8 \cdot 7}{3 \cdot 2 \cdot 1} = \dfrac{504}{6} = 84$

(h) $C(9, 9) = \dfrac{9 \cdot 8 \cdot 7 \cdot 6 \cdot 5 \cdot 4 \cdot 3 \cdot 2 \cdot 1}{9 \cdot 8 \cdot 7 \cdot 6 \cdot 5 \cdot 4 \cdot 3 \cdot 2 \cdot 1} = \dfrac{9!}{9!} = 1$

17. $S = \{Y_1, Y_2, Y_3, Y_4, Y_5, B_1, B_2, B_3, B_4,$
$G_1, G_2, G_3, G_4, G_5, G_6, G_7, G_8\}$

(a) There are

$$C(8, 5) = \frac{8 \cdot 7 \cdot 6 \cdot 5 \cdot 4}{5 \cdot 4 \cdot 3 \cdot 2 \cdot 1} = \frac{6720}{120} = 56 \text{ ways}$$

to select five green marbles.

(b) There are

$$C(5, 5) \cdot C(8, 5) = \frac{5 \cdot 4 \cdot 3 \cdot 2 \cdot 1}{5 \cdot 4 \cdot 3 \cdot 2 \cdot 1} \cdot \frac{8 \cdot 7 \cdot 6 \cdot 5 \cdot 4}{5 \cdot 4 \cdot 3 \cdot 2 \cdot 1}$$

$$= \frac{6720}{120} = 56$$

ways to select five yellow and five of the eight green marbles.

(c) From (a), there are 56 ways to select five green marbles. There is only

$$C(5, 5) = \frac{5 \cdot 4 \cdot 3 \cdot 2 \cdot 1}{5 \cdot 4 \cdot 3 \cdot 2 \cdot 1} = \frac{5!}{5!} = 1 \text{ way to}$$

select five yellow marbles. Therefore, there are 56 + 1 = 57 ways to choose either all five of the five yellow or five of the eight green marbles.

---

19. $S = \{5 \text{ white, } 6 \text{ red, } 4 \text{ black}\}$

There are $C(5 + 6 + 4, 2) = C(15, 2) = \dfrac{15 \cdot 14}{2 \cdot 1} = \dfrac{210}{2} = 105$ ways to randomly select two balls.

(a) There are $C(5, 2) = \dfrac{5 \cdot 4}{2 \cdot 1} = \dfrac{20}{2} = 10$ ways to select two white balls, $C(6, 2) = \dfrac{6 \cdot 5}{2 \cdot 1} = \dfrac{30}{2} = 15$ ways to

select two red balls, and $C(4, 2) = \dfrac{4 \cdot 3}{2 \cdot 1} = \dfrac{12}{2} = 6$ ways to select two black balls. Therefore,

$$P(2 \text{ balls the same color}) = \frac{C(5, 2) + C(6, 2) + C(4, 2)}{C(15, 2)} = \frac{10 + 15 + 6}{105} = \frac{31}{105} \approx 0.30.$$

(b) From (a), there are $C(5, 2) = 10$ ways to select two white balls, thus $P(2 \text{ white}) = \dfrac{C(5, 2)}{C(15, 2)} = \dfrac{10}{105} \approx 0.10$.

Copyright © 2015 Pearson Education, Inc.

(c) From (a), there are $C(5, 2) + C(6, 2) +$
$C(4, 2) = 10 + 15 + 6 = 31$ ways to select two balls the same color, and $C(5, 2) = 10$ ways to select two white balls. Therefore, $P(2 \text{ white}|2 \text{ same color}) = \dfrac{C(5, 2)}{C(5, 2) + C(6, 2) + C(4, 2)} = \dfrac{10}{31} \approx 0.32$.

(d)

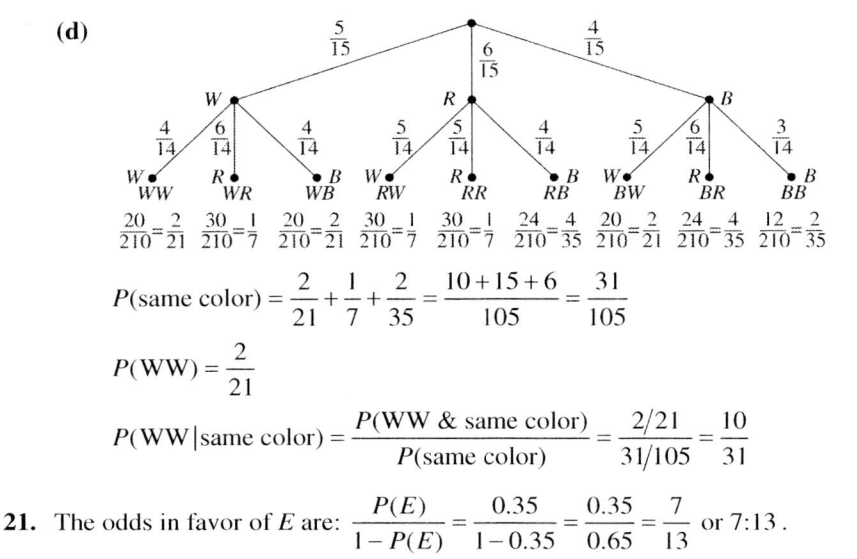

$P(\text{same color}) = \dfrac{2}{21} + \dfrac{1}{7} + \dfrac{2}{35} = \dfrac{10 + 15 + 6}{105} = \dfrac{31}{105}$

$P(\text{WW}) = \dfrac{2}{21}$

$P(\text{WW}|\text{same color}) = \dfrac{P(\text{WW \& same color})}{P(\text{same color})} = \dfrac{2/21}{31/105} = \dfrac{10}{31}$

21. The odds in favor of E are: $\dfrac{P(E)}{1 - P(E)} = \dfrac{0.35}{1 - 0.35} = \dfrac{0.35}{0.65} = \dfrac{7}{13}$ or $7{:}13$.

23. (a) Expected value $= \$0(0.15) + \$5(0.50) + \$10(0.25) + \$20(0.10) = \$2.50 + \$2.50 + \$2.00 = \$7.00$
The expected value of the game is $\$7.00$.

(b) You expect to collect $\$7.00$ per play but you must pay $\$10.00$ per play for an average loss of $\$3.00$ over the long run. If you want to make money, it is not wise to play. If your enjoyment during the play of the game is worth at least $\$3.00$ per play, then you can play for fun.